The Foundations of Phenomenological Psychotherapy

Giampiero Arciero • Guido Bondolfi
Viridiana Mazzola

The Foundations of Phenomenological Psychotherapy

Springer

Giampiero Arciero
IPRA
Rome
Italy

Department of Psychiatry
University Hospitals of Geneva
Department of Psychiatry
Genève
Switzerland

Viridiana Mazzola
Liaison Psychiatry
Department of Psychiatry
University Hospitals of Geneva
Department of Psychiatry
Genève
Switzerland

Guido Bondolfi
Department of Psychiatry
University Hospitals of Geneva
Department of Psychiatry
Genève
Switzerland

ISBN 978-3-319-78086-3 ISBN 978-3-319-78087-0 (eBook)
https://doi.org/10.1007/978-3-319-78087-0

Library of Congress Control Number: 2018940445

Printed on acid-free paper

This Springer imprint is published by the registered company Springer International Publishing AG part of Springer Nature.
The registered company address is: Gewerbestrasse 11, 6330 Cham, Switzerland

In memory of Prof. Pasquale Pellecchia,
a master of natural philosophy, who gave us
the keys to ancient thought

Introductory Quotation

The transformation of science is always accomplished only through itself. But science itself thereby has a twofold foundation: (1) work experiences, i.e., the direction and the mode of mastering and using what is; (2) metaphysics, i.e., the projection of the fundamental knowledge of being, out of which what is knowledgeably develops (Heidegger, pp. 65–66, GA 41).

Prologue: A User's Manual

Why entitle the prologue to a text—a text designed to establish psychotherapy on phenomenological foundations—as though it were the booklet that comes with an object and explains how to use it? Is it, perhaps, because a book too is a finished product?

This title does not actually refer to the book as the outcome of a process of production and hence to its usage; rather, it is a reference to a famous novel from the 1970s: Georges Perec's *Life, A User's Manual.* The subtitle is *Novels* (*Romans*), significantly in the plural. As this subtitle suggests, Perec's book is made up of individual stories that intersect, merging with one another. The common theme is "life in a Parisian apartment block: the tenants, objects, actions, memories, feelings, reveries… Each apartment in the block conceals a mystery, a drama, an adventure that branches off into the most exotic areas of the world or finds its origin in remote ages. The hundred rooms turn into the 'Thousand and One Nights'." This is how the book is summed up by Italo Calvino, who regards it as "the last real event in the history of the novel" (p. 88).

The plot of the book, then, finds its model in this building teeming with life and enclosing a "range of possible narratives" (p. 88), a building which for Perec—as much as for the Italo Calvino of *Six Memos for the Next Millennium*—requires us to apply rigorous rules if we wish to make our way through it over and over again—as though we were arranging and rearranging a web of meanings by leafing through a deck of tarot cards.

Our work falls within the context of this sort of narrative composition and therefore requires a user's manual. The text, while unitary, may be read as three heterogeneous narratives that may be interpreted on different levels even though they all follow the same thread.

The first narrative, "The Crisis. The Natural Sciences and the Unthought Debt," concerns the crisis of a paradigm: the paradigm of production, which grasps things as products. This paradigm has supported the ancient Greek as well as the contemporary vision of man that constitutes the foundation of therapeutic practice. Via Kant, it has slipped into the life sciences—hence the "unthought debt"—only to clearly resurface through the devastating impact of the neurosciences on natural-scientific psychology. The protagonist of this first narrative is the self, conceived as a natural entity, a thing among other things, the reality of which—being—is given by the enduring presence of its organization—just as in the case of the table, which, once crafted by the carpenter, stands on its own, as a finished object that is on hand. Just like the table, then, the self may be studied as what is complete and remains unchanged with respect to any transformation, offering itself up to analysis as a stable and intelligible manifestation. We thus gain access to the self—a presence enduring in self-identity—by grasping its immutable features through the theoretical mode of knowledge (be it of the empiricist or rationalist sort). A certain way of conceiving man is here accompanied by a particular conduct—the theoretical conduct—to grasp its reality.

The second narrative is "A New Beginning. Formal Indication, Non-rationalist Psychology, and Phenomenological Psychotherapy." While linked to the previous one, which is to say, to the failure of a way of conceiving and grasping man's being starting from the question of what that being is exactly (as though the "whatness" of a thing could define its truth!), this narrative illustrates the need for a critique of the foundations of psychology and hence psychotherapy. The urgency it reflects is not only due to the question of whether psychology—understood as a natural science—is capable of surviving the age of "Brain-mapping technologies"; rather—and at a deeper level—it springs from the need to call into question the guiding idea of man that has shaped the development of psychology.

What is suggested, then, is a radical critique of the idea of man as a finite being and hence of the paradigm of production that lies at the basis of such an idea. In turn, this leads to the outlining of a new perspective on man's being, which is to say, of an ontology capable of conceptualizing the precariousness and incompleteness that characterize the movement of each person's existence without the need to erase the latter in the light of what endures.

The new beginning may therefore be taken to refer to an ontological change that places—and preserves—at the center of the enquiry "that which happens or may happen to someone" at any moment. This new view assigns a new meaning to the motility of existence, to its incompleteness, which the paradigm of production interpreted as a deficiency in the light of the idea of the finished product—that is, by understanding the movement of life in terms of what the log carved by the carpenter, for instance, is still missing in order to become the table which he has in mind. The constitutive temporariness of human existence, its motility, is seen at each moment as a disclosure of meaning, as an intelligible space that opens up with respect to the meaningful things we encounter. All of a sudden, the reality of things, their being available as this or that, manifests itself together with our possible position with respect to them. In other words, the existence of things at the same time coincides with the possibility for us to access their specific content. Human experience thus emerges as the unity of (1) our interpretative position with respect to the ongoing situation (*Bezugssinn*); (2) its concrete enactment (*Vollzugssinn*); and (3) the content with respect to which we situate ourselves (*Gehaltssinn*). Such is the structure of the phenomenon, and an ontology of this sort is therefore phenomenological: for phenomenology is its method of enquiry. Thomas Sheehan sums it up as follows: "Meaningfulness is a thing's relatedness-to-oneself (*Mich-Bezogenheit*), and the phenomenologist studies this relatedness of the subject matter and the person involved with it" (2014, p. 121).

We thus come to the great theme of ipseity. Rather than as a produced thing, the self appears as an event, or rather as a phenomenon whereby ipseity, in contributing to the appearance of the world, relates to itself and comes to manifestation by virtue of that appearance it conditions. Clearly, ipseity does not create the world, nor does it "constitute" it; rather, in each case it comes to itself, it *finds itself*, while at the same time enabling the appearance of the world as something meaningful.

If, in taking charge of the precariousness of man, phenomenological ontology brings the structure of experience in general to light, the psychology which is re-founded on it is forced to call its own positum into question, along with the method used to access it. The psychological sphere can no longer be grasped as an object among other objects under the theoretical gaze of natural-scientific psychology; otherwise the domain itself will vanish. The fact of being oneself, ipseity, takes shape as a feeling oriented toward something (*Bezugssinn*), which in each case concretely emerges (*Vollzugssinn*) through the actual encounter with an occurrence (*Gehaltssinn*) and at the same time, and most crucially, as a form of self-intimacy. This is the fundamental indication that psychology, a necessarily a-theoretical and hence non-rationalist psychology, receives from phenomenological ontology. Related to this is a methodological problem: for, if the generation of meaning which a non-rationalist psychology focuses on always concerns a well-defined meaning, an individual story, a unique life which takes shape again and again as a concrete determination of one's encounter with the world in factual situations, then it is this relation that we must examine if we are to grasp personal experience in its making.

This kind of psychology, which cannot access the meaning of experience through reflective objectification, must do so by grasping it in its enactment: by grasping the taking shape of lived experience in relation to this or that circumstance, as it actually occurs in everyday life—in different ways for different people. What emerges here is the prereflective domain of life: a domain which can only be grasped through a methodology capable of understanding the experience of life by abandoning any a priori point of view and direction.

The specifying of what method guides us in grasping the way in which each person prereflectively *experiences* this or that situation is the cornerstone of this second narrative. In the footsteps of Heidegger's phenomenology, we have referred to this method as formal indication. In order to account for the incompleteness of each person's being there, this indication focuses on the intentional involvement without specifying any content. This method, which accompanies the taking shape of other people's experience of this or that, enables us to study the way in which ipseity relates to the world in different situations, in a way that carries different meanings for different people. Deriving from this methodological approach is the general indication that guides what we refer to as phenomenological psychotherapy. It consists in an interpretative understanding faithfully focusing on the actual occurrence of living experience, as this manifests itself to the individual, with no theoretical mediation.

The epilogue of this narrative takes shape as a destructive critique of theoretically founded psychotherapies, although it makes no claim to be judging or condemning psychotherapy as a whole, or the kinds of knowledge it has produced.

The engagement with the history of "scientific psychotherapy," and the radical critique marking this final point, enables us to grasp the stages through which the scientific psychotherapy has come to touch upon the experience of life, and hence things themselves: the very things that are the constant focus of phenomenology.

Finally, we have the closing narrative: "The Renewed Pact. Corporeality, Experimentation, and the Care of Self." It springs from a question implicitly raised by the previous account, picking up a theme already explored in the first: what possible new relations with the natural sciences are engendered by the ontological and methodological release of scientific psychology from the paradigm of production on which it rests? In other words, how can a phenomenological psychology and psychotherapy open up to the natural sciences by cooperating with them in a relationship of mutual enlightenment? On the other hand, how do non-rationalist psychology and psychotherapy stand in relation to that tradition of the care of self, which ever since Pythagoras' day has thematized the relation between one's way of living and the disclosing and transforming of oneself?

These comparative analyses—developed on the basis of the refounding of the position of psychology as a science, and hence of both its positum and method—shape the renewed pact that constitutes the focus of the final narrative. This new alliance between phenomenological psychology and the natural sciences most certainly revolves around corporeality, which is in turn intertwined with the debate on theoretical biology, the neurosciences, psychiatry, cardiology, and finally the care of self.

The central role played by corporeality finds its starting point in a rather self-evident fact: if each person relates to him or herself by relating to this or that concrete situation, this necessarily implies the activity of a corporeal being. This corporeality is understood as a phenomenon, and hence something that is constantly in the process of being actualized and is not limited to the mere presence of the body as a material entity that ends with the skin. Corporeality rather takes shape as a form of engagement, as a way of comporting oneself solicited by the ongoing situation: a mode of conduct that in its enactment discloses the historicity which lies within corporeality, the traces it conceals. The corporeal thus only emerges within the context of our relationship with the world and its limits coincide with those of our openness to the world: just as, when I listen to a concert, the limits of my corporeality coincide with those of the concert hall.

In this respect, the living body, the flesh, takes shape at each moment; in other words, the body becomes such starting from the openness to the world—the motility of life—according to which each person's existence unfolds at each instant. This perspective—envisaging being alive as the fact of *finding oneself* with respect to existing circumstances as the body becomes such over and over again—allows us to easily approach the problem of disease (and hence of treatment), as it allows us to understand the body according to the perspective of the unique history of the person whose traces it bears. The physiopathological mechanism, therefore, may be read in the light of the person's historicity and uniqueness, in such a way as to disclose those possibilities of engaging with the natural sciences and various branches of medicine that make up the various episodes of this narrative. What is new for empirical and clinical research is the possibility of being guided by the categories offered by life itself, thereby opening a new window onto the vast domain of individual differences.

If we now focus our gaze on corporeality, starting from the motility of life, it will enable us to account for another theme dear to theoretical biology and physiology—which has always gone hand in hand with comparative physiology—in addition to phenomenology: the difference between human beings and animals. On the one hand, this difference has to do with an unequal relation with the world, namely the ability to access the meaningfulness of the world that distinguishes the human being from all other life forms; on the other hand, this difference emerges from an original unity of life, understood in terms of motility. Man and animal (and, more generally, every living organism) are therefore different modes of enacting the motility of life. Indeed, the motility of life establishes again and again that relation which for each living creature takes shape as an act of individuation pertaining to the choice of life's possibilities. But whereas an animal remains chained to things according to its instinct in the enactment of the relation with the environment in which it lives, a human being leaves the meaningful things he encounters. In relating to them, he brings them to manifestation and at the same time comes to himself, *finds himself*, precisely through that which he causes to appear. Unlike in the case of an animal, this relation is not a necessity: man is free. In other words, in relating to things in view of the world, which co-appears as a background beyond all objects yet always together with them—as a constant presence on the basis of which a thing is grasped as this or that—man in each case experiences things by grasping their reality in terms of intelligible meaning.

What we have here is the experience of freedom, understood as the open possibility of comprehending that which addresses us and which belongs to man by virtue of his condition of having come into the world, of having been born. From this perspective, ipseity manifests itself as the very possibility of the experience of freedom. It appears as an opening up to *(Bezugssinn)* that which addresses us (*Gehaltssinn*) and which actually realizes itself again and again (*Volzugssinn*) as the manifestation of oneself and the appearance of the world. The enactment of man's life and hence the experience of freedom are founded on corporeality, understood as motility. Therefore, the difference between men is founded on the choice and responsibility of the orientation of one's way of living.

The narrative at this stage touches upon the great tradition of the care of self which, from Pythagoras onward, has focused its enquiry on the motility of life and on the care of the soul as the transformation and reorienting of that motility so as to open up to the truth about oneself.

We thus reach the conclusion of our account: a grand ending that unfolds through an itinerary that centers on the relation between the search for truth within oneself and the relation with a master capable of guiding this search. This itinerary starts from Classical Antiquity, passes through the practice of confession in Eastern monasticism, and reaches the Cartesian turn of the self-foundation of the thinking I. Resting on this new basis, which marks the beginning of modern thought, is the link between the truth about oneself and the truth of reason, mediated by the act of reflection and the theory guiding it, which has endured as the foundation of contemporary psychology and psychotherapy. This "Cartesian moment" thus marks the dramatic transition from a mode of accessing the truth—typical of the care of self—that requires the subject to make some changes to himself, to put himself to the test so as to ensure the possibility of grasping this truth, and a different mode, according to which the accessing of the truth about oneself actualized in the act of knowing is fixed by those principles that guide the process of knowledge acquisition.

The first method requires the person to implement some changes and change his position in order to gain access to himself, so as to engender through this work on himself a new disclosure of the truth about himself. As a consequence of the individual's change of position with respect to himself, the new access in turn produces some rebound effects that modify the current being of the subject, establishing a sort of virtuous circle. The other mode of grasping the truth about oneself simply requires the subject to reflect upon himself, guided by rational principles that establish the very conditions for knowledge, with no need for the person to call himself into question.

Within the framework of this contrast, it is possible to appreciate the difference between phenomenological psychotherapy and all forms of theoretically oriented contemporary psychotherapy. The course it follows is a path toward the transformation of one's own mode of living: a transformation for which the individual himself takes responsibility by carrying out some extended and taxing work on himself with the support of a therapist, a "well-meaning friend with a passion for healing."

By reclaiming the tradition of the care of the soul and in keeping with its underlying theme, namely, the idea that man takes care of himself through the way in which he relates to the world, this last narrative merges into the first, referring the reader back to the beginning of the text and its interpretative retrieval. Starting from the new mode of access ensured by the interlocking of the various narratives, this final account ends with the encounter between Leon and Pythagoras that opens the first narrative—an ancient tale illustrating the differences between men and their ways of life—thereby establishing the text as a unitary composition with neither beginning nor end. The book is thus open to different reading paths, marked by different modes of access corresponding to the three narratives which it comprises and which address the reader according to distinct modes of knowledge: the first according to philosophical knowledge and the second with respect to psychological and psychotherapeutic knowledge, while the third and final one implies a degree of familiarity with the natural sciences and phenomenology.

This threefold division that is only visible from outside the text, and which therefore enables the reader to access it from one of the three narratives, according to the form of knowledge he or she favors, loses its significance once we plunge into its living current. Little does it matter by what access we choose to delve into the book, for we will always be making our way through it by following its development and plot according to a circular pattern. In other words, the reader who reaches the last page is led to return to the beginning, making his or her way through its narrative paths, according to the kind of appropriating repetition that through reading picks up what the previous understanding of the text had left behind. Therefore, a new access key emerges from within the text itself, so to speak, and within its flow, brings the theme of the book to light: the motility of life, which is referred to in the text in various ways and constitutes the foundation of phenomenological psychology and psychotherapy.

The text consists of 11 chapters spread across three parts—the three narratives—that constitute as many epistemological links. Chapter 1 introduces the theme of healing in the light of the paradigm of production and hence of a way of accessing the motility of life according to a form of knowledge, *theorein*, capable of ensuring a stable and enduring form of knowledge. Chapter 2 follows the various stages in the development of this paradigm in the modern age, with a particular attention to how, first with Kant and then—and especially—with Fichte, it came to establish itself as the foundation of the life sciences. Helmholtz grafted Fichte's philosophy onto the very heart of physiology, producing two remarkable consequences: (a) he disclosed a new method of investigation that via Magnus, DeBarenne, and McCulloch ushered in a new—and still contemporary—era in the mathematical modeling of natural phenomena; (b) he laid the physiological foundations for the development of technologies in the image and likeness of man. Chapter 3 continues the narrative of the previous chapter by tracing the conceptual developments of cybernetics down to the contemporary neurosciences (in particular neurophenomenology) and psychologies that stem from them. The chapter then touches upon the debate between the easy problem and the hard problem that monopolized the attention of scientists, philosophers, researchers, and thinkers in the late twentieth century, also focusing on the deconstructive criticism delivered by Varela's suggested way of approaching this dialectic.

At this point in the book, a second part begins, revolving around the definition of a new framework for psychology and hence of a new therapeutic method. Chapter 4 explores the problem of making a new beginning in the light not of the paradigm of production but of the need for an ontology capable of thematizing the incompleteness of existence: "that which happens or may happen to someone." Phenomenological ontology thus presents psychology and psychotherapy with a new positum, ipseity, and hence a method to study it. Chapter 5 elucidates the structure of ipseity in relation to the world, everyday life, one's own history, the changes occurring over the course of one's life, and the creation of a symptom. Chapter 6 develops the new method, formal indication, capable of grasping living experience as it actually manifests itself to those who have it, with no theoretical mediation. The analysis of a clinical case helps further explicate the new method. Chapter 7, which ends this part, focuses on the definition of the margins, themes, and disclosures offered by phenomenological psychotherapy, through a close engagement with the hot topics in psychotherapy research and with the answers that theoretical forms of psychotherapy have to offer with regard to these topics.

Chapter 8, which opens the last part, introduces the phenomenon of corporeality and its intertwining with existence, the relation between corporeality and the body, and the topic of pathology—of the trauma of violence and its healing. The core of Chap. 9 consists in an engagement with theoretical biology. Touching upon Heidegger's dialogue with von Uexküll's work, by developing a phenomenological reappraisal of facticity, it reaches a new understanding of the difference between human beings and animals. This last topic is more fully explored in Chap. 10, which elucidates the difference in question in the light of the contrasting modes of enactment of the motility of life among living beings. These phenomenological analyses ultimately trace the humanity of man—and hence the difference between human beings and animals—back to the freedom of man to understand the things that meaningfully address him. The analyses conducted enable an engagement with some of the key points of extended evolutionary synthesis, and outline a range of new spheres in which psychology and the natural sciences can cooperate, according to a phenomenologically oriented methodology. In particular, the focus becomes the co-productivity of psychology and cardiology, the neurosciences, and psychiatry. Chapter 11 outlines a path that allows us to grasp phenomenological psychotherapy within the framework of the care of self, as the reappropriation and renewal of this tradition. The chapter unfolds along a trajectory that links the end of this part with the theme of healing in Chap. 1, fading into it.

Acknowledgments

The present study owes its origin to a series of seminars delivered at IPRA Institutes of Rome, Bari, Tenerife, Santiago del Chile, Buenos Aires, and at the Department of Psychiatry of the University Hospitals of Geneva over a period of 10 years.

We warmly wish to thank our Ipra colleagues and the members of the Department of Psychiatry for their constant support. We particularly are grateful to Jean-Michel Aubry for his assistance toward the fulfillment of this project, to Vittorio Conti for the lively exchange of ideas during the preparation of the manuscript, to Sergio Knipe for his professional competence and his hermeneutical sensitivity in helping us to put the work in the final English form, and to Steve McKend for reviewing the manuscript.

Giampiero Arciero
Guido Bondolfi
Viridiana Mazzola

Contents

Part I

The Crisis: The Natural Sciences and the Unthought Debt

On the Care Path

1

When people enter into therapy, they bring a story that usually corresponds to the way in which they account for a series of events, to which they assign causative value, with respect to the condition for which care is being sought. The narrative in question, a particular way of configuring events, is often accompanied by a certain malaise. At times this corresponds to specific symptoms; at other times, it amounts to an incapacity to gain access to oneself or to a sense of inadequacy in the way of relating to oneself. In all cases, it is a form of personal disunity that is suffered and experienced as a disorder. Usually, therefore, the story which the person tells is also accompanied by suffering, and it is this which leads the patient to seek treatment. So, in order to explore the topic of care, let us start by addressing the question of how this need of the patient emerges from the cracks of a narrative developed by Maria, with the aid of her therapist, during her first session.

Maria, a separated 40-year-old woman with a son just over one, starts off by saying that she is visiting our practice on account of the problems she has with the father of the child, with whom she would like to build a peaceful relationship. Maria had first met this man, whom we will call Carlo, 3 years earlier, at the end of a long and extremely fraught process of separation, which she experienced in a condition of complete sentimental and sexual isolation. The origins of the difficult nature and duration of this separation (3 years) are explained by Maria by referring to the episode that triggered it. By chance, her husband had unexpectedly learned about an affair which had been going on for a few months. This discovery was followed by a de facto separation—in the sense that the two spouses now no longer live together—but not by a legal separation. This is an interesting detail, because Maria will interpret this failure to lend official acknowledgment to an existing condition as a sign of indecision on the part of her husband and therefore as a possibility for them to recover their relationship. A few days after leaving their shared home, Maria realizes what a serious mistake she has made: the freedom she had yearned for in the last months of her relationship with her husband is of no use to her now. Her affair is soon over and Maria finds herself alone. This loneliness is marked by terrible feelings of guilt about her infidelity, combined with a hope for reconciliation, which is

G. Arciero et al., *The Foundations of Phenomenological Psychotherapy*,
https://doi.org/10.1007/978-3-319-78087-0_1

left open by the lack of any request for divorce on her husband's part. Maria spends 3 years in this condition, without as much as a one-night stand. Positive that she could never find a man to match her husband, she is caught in a limbo, as she waits to return to him.

Three years after learning about his wife's affair, however, the husband decides to file for divorce. Maria's reaction is ambiguous: on the one hand, she is stunned, bewildered, and painfully surprised; on the other, she feels free because her guilt has finally lifted. It is against this background that the relationship between Maria and Carlo blossoms. One day, a few months later, Maria is suddenly struck by Carlo's gaze in her office. The same day, the two brush by each other during lunch break—an erotically charged encounter. Carlo looks younger than his age. He is elegant and ready to pick up any woman's gaze; in other words, "he looks like a man who is after some fun rather than a stable relationship."

Maria, however, is soon forced to change her mind: after a few months, Carlo asks her if she would like a child from him. Maria can hardly believe her ears. In the grip of a deep and unexpected passion, she immediately accepts the suggestion. She envisages a new horizon with a man who, unlike her ex-husband, has asked her to give him a child, thereby setting their relationship within the framework of a shared life project. Strangely, though, in the following 4 months of cohabitation, Maria does not become pregnant. In fact, the relationship grows strained. Maria is wary and throws jealous fits. She becomes intrusive and is easily irritated by the flirtatious attitude which Carlo continues to display toward his female colleagues at work. The situation becomes so critical that Maria ends up doing the house move they had planned together 4 months earlier all on her own. Carlo, stifled by her jealousy and aggressiveness, explains that he needs a break: he takes a 2-month vacation, interrupting all regular contact. Maria realizes that she has gone too far and awaits his return, swallowing the bitter pill of his summer pause for reflection. During his 2 months of leave, Carlo only phones now and then, but his behavior is erratic: he buzzes Maria's door out of the blue, then phones her three times in one day, and then writes to her. In any case, he avoids meeting her.

In this period, however, something peculiar happens. One August afternoon, Maria receives a strange phone call; on the other end of the line, she hears a woman's voice: "My name is Cecilia Smith and I'm calling you to tell you that I'm 10 years younger than you and have been seeing Carlo for some time. We have a shared dream and some problems that unfortunately are preventing us from fulfilling it…." Maria is speechless, dumbstruck, and stunned.

A few days after this episode Carlo shows up again, as if by magic. He repeatedly phones Maria, but she does not answer his calls. He comes round to her place and rings the doorbell. She ignores him at first but then finally opens the door. He asks her to listen to him. The story Carlo tells is a rather unlikely one, but Maria chooses to believe it anyway. Carlo explains that his ex-lover—Cecilia Smith—has been stalking him for a year, trying to sabotage all his relationships. A few days earlier that woman had entered his office and stolen his mobile phone. Knowing Maria's full name, she had stolen his phonebook to call her. This is the explanation Carlo wanted to give her. But he goes even further by telling Maria that he has finally made his choice: "From tomorrow I'll be moving in with you." The following day Carlo indeed moves into Maria's house. Four months later Maria is pregnant.

At this point, their relationship takes a radical new turn. During her pregnancy Maria gains more and more weight. As her body image changes, she become more insecure and hence more distrustful and jealous. She clearly feels she cannot live up to other women, and this leads her to hate Carlo: she attacks, provokes, and exasperates him and cannot bear the sight of him. The pregnancy turns into a nightmare. At last, Maria gives birth. But Carlo shows very little enthusiasm for the child. In fact, he starts getting home increasingly late. He joins a fitness center far from home where he goes every evening, and when he gets back he mistreats and insults Maria. A month goes by, and one morning Carlo wakes up and tells Maria: "I can't bear you any longer. I don't love you anymore. Don't stand in my way, or in the way of my choices—I'm leaving. I don't think this will come as a surprise to you: we haven't had intercourse for months. I've been wanting to leave you for months and you know this full well, because you chose to ignore it and... I don't think it's right to raise the child in a situation like this." In that period—although Maria only discovered it later—Carlo had just begun a relationship with a colleague of theirs, a woman about 15 years younger than her. This event triggered what Maria refers to as her obsession with women in general, an obsession which flares up—in an increasingly intense way—each time she meets Carlo or her rival in the office. Carlo is a handsome 36-year-old man who seduces 20-year-olds and is involved with a woman 15 years his junior, a young woman with whom Maria cannot compete. And here is where Maria's obsession lies: the only way she has found to survive the separation is to constantly engage with Carlo's real or imaginary women—a soul-shattering film that is constantly playing in her head! Even the love she feels for her child is not enough to get Carlo out of her mind and free her from all the resentment, rage, but also desire she is experiencing. This is the affliction for which Maria has entered into therapy: she would like to change her life condition but is having trouble doing so. She is looking therefore for some care that would promote, engender, or bring about the end of the affliction she is suffering from. But what does it mean to bring someone into care? What is healing based on and how is it accomplished?

1.1 On Healing

There is a principle, an echo from the ancient world, which still underlies the praxis and research of all the various schools of psychotherapy. Aristotle formulates it in chapter 12 of *Metaphysics* book 5: "Potency (*dynamis*) means a source of movement or change, which is in another thing than the thing moved or in the same thing qua other. The art of building, for example, is a potency which is not in the thing built, while the art of healing, which is a potency, may be in the man healed, but not in him qua healed" (*Met.* book Δ 12, 1019 a 15 ff.).

Aristotle's definition first of all establishes an equivalence between movement and change. Potency therefore describes a principle that is the source of all change. Aristotle then states its location: this principle, potency, may reside either in another thing than the thing moved or in the same thing, which, however, is moved qua other. In order to explain these two aspects of potency, he uses two examples: the art of building and that of healing.

When we build a house, for example, in order for a range of materials such as stones, bricks, mortar, and tiles to turn into the house, it is necessary for these elements to be arranged according to a specific plan. The idea of the house and, more generally, the art of building it do not lie in the materials themselves: the starting point of this change is, rather, the architect who foresees it and hence another thing than the changing one. Just as the principle of movement lies in something other (the architect) than what is moved (the building material), so the principle of change originates from something other than what changes.

The second example concerns the art of healing. It is similar to the first, but more complex: for the principle of transformation here may reside not just in something else (as in the case of the architect with respect to the house) but also in the changing thing itself. A doctor will heal his patients but may also heal himself when he is ill; in this case, he can alter his health condition because, apart from being an ill person, he is also something else, namely, a doctor. In other words, in order for a patient to be healed, it is necessary for him either to undergo a change at the hands of the person capable of exercising the art of healing or, if the patient himself is a doctor and hence capable of providing treatment, to become the object of his own treatment. The principle of healing, therefore, lies either in another or in oneself as other; therefore, change brought about by another lies at the heart of the healing process. It is evident here that the Greeks were already familiar with the notion of the being of a person qua object (as was bound to be the case, given the primacy assigned to theory). Immediately afterward, however, Aristotle adds: "Potency means ... the source of a thing's being moved by another thing or by itself qua other: for in virtue of that principle by which a patient undergoes some change, we say that the patient himself is capable of undergoing changes" (*Met.* book Δ 12, 1019 a 20 ff.).

Aristotle better articulates the first definition of potency by reaffirming its defining features, this time in passive terms. This perspective, complementary to the previous one, reverses the examples: potency is now seen to lie in the materials which determine the possibility for the builder to employ them and set limits to this operation. We cannot build a house by arranging prime numbers one on top of the other, for instance, because they do not possess the capacity to be used in construction work. Likewise, the potency to heal, the disposition toward recovery, pertains to the patient, who may or may not be capable of enduring and reacting to the stimuli provided by the person treating him. In other words, in order for care to prove effective, the healing method must conform to the nature of the patient rather than seek to replace it as the foundation for health.

Aristotle thus furnishes a series of specific indications with regard to the art of healing. First of all, this is modelled after the art of building and hence regarded as a *techne*: as the kind of knowledge, that is, which governs those activities which have the crafting of objects or the construction of a house as their aim. Moreover, the art of healing corresponds to a principle of movement or change: healing implies the transition from one condition to another, from illness to health. Finally, the capacity to heal lies either in something else or in oneself qua other: the source of the kind of movement with which healing is identified is to be found in alterity vis-à-vis the person to be healed. On the other hand, the healer's action is accepted and received according to the patient's pre-existing possibilities.[1] This view of healing opens up a range of thematic areas which have been shaping therapeutic knowledge since Antiquity, without ever having been disputed.

If healing is governed by the same fundamental relationship which enables the potter to mold his vase and the architect to design his house, if—that is—healing stems from the application of knowledge in the light of which the being of a person may be transformed, what way of understanding man makes the art of healing accessible?

Healing, envisaged as a transformation leading from one existential condition to another, concerns the life of the patient, even though it occurs through a relationship with the healer or, rather, the person providing care. In other words, to every change there are two complementary sides, which correspond to the two Aristotelian determinations of potency: the active and the passive. On the one hand, the person who provides care must possess the knowledge required to trigger a transformation; on the other hand, the person who is being treated must undergo change. The healing process, therefore, takes shape through a relationship. On what way of envisaging this "interpersonal" foundation is the healing relationship based?

Finally, if the origin of healing lies in the person who provides treatment, while the patient receives his action according to pre-existing dispositions which determine its meaning, is it possible to identify a method to govern both the healing process and the processes of change of the patient?

These three questions, around which the present work will revolve, concern the basic themes, and hence issues and empirical data, which have defined and lent support to studies on psychotherapy and its praxis so far. What are at stake are three underlying aspects of every form of psychotherapy: (1) What view of man is care based on? (2) What way of envisaging one's relationship with the other is implicit in a therapeutic relationship? (3) By what path can care take shape?

Evidently, addressing the issue of the horizons of psychotherapeutic practice and research requires us to rethink the experiences which the above principles are meant to account for. These principles have provided a basis and context for developments which began in Antiquity and which, in the nineteenth century, led to the establishment of psychology as a branch of natural science and to the birth of the science of the care of the psyche.

---

[1] In his commentary on this passage, Reale (2004) aptly quotes the Scholastic formula *Nihil potest recipi, nisi praeexistat in patiente ordinatio ad finem, et ideo recipitur actio agentis ad modum recipientis* ("Nothing may be received, unless a disposition toward an end is already present in the patient, and therefore the agent's action is received in the patient's own manner").

1.2 The Man of Psychotherapies

Psychotherapy, in all its various forms, has been determined by the way in which psychology has thematized human life. Starting from the second half of the nineteenth century, psychology has conceived man on the basis of the suggestions provided by the natural sciences, using the conceptual tools and categories favored by them. The definition of the psyche as a natural entity, a thing among other things, is what has determined the limits and horizons of the field of inquiry, as well as of all praxis and research; it has fixed the boundaries of the subject matter and hence the way in which the fundamental structures related to it are to be investigated.

This preestablished mode of approaching the realm of the psyche, which has never really been brought into focus, has determined the path (*methodos*) of psychology and therefore the way in which its research and experiments are to be conducted. In other words, if the psyche is conceived from the perspective of the natural sciences, what psychology employs in order to convey its distinctive essence on the conceptual level are not just categories appropriate for grasping natural entities but also the appropriate method for studying them. It is therefore possible to look at one's psyche (or that of another person) as though it were an object, just as I might gaze at the car parked opposite my house, observing its shape, seats, wheels, color, etc. What I grasp of the psyche, understood as a thing that meets my gaze, are those properties that fall within the field of my disinterested vision—be they isolated or interconnected.

In this respect, it makes little difference what aspects the various currents of psychology may focus on: inner states, behavior, cognitions, emotions, feelings, narratives, relations, the flow of consciousness, or dreams. Different schools of psychology are distinguished by different focuses of observation, more or less articulated within given heuristic models—comparable to different car models in the previous example. Ultimately, however, they all share the same ontological foundation and the same method of inquiry.

All the various forms of psychotherapy deriving from this way of understanding psychology emerged over the course of the twentieth century as a sort of testing ground, or justification, for the psychological theories which had inspired them. So whereas psychic suffering and related therapies have been explained in the light of psychological models of reference, at best therapeutic practice has tended to prescribe a series of rules and patterns—more or less consistent with those of the model of reference—for structuring therapeutic action. In most cases, these rules have become methods of intervention. In recent years, starting from these premises and under the influence of insurance companies requiring an increasing degree of certainty in the choice of effective treatments, there is a growing pretension to measure the effectiveness of therapeutic intervention: evidence-based psychotherapy. This quality label for treatments that have been proven demonstrably effective, according to the empirical evidence, both certifies their scientific standing and ensures insurance reimbursement. But this claim to a scientific standing—as we shall see—is comparable to collecting data to prove the usefulness of prayer!

This state of affairs, which, on the whole, concerns these disciplines both in the professional sphere and in the academic and educational one, has to do with two foundational aspects that are closely interconnected and reflect the Western tradition to which they belong. On the one hand, the interpretation of the "soul" (*psyche*), which predetermines the course of this research, is based on an ontology which has remained the same ever since Greek Antiquity and which envisages human life according to the essential category of production (*poiesis*): a category implying a relation with a creator, a demiurge. The art of healing, therefore, takes shape in the light of such knowledge. This way of conceiving man understands human existence according to the model of the production of objects and of the way we experience objects produced—as things that stand on their own and are available to us. Starting from this vision, the object of psychology, its *positum*, becomes accessible: the self as a subsisting entity, as a thing among other things.

On the other hand, grasping the self as an object, envisaging it as something complete that resists all transformation, as "that which already was" (*to ti en einai*), to quote Aristotle—in other words, its remaining ever the same—defines the way in which it is possible to study the self but also the limits to every possible cure. This immutability is precisely what shapes one's observational relation to this object. The self can only be grasped through the regularity of its existing and occurring through the kind of theoretical knowledge which, freed from all prospectivism, makes it possible to identify its objective properties. Clearly, then, any form of change too can only be grasped by reducing it to the stability of the entity under consideration.

1.3 A Gaze Fixed on the Self?

Psychology (and consequently psychotherapy) may be regarded as a theoretical science, on account of both the ontological characteristics of the *positum*—the self as an unchanging substantial unity that constitutes the foundation of any transformation—and of the mode of observation which distinguishes its scientific method. This way of grasping things, i.e., theoretical knowledge (*theorein*), shaped the various currents of psychology and psychotherapy over the course of the twentieth century, down to its contemporary expressions.

On a practical level, the same kind of knowledge provides a range of interpretative rules to be followed in the care of self, which is to say, in psychotherapy. The idea is to treat this self—by promoting self-knowledge—according to a theoretical approach whose object is to be found within the so-called inner experience: within consciousness.

From being a science of the spirit, psychology therefore becomes a science of psychic phenomena to be investigated in the light of the knowledge which distinguishes the natural sciences. Natural-scientific psychology thus irrupts with its methodology into the sphere of awareness: while searching for invariants, it devitalizes lived experience and dehistoricizes factical experience. The development of physiological psychology is the first major achievement of this epistemological transformation, which finds its origin in Descartes and its culmination today in cognitive psychology.

At this stage we are bound to follow a path that is usually of little interest to psychotherapists, as it apparently adds nothing to the knowledge of practices which might make therapeutic intervention more effective. Therapists who share the everyday problems, dramas, events, and troubles of their patients would certainly prefer to be presented with the analysis of a noteworthy session within a given course of treatment, with a new technique or with some intervention guidelines. What would appear to shape their practice is especially the need to know in advance how to deal with this or that problem, how to react to this or that situation or disorder. It is baffling how the requests advanced by therapists in training coincide with those of many professionals who have long completed their education: both are searching for reliable tools, categories that may be applied, manuals telling them what to do, and preestablished patterns on which to model their treatment. The simpler the guidelines, the more likely they are to be followed.

Faced with patients who feel at a complete loss and seek help, novices and established professionals alike claim to possess a priori a degree of certainty ensured by the model they subscribe to and in the light of which they interpret all evidence—a model which provides a key for them to access the workings of the human mind and to operate upon it. Once acquired, this certainty often becomes dogma: it is no longer thematized, but is rather confirmed through the constant application of patterns deriving from one's model of reference. Thus psychotherapy "parishes" are established—of the sort that sparked the holy wars of the 1970s and 1980s. Each has its own truth and liturgies, its own shrines and saints, and its own gospel, to be applied and adapted over and over again to the lives of patients seeking help.

1.4 The Theory and the Philosophers

While the ontology on which psychology is based (and which has never been thematized) reflects a conception of being as a single whole, which remains itself unchanged but provides the basis for all change, theoretical knowledge is the method that ensures a firm understanding "of divine and human things." Both of these are founded on the same nature (*physis*), in an original and fundamental sense. Philosophy has laid claim to this detached form of knowledge from its outset: the Greeks called it *theorein*. According to Chantraine (1977), the word *theoros* (spectator) describes someone who is sent somewhere to consult an oracle or to take part in a religious celebration but also a traveler visiting unknown lands.

From the very beginning, then, this term must have been used to describe "he who goes as a spectator" to certain places and is engrossed by the spectacle before his eyes. Starting with Plato, the word *theoria* acquired the meaning of "contemplation, consideration," which it retained in Hellenistic Greek, where it is used to indicate "theory and speculation," as opposed to practice.

These different meanings, which Chantraine traces in the etymology and history of the word, provide the background to a testimony on the life of Pythagoras (Malingrey 1961) that sheds light on the origin of the term. This account (fr. 88 Wehrli) comes from Heraclides Ponticus (c. 390–320 BC), a disciple of Plato's; it was quoted by Cicero and then taken up by Iamblichus in his *Life of Pythagoras*, written roughly seven centuries later.[2] According to this account, when Pythagoras was visiting the city of Phlius, he discussed certain questions with the local tyrant Leon. Struck by Pythagoras' sharp mind, the latter asked him what art he practiced: Pythagoras replied that he specialized in no art but was a philosopher.[3] Leon, who had never heard that term before, asked Pythagoras what a philosopher might be and how he differs from other men. Pythagoras then told him that there is an analogy between the way in which men come into life and the way in which crowds attend solemn celebrations.[4] All sorts of people would flock to these celebrations, for a variety of reasons. Some would go there to sell their wares and earn a little money, others to compete and seek glory. Then there was a third category of people, comprising the noblest individuals, "who seek neither approval nor profit, but simply go there to see and carefully observe what takes place and how it takes place." So just as people will set out from different cities to gather at the same crowded celebration, by setting out from different natures in life, we find ourselves driven by monetary greed and ambitions of glory; only a few people—rare types—devote themselves to the study of nature. These, Pythagoras explained, are the philosophers, who investigate and learn about nature as disinterested observers (*theoroi*). It is worth noting that what is meant by nature here is the unity of the whole.

Iamblichus adds the following words to Cicero's account, lending it a distinctly Neoplatonist twist: "It is beautiful to contemplate the whole heavenly vault and to recognize the order of the stars moving across it ..." philosophy for him (i.e. Pythagoras) was the search for this sort of contemplation (*philosophia e zelosis tes toiautes theorias*).

We do not know within what narrative context Heraclides Ponticus presented the above anecdote, given that the work *De mortua*, which reportedly included it, is now lost. What emerges from the various accounts of the episode, spanning several centuries, are two aspects which raise some interesting questions even today.

[2] This testimony was first quoted by Diogenes Laertius (*Lives of Eminent Philosophers*, 1.12); it was then recorded by Cicero in *Tusculan Disputations* 5.8-9 (45 BC); finally, it came to be included in Iamblichus' *Life of Pythagoras* 12.58-59 (early fourth cent. AD).

[3] The word *sophia* originally referred to craftsmanship and to an artisan's know-how but also to the sphere of musical proficiency. Hesiod instead also uses it to describe sailing expertise. The field of use of the term was later widened to include poetry and, more generally, anything that may be learned and put into practice (see Heidegger, GA 27 1996; Hadot 1995).

[4] Pythagoras here may have been referring to the Panhellenic Games which, from 776 BC onward, were held in Olympia every 4 years and which drew vast crowds from every corner of the Greek world. Moreover, many Olympic champions hailed from Croton, the principle headquarters of the Pythagorean school (Riedweg 2005).

The first aspect is the close connection between philosophy and *theoria*. The difference between men and philosophers—the rarest and most noble of men—lies in the fact that the latter are capable of grasping the real being of things through a disinterested mode of observation. Understanding the being of things known to man (*sophia*) requires a particular inclination (*philia*) toward things, an inclination sustained by prolonged and intense exercise: contemplation, *theorein*.[5] From the outset, then, philosophy has been conceived as an activity aimed at the attainment of firm knowledge, as a practice that, while leading one to understand (*sophos*) those things which cannot be in any other way, brings about a transformation of one's way of life.

Yet this episode also presents a second striking aspect: it draws a contrast between the mode of acting of the few—the philosophers—who "see and carefully observe what takes place and how it takes place" and that of the majority of people, who are driven by their desires and passions.

1.5 The Theoretical Way of Life and the Care of Self

Where does this difference between men stem from? The most explicit answer provided in Antiquity comes from Socrates in the *First Alcibiades*. The young Alcibiades, a youth who has spent his best years indulging in soft living, now realizes that his beauty is fading. At a loss as to what he should do with his life, he dreams of becoming a politician, but actually has no means of taking his life into his own hands. Socrates makes an explicit suggestion to him. Most men, he explains, believe that taking care of oneself means taking care of one's business; but the art by which we take care of ourselves is not the same as that by which we take care of what belongs to us.[6] In order to know what art (*techne*) makes us better, we must first know ourselves, which is to say, our soul. This notion is reflected by the famous saying *gnothi seauton*, the injunction which the Delphic oracle would always address to those who consulted her.

If the soul must know itself, what must it turn toward in order to see at the same time itself and what it really is? To answer this question, the well-known metaphor of reflexiveness is introduced. Just as an eye gazing at itself in the pupil of another eye facing it can see, in addition to this pupil and reflected within it, the part in which its own visual capacity (*arete*) lies, so the soul, in order to know itself, must reflect itself in the brightest part of another soul, the part where the virtue of the soul lies: its *sophia*. "He who gazes into that region sees all that is divine" (Friedländer

[5] Reflecting on the beginning of philosophy, Heidegger wrote: "Philosophie nennt nicht das, was da behandelt, erkannt werden solle, sondern das Wie, die Grundart des Verhaltens. Daher sagen wir: Philosophie ist Philosophieren" (p. 25 GA 27 1996: "Philosophy does not name what must be recognized and treated through it, but rather the 'how', the fundamental mode of operating. We therefore affirm: philosophy is philosophizing.").

[6] These words may be seen to foreshadow the Stoic maxim "Of things, some are in our power, and others are not" (Epictetus, *A Manual for Living* 1.1).

et al. 2004, p. 656). This means that the act of reflexion, by which the soul can simultaneously know itself and what it is, only takes place provided one's gaze is turned toward another soul engaged in the same quest. Patočka notes: "Self-awareness requires two elements, namely the power to see and to seek, reflected in another power to seek: it corresponds to the power of questions and answers, the power to remain in dialogue; and this, clearly, means remaining within the problem, within the question" (p. 383, Patočka 2003). In other words, the movement by which the soul knows itself, while at the same time gaining access to the being of things—what endures in an unbroken present—is necessarily associated with a constant questioning, jointly conducted with others engaged in the same quest. The care of self, therefore, entails the establishment of a relationship with another person. In Greek Antiquity, this person was one's philosophical master; in early Christianity, one's spiritual father. Then, in the modern age, the idea emerged that access to the truth could be ensured by acts of conscience alone (Foucault 2001).

Socrates' advice to Alcibiades, who seeks to administer the city and therefore claim responsibility for the public good, is that, in order to correctly and honestly engage in politics, he must first of all be capable of seeing things for what they really are, rather than what they appear to be. "And you will act with your eyes turned on the divine light," Socrates states (134d). By following the divine light, the soul finds stable guidance in its relations with things, people, and the ever-changing affairs of the world. Only by looking to what is bright and divine can the person wishing to take care of the common good act according to justice and wisdom. Besides, self-knowledge corresponds to a way of living that acknowledges the value of truth above all, and entails a constant attempt to cultivate one's relationship with it, and therefore a constant testing of our capacity to gain access to things in themselves. The difference between men emerges and takes shape starting from a different care of the soul. The practice of self-knowledge as the contemplation of the being of things, *theorein*, is thus connected to the knowledge of the care of self and hence to a tendency to question one's own opinions, to the suspension of judgment, and to the ongoing quest to uncover the meaning of the famous Socratic notion of "knowing that you know nothing."[7]

Patočka is right in stating that what makes the care of the soul necessary is the dualism distinguishing its movement: on the one hand, we find the immediacy of things, which by their very nature can only show us indeterminate shadows of the real forms; on the other, the reflexive quest for what is not subject to change, for unalterable principles, and for the ultimate sources of all things. Self-knowledge and human finiteness therefore coincide with the knowledge that one does not know; and it is only in the light of this awareness of existing limits that consulting an oracle makes any sense.

Plato's answer is clear, then: the rift between those who live according to wisdom (*sophia*), i.e., philosophers, and the multitude of people who are not philosophers, is reflected in the way in which the former lead their lives, day after day, on a quest for the being of things. What distinguishes the philosopher's way of life is this stable relation between knowledge of the self and care of the self. It is only by virtue

[7] "If we know this, we can also know how to take care of ourselves" (129 a).

of the capacity for permanent reflection, which guides his investigation, that the philosopher can keep his gaze fixed on ever-changing things, acquiring an enduring knowledge which resists all changes of opinion: care of the soul (*tes psyches epimeleisthai*). This constant practice regulates one's explicit relationship with truth and hence with the good. It is developed through dialogue, which challenges the interlocutor and calls his beliefs into question. This dialogue may also be one which the soul entertains with itself. In all cases, it reveals one's limits, lack of knowledge and lack of wisdom; and through this experience of not knowing, it makes the soul aware of its own being. In this respect, the care of self is a self-oriented praxis, a practice oriented toward one's self-determination, one's thinking and acting well. This is why we can only grasp the nature of the soul through its care. "The care of the soul discovers this dualism ... discovers the splitting, both what is changeless and stable, and what flows—and both discoveries are equally crucial. The care of the soul therefore coincides with the discovery of the two fundamental possibilities of the soul, the discovery of the double path along which the soul moves" (Patočka p. 123, 1997). Indeed, this dual, dialogue-based structure is what provided the matrix for the examination of conscience and confession, understood as modes of exposition "of the soul and its errors" to oneself or others, in Hellenistic philosophy first and then—with some nuances—in Christianity. From our perspective, the archaeology of the subject which Foucault explored in his final years constitutes precisely a development in the ethico-political sphere of this underlying ontological theme, which has shaped the course of European history (Patočka 1976, 1977). Foucault argued that the mutual implication between knowledge and care of the self represents a problem which varies considerably "with different balances and relations, with different emphases on one or the other, and with a different distribution of the moments of self-knowledge and care of the self in the various systems of thought" (Foucault 2001, p. 60). As we shall see, while grasping the remarkable richness of this mutual relation, Foucault somewhat overlooked the fact that every interpretative configuration of the self—be it the Stoic, the Epicurean, or that of the Greek Fathers or of the cenobitic tradition, down to John Cassian and the Christian technologies of the self—is always defined "from the point of view of the gaze directed toward the eternal truths." What Foucault never discusses is the fact that theoretical knowledge, understood as a life praxis, underpins the ethico-political sphere and hence both Classical techniques and Christian interpretative analyses. The ethical distinction between good and evil, in other words, presupposes a more fundamental ontological knowledge—an enduring focus—on the basis of which to grasp the distinction between the being of things and their appearance; and the care of the soul, the exercising of which has evolved over the centuries, giving rise to different technologies of the self, remains oriented on the same ontology.

1.6 Theoretical Knowledge and the Essence of Movement

In chapter 8 of *Metaphysics* book 9, Aristotle offers what is arguably one of the most significant passages from Greek Antiquity, a passage which also sheds light on the later development of the relation between self-knowledge and the care of self in the

Hellenistic period. Within the context of a discussion of the relation between corruptible things and eternal beings, Aristotle provides a surprising, yet at the same time enlightening, indication as to the emotional climate of the earliest philosophy: "And so the sun and the stars and the whole heaven are ever active, and there is no fear that they may sometime stand still, as the natural philosophers fear they may" (1050 b 22 ff.).[8]

The ancient physiologists, the first to have thought about the being of nature, developed their considerations out of fear that the sun, the rotation of the planets, and the whole sky might suddenly cease to be. This fear led them to focus their observations on the notion of what endures unchanged. "The fear that, here, leads the analysis of being, lives from the hope or conviction that beings, genuinely speaking, may and should have to be being-there-always," Heidegger notes (GA 18 2002, p. 289).[9] By practicing a new form of knowledge, these early thinkers succeeded in grasping the unity of an indivisible nature which endures as the foundation of all transformation. At the same time, they discovered that they were able to use this knowledge to govern all forms of emotional immediacy.[10]

Theoretical knowledge thus started shaping education, the care of self, and, more generally, the ethico-political sphere. Not only that, but on the basis of the above fear, the Ancients started cultivating a fundamental attitude which endured throughout Western history: they learned how to grasp the meaning of being as presence, as the present time. They were the first to bear witness to the fact that man understands being starting from time, from the enduring present that is always "on hand."

Setting out from the fear that everything might vanish, the Greek world first and, later, Western man came to value sight above all and, through it, the ideal of direct knowledge, which is chiefly defined by the fact of being present, of being an eyewitness (Brague 2009). From the very outset, therefore, privilege has been assigned to *theoria*, the meaning of which is essentially connected to the action of seeing. The privilege assigned to it is a double one: for, on the one hand, contemplation directly partakes of what subsists in a stable form, as sheer presence—through a "spiritual contact," as Festugière (1950) phrased it; but, on the other hand, this "hope of salvation" silences all fear, neutralizes all emotional excesses, and frees one from the worries of everyday life. A famous passage of the *Iliad* illustrates the sense of wonder which the Greeks experienced at the sight of the heavens: "These then with high hearts stayed the whole night through along the lines of war, and their fires burned in multitudes. Just as in the sky about the gleaming moon the stars shine

---

[8] These words seem to echo the anxiety of these ancient (*palaioi*) men who, according to the *Cratylus*, "always get dizzy as they turn round and round in their search for the nature of things, and then the things seem to them to turn round and round and be in motion" (411b-c).

[9] The fickleness of fortune and happiness, like the instability of destiny, made its appearance as a theme in Ionic lyric poetry very early on.

[10] A few centuries later—as Hadot notes—while conceiving wisdom in different terms, Hellenistic schools all share the (ancient) view of philosophy as the pursuit of perfect tranquility. Hence, philosophy is envisaged as a "remedy for human worries, anguish, and misery brought about for the Cynics, by social constraints and conventions; for the Epicureans, by the quest for false pleasures; for the Stoics, by the pursuit of pleasure and egoistic self-interest; and for the skeptics, by false opinions" (p. 100, 1995–1998).

clear when the air is windless, and into view come all mountain peaks and high headlands and glades, and from heaven breaks open the infinite air, and all the stars are seen, and the shepherd rejoices in his heart" (VIII, 553–559). This wonder also brought a sense of security on earth, since the light of the sky illumines all things and gives each its place, as well as providing a sense of certainty with regard to the alternation of light and darkness, works and days, and—last but not least—an unshakable faith in everlasting time, in the intangible eternity of the motion of the stars, a movement which unfolds before the eyes of those capable of grasping it.

With theoretical knowledge life forgoes its circumspection and practical engagement and is instead absorbed, through sheer contemplation, by those entities which cannot be other than what they are, those entities which preserve a degree of constancy in their being and occurring—"and the shepherd rejoices in his heart."[11]

An entity of this sort, whose mode of being consists in enduring as a whole (*entelekeia monon*), was never generated or produced and cannot cease to exist or become other than what it is: it is simply present. Hence the movement of the stars, which is always already completed,which is neither thought of nor simply observed, but is rather experienced through fear and grasped as being-there-always (as a presence) even when it vanishes from sight. Here is the god of the philosophers (Rm 8:19) that Luther thundered against!

1.7 Above and Below the Moon

Being, therefore, primarily coincides with that being whose movement is always already completed and knows neither beginning nor end. This is the crucial point which Aristotle—drawing upon the legacy of the ancient physiologists—turned into the centerpiece of his *Physics*: he sought to access the meaning of being by interpreting movement. His investigation would still appear to proceed from the key experience of the fear of the end, which distinguishes the discussion on nature (*physis*) conducted by the ancient philosophers, precisely because Aristotle provides a single, certain answer to this concern: movement is eternal and incorruptible and represents a constant presence free of all change.

Alongside those entities which, being complete, manifest themselves as sheer presence, we find other entities whose character consists in being incomplete and, therefore, open to the possibility of being other than what they are. This is the case with human life, for example, which in its movement of maturation, in its transformation (*metabole*) and constant potential for completeness (*energeia*), is negatively defined by contrast to those entities which are ever perfect (*entelekeia monon*). This

[11] Besides, even though *theorein* is conceived as a religious and political practice—i.e., as a mandate issued to some citizens to represent their city at Panhellenic festivals (Wilson Nightingale 2004)—it implies one's participation as a witness in a series of events that take shape within a "space" which transcends political differences and which therefore entails a kind of suspension of one's belonging to an individual polis. Wilson Nightingale sums it up as follows: "the defining feature of *theoria* in its traditional forms is a journey to a region outside the boundaries of one's own city for the purpose of witnessing some sort of spectacle or learning about the world" (p. 68).

incorruptible presence becomes the precondition for conceiving the precariousness, emotivity, and changeability of human life, whereas movement in all its various forms represents the common matrix which enables one to account for every entity.

"From what has been said, then—Aristotle concludes in chapter 4 of *Metaphysics* book 5—the original and fundamental sense of nature (*physis*) is the substantiality (*ousia*) of those things which contain in themselves, as such, a source of movement" (1015 a 13–15). Anything that changes is grasped, therefore, on the basis of a certain and ever-present foundation which always remains the same: nature (*physis*) as the unity of all, as the substantiality determining all beings both above and below the moon qua substances.

Heidegger always maintained that Aristotle's *Physics* is the book that secretly underpins all Western philosophy, which is to say, Aristotelian ontology, and therefore also all the theoretical, practical, and poietic disciplines which rest on an interpretation of the being of nature (*physis*). The principle of movement, the foundation of all things divine as well as of the sublunary world, thus coincides with substantiality (*ousia*), which pre-structures the mode of being of God, the world, and things (Marx 1961). In this respect, asking what being is means asking what its substance is. This view finds support in the first chapter of *Metaphysics* book 7, which states: "Indeed, the question which was raised of old, as it still is and always will be, and which is always an object of enquiry, viz. 'What is Being?', is in other words 'What is substance?' (1028 b 2–4)." It is chiefly the theoretical faculty which enables the contemplation of substance understood as sheer presence (substantiality (*ousia*), movement (*kinesis*), nature (*physis*))—and this is precisely what makes such faculty divine.

1.8 Different Men, Different Forms of Knowledge

On the one hand, we have the pursuit of firm knowledge, of theoretical wisdom (*theorein*); on the other, we have praxis, the exercising of this attitude of detachment as a mode of life, and the permanent relation and intertwining between self-knowledge and the care of self: philosophy. Philosophers integrate theoretical research within a mode of living, which from Socrates onward has chiefly concerned the ethico-political sphere.

But what about all other people, the multitude of people who are driven by the desire to acquire wealth or who act in the pursuit of glory and success?[12] If it is different ways of living which engender the differences between men, why do men enter life with these different inclinations, as illustrated by the crowds at celebrations? Evidently, that kind of knowledge, theoretical knowledge, which finds its object in the kind of beings which cannot be other than what they are, is ill suited to understanding men's praxis, which is unstable, variable, and dependent upon changing circumstances.

[12] The distinction between judgment, desire, and action runs throughout the Classical and Hellenistic world and into the Christian era.

The ancient world was therefore faced with the question of how concrete life gives rise to two distinct kinds of existence and of how men may attain happiness by choosing different modes of living. Aristotle discusses the main aspects of this problem in a way that illustrates the defining features of his method on a broader, epistemological level.[13] The issue which Aristotle addresses is the precariousness of human life, its fragility: the fact that it is subject to constant transformation depending on the circumstances at hand.[14] The crucial step he takes lies in the discovery that the human soul has different ways of revealing things and hence making them available. By "adopting as his starting point" the wide range of attitudes displayed by the human soul, Aristotle refocuses his inquiry on that original movedness of life (*kinesis tou biou*), whose articulation produces all the various dispositions. The inquiry cannot be restricted to an attempt to grasp the kind of entity which endures unchanged—that which necessarily is what it is—but must take account of human life in its different concrete expressions, i.e., of that kind of entity which may become other than what it is.

The philosopher's contemplation, the profit-oriented transactions of someone attending a celebration in order to sell his goods, and the Olympic athlete's assessment of his own prowess are three different dispositions which reflect three different ways of articulating the movement of life: for the philosopher, non-prospective knowledge, *theorein*; for the tradesman, the performance of an action (trading) aimed at producing a profit, *poiein*; and for the athlete, an evaluation of how to deliver the highest performance, *prattein*.

Unlike *theorein*, however, doing (*poiein*) and acting (*prattein*) refer to two determinations of the movement of life pertaining to "what can become different from what it is and therefore can be an object of both production and action" (*Et. Nic.* VI, 4, 1140a 1–5). In the case of the tradesman, in other words, as in any other context of production, the aim coincides with what is produced (i.e., money). The aim therefore lies outside the action itself.

In the athlete's case, by contrast, as in any other circumstance in which the aim of the action is intrinsic to the action itself (as in the case of acting well or making the right choice), the movement of life is co-determined by the concrete situation in which the individual finds himself acting and in which he actualizes his existence.[15]

These three modes of actualization of human life, corresponding to as many ways of accessing and discovering the world, go hand in hand with different forms of knowledge which shape the care of self. Contemplation, *theorein*, the highest human activity, is guided by wisdom (*sophia*); doing, *poiein*, which has production

---

[13] As Düring notes, Aristotle's method of inquiry consists in approaching the same problem from different angles. The formula he uses is "Adopting a different starting point now..." (p. 247).

[14] This fragility of human action, which was considered the cause of the weakness of the city state (*polis*), leads Plato to observe that: "Human things are not worthy of great preoccupation, and to have to take an interest in them, is an unfortunate thing" (*Laws* 803 b).

[15] It is evident, therefore, that any kind of action may be seen to reflect either one of these two modes depending on the context and aim pursued. One example would be competing in a race for the sheer pleasure of doing so and competing in the pursuit of glory and success, as Iamblichus suggests (...).

as its aim, is directed by technical knowledge (*techne*)—the kind of knowledge which, as we have seen, will guide the healing of the sufferer (therapy) as much as the building of a house; and acting, *prattein*, where the self is always at stake, is oriented by the kind of reasonableness or practical wisdom (*phronesis*)[16] which changes depending on the context and whose exercising implies the search "for a mean between excess and deficiency." The division of knowledge into theoretical, poietic, and practical disciplines therefore reflects the view of man on which they are based.

1.9 Movement and Desire

Research on human life—the defining feature of which is the fact of being in movement or, rather, of orienting itself by moving toward something that is meaningful for life itself—thematizes not just the formation of the three abovementioned inclinations but the very essence of movedness. What is it that drives man to choose one kind of life over another? "It is apparent to be sure that these two are movers, either desire or mind ... both these therefore have the power to move according to place, mind and desire" (Arist., *De anima* 3, 433a9-15). What is meant by mind here is the intellect which reasons with a practical end in sight and which therefore differs from the theoretical mind with respect to its aim.[17]

According to Aristotle, therefore, what underpins the various expressions of the movement of life is the integration between desire and discernment. Life moves by approaching one thing but avoiding another. The structure of desire determines not only the possible sphere of action, pursuit, or avoidance but also one's perceptual orientation. The object of desire, in other words, reveals itself in relation to desire itself; at the same time, the life which craves for, or shuns, something finds itself either pleasantly or unpleasantly disposed toward what it perceives through its various emotional forms (*Befindlichkeit*).

Whereas an animal realizes its existence by nature, and it is only by nature that it becomes what it can be, man is not bound to the sphere of natural instincts. Since man can speak, he can grasp through language the object of his desire, not simply as something to be craved for or avoided, but as the reason for his way of acting. Man, that is, is capable of accounting for his actions because, in accordance with his mode of being, he is capable of discursively justifying his choices (*Rede*).

While desire is what determines one's conduct in relation to existing circumstances, the human capacity for discernment modulates the intensity of desire and the ability to control it, to the point of altering its structure. Human beings are capable of anticipating (*prohairesis*, *verstehen*) things as the reason for a certain action or choice with respect to the existing circumstances. This means that,

[16] Gadamer translates this as *Vernünftigkeit*.

[17] These sentences still echo the religious thought of the archaic age, which envisaged the struggle between blind passion and discrimination for control over the human heart as the key problem in the education of youth (Jaeger 1944).

although human beings may fall under the spell of desire, in every situation they can still choose between one thing and another, as well as determine the direction of their lives—so as to transcend mere craving—through the possibility of anticipating a given thing as the basis of their actions. This is due to the distinctly human quality of lending meaning to one's existence, even by going against one's immediate desires, based on the perception of time: "for while the mind bids us to hold back because of what is future, desire is influenced by what is just at hand: a pleasant object which is just at hand presents itself as both pleasant and good in an absolute sense because of want of foresight into what is farther away in time" (Arist. *De anima* 3, 433 b5-11).[18] Taking this fragile balance between the passions, the mind and existing circumstances as his starting point, Aristotle attempts to account for how human life, in its various forms of realization, can express movement, thereby making individuals responsible for the way in which they realize their own lives and for the content they lend to their own happiness (Volpi 2008).

If life is movement, and if this movement takes shape over time as an articulation of passion and thought, then the problem which emerges is how to regulate the combination and development of these elements, not just in the formative period, but throughout one's life. Thematizing the issue of the fragility of human life inevitably extends the care of self—which in the *Alcibiades* concerns the stages of development before adulthood—to the whole span of one's life and regulates it according to forms of knowledge other than theoretical knowledge. It is certainly significant that in one of the most famous books of the *Nicomachean Ethics*, book 6, Aristotle mentions Pericles as an example of a wise man, whereas in the *Alcibiades* Socrates had belittled his role as a leader (*Et. Nic.* 1140B 5–10). Besides, if practical life—understood as praxis—tends toward the attainment of a possible good through action, as a way of pursuing personal happiness within specific, contingent circumstances (*kairos*), the acting subject is accountable throughout the course of his/her life. Agents are responsible for their choices—after all, they develop their decision-making competence precisely through the choices they take and the knowledge they derive from this.

The transformation of the care of self, which Foucault traces from the *Alcibiades* to first- and second-century AD texts, highlighting its key aspects in relation to politics, education, and self-knowledge, becomes particularly significant—although this is generally overlooked—within this new horizon which concerns the realization of human life both as a whole and its uniqueness.[19]

---

[18] The issue of the relation between this possibility of envisaging the future and the capacity for discursive reasoning (*logos*) is a puzzle which arguably has only been solved by Ricoeur's work.

[19] That this was indeed the case would appear to be confirmed by the reception of Aristotle in the fourth and third century BC. The central role assigned to activity, understood as political rather than theoretical activity, was one of the reasons why the Epicureans—and probably Hellenistic schools in general—were so critical of the philosopher. One highly interesting testimony is provided by Philodemus' anti-Aristotelian polemic (see Berti 2009).

1.10 Movedness and the Incompleteness of Everyday Life

If sheer presence is what defines the authentic nature of being, how are we to understand the movement of those entities, such as human life, which have not yet attained completeness?

It is through this incompleteness, understood as a deficiency, as an absence perceived as a "not yet," that movement becomes intelligible. Transformation becomes decipherable precisely in the light of this privation (*steresis*), which emerges as a fundamental category of Aristotelian ontology.

Let us take the example of a log of a certain kind and a certain size. This wood carries the possibility of being employed in a certain way: it may be turned into a table, for instance. The wood has the availability to become a table: the carpenter actualizes this availability when he starts processing the log. The fact of not being a table yet, in other words, becomes present within the workshop, as the log is ready to be transformed into a table by the carpenter. This indeterminateness (*aoriston*) pertains to movement.

Let us suppose that at the end of his day's work, the carpenter leaves that piece of wood, which is neither a log anymore nor yet a table, in his workshop. The wood is no longer present as it was when the carpenter first laid his hands on it: it is present at rest, so to speak; and rest is an extreme instance of the movement initiated by the carpenter. The entity present there, in that workshop (spatially), at the end of a work day (temporally), is an entity which is neither exclusively a log nor exclusively a table; and this way of being-there of the entity coincides precisely with movement. It is clear, then, that when he refers to this phenomenon, what Aristotle has in mind are not the carpenter's "movements", but rather the movedness of the table itself, which is about to be born as such, which is about to acquire a certain appearance (GA 9 1976). Movement is not a being but a how of being-there, as Heidegger argued; it is not added from the outside, but is intrinsic to the entity itself.

This unique availability (*dynamis*) of the log to become a table is therefore a presence which is already at work. Movement consists in this change leading from the log to the table: in the transformation of an availability already present which, by being mobilized, acquires an even higher degree of presence (*energeia*),[20] until it attains rest in the thing produced. Movement is the presence of an ability to be (Arist. *phys.* Γ 1, 201 a 10 ff.).

At this stage, the problem becomes: where does movement lie? In that which moves or in what is moved? Does movement lie in the productive attitude (*poiesis*) or in the entity produced (*pathesis*)? In the mover or in the moved? In the therapist or in the patient? Aristotle's answer is that the productive attitude and the entity produced—just like what we teach and what we are taught or the therapy we practice and the one we benefit from—are but two ways of looking at the same phenomenon of the entity-in-movement: "the potentiality for acting and being acted upon is one ... Hence, in so far as it is a natural unity, nothing is acted upon by itself; because it is one, and not a separate thing." (*Metaph.* Θ 1, 1046 a 19–29).

[20] *Energeia* here describes a being that is not yet perfect or complete but is about to become so.

1.11 Care and Movement

In these two determinations of movement, Heidegger sees a guiding thread which sheds light on over two thousand years of philosophical history.[21] "Significantly—Heidegger writes in *Natorp-Bericht* (GA 62 2005)—the coming-to-be of the statue starting from shapeless bronze (through the practical movedness of production) serves as a guiding example for the problem of movement" (p. 395). According to this perspective, therefore, the practical mode of production is the essential paradigm enabling one to access the problem of movedness, which represents the foundation and hence the basis for understanding the nature of natural beings.

As already mentioned, what guides the productive attitude is a kind of knowledge which leads to the realization of a product starting from the idea which the sculptor has about his statue, the carpenter about his table, and the architect about his house. This idea is the source of movement. Healing is actualized according to the same paradigm.

"Health is the form present in the soul [of he who heals] and [therefore is] knowledge. Now, the healthy subject is produced as the result of this reasoning: since health consists in a given thing, in order for healing to take place, this thing must be realized, for example a certain equilibrium ... the movement which is effected by the healer at this stage, which is to say the healing movement, is called production. Therefore it follows in a sense that health comes from health and a house from a house, which is to say: that which is material from that which is immaterial" (*Metaph.* H, 1032 b 5–14). So just as the sculptor brings forth into appearance the statue by gathering together (*techne*) a certain material (*hyle*), to which he must lend shape (*eidos*) in view of a given end (*telos*), the healer, starting from the idea of health, brings about a series of transformations the principle of which is present in his soul, "until he ultimately reaches that which he has the power to restore to good health" (*ibid.*). A few lines later, however, Aristotle notes that health may also be restored spontaneously, precisely because it is constitutive of man: human nature is intrinsically capable of triggering a healing process even without the external intervention of a healer. Bricks too are preexistent compared to the house, yet they can only become part of the house if the architect chooses to include them in his building plan. The difference between natural and produced entities, then, lies in the origin of the movement, which in natural entities is to be found in their very being. Nevertheless, in all cases, that form of knowledge (*techne*) which directs the productive attitude (*poiesis*) leads one to approach even natural entities—be they bronze, clay, or health—in terms of the possibility for them of being used as a

---

[21] Movement, which in this perspective refers to an entity moved by a mover, extends all the way down to that entity which is pure act, sheer actuality devoid of any potentiality and lying beyond any possible change: the unmoved prime mover. The eternity of movement is demonstrated by Aristotle starting from the idea of motion and the phenomenon of time (*Phys.*, IV 219 b 1–7, b 23–25). Complementing this determination of the essence of movement, as a counterpart to the power to act, is the lack of power, *steresis*. With reference to the power of generation, for example, Aristotle states: "For we cannot claim that the boy, the man and the eunuch are powerless to procreate in the same sense."

material to create a given product.[22] It is precisely this relationship of man with his work—which is accomplished so as to be made available—that determines the Greek concept of knowledge.

The model of production, with the conceptuality it entails, is transferred from what is put into motion through production—the world of artifacts—to what is self-moved: life (*zoe*). In other words, ancient ontology interprets life starting from ontological categories pertaining to the practical experience of the production and manipulation of things in the world, giving rise to the technological view of man which has distinguished the history of the West down to the present day.[23] According to this view, natural entities exist in the world, not as things produced by someone, but as entities coming into presence, as self-producing entities. A natural entity may thus be seen to emerge into presence, to gather itself in a stand and to endure in its selfsameness. Following in this tradition, 2400 years later, cybernetics was to conceive of the living organism as a technological-computational artifact which produces itself, according to the notion of self-organization. In doing so, it unwittingly inherited—by way of neo-Kantianism—the conception of being tied to the notion of permanence and endurance.

1.12 Degrees of Movement

Life is therefore envisaged as a presence in the world: a perfect presence, since it is complete in its being, yet one that is grasped on the basis of the reality of objects. There is one difference, however: in the case of artifacts, the principle of production lies in something else, and the movement comes to a halt when the external end is reached—when the table has been completed.[24] At that moment, the object, which has come into presence, subsists in itself as complete and hence independent of its producer. The table has become available.

In the case of a living being, by contrast, the end exists as something internal, as the mode of its movement. A living being does not have a form it must reach outside itself, but possesses the principle of movement and rest within. The completeness of the movement is already within the movement itself, in the very act of moving: the movement of life is completed at every moment—it does not result in any work. In

[22] The concepts of matter and material are rooted in an understanding of the soul oriented toward production. For otherwise the idea of a material as that starting from which something is produced would remain concealed. The concepts of matter and material—the concept of *hyle*, which has the notion of *morphe*, or imprint, as its counterpart—therefore acquire a crucial role in ancient philosophy, not because the Greeks were materialists, but because that of matter is a fundamental ontological notion which necessarily emerges when an entity, be it a product or something which must not be produced, is interpreted within the horizon of the understanding of being inherent to the productive attitude as such (see GA 24 1975 pag 164).

[23] The same view has also shaped the perception of man's being according to those categories which enable us to grasp the meaning of objects, of things that are present and within our reach.

[24] Another essential difference concerns generation, since human beings beget human beings but tables do not beget other tables (see Arist. *Phys*. B1 193 b 8–9).

this respect, it therefore differs both from the perfection of sheer, ungenerated and incorruptible presence, which characterizes the divine, and from the realization of an artifact, where movedness gives way to rest through the completion of the product.

If life is movement and all movement is change, then change is a mode of being-there of life: its way of being at work in order to become complete (*energeia*), even though it is not yet complete (*dynamis*). In living beings change is understood on the basis of two principles: what remains unchanged, the substrate, which remains constant as transformation takes place, and form (*eidos*)/the privation of form (*steresis*), which instead constitutes that which changes. Hence the transition from one thing to another, while the selfsameness of the living creature endures in its being. Every transformation, therefore, can only be conceived as a transition from a given form to its privation, while that which endures beneath it and through it remains unchanged (substance, subject, organization).

So although human life realizes its perfection at every moment, ancient ontology grasps it as a self-subsisting entity: as an entity present within the world. The living being, in other words, is understood within the horizon of production and hence as a thing which is spatiotemporally situated and whose selfsameness also defines the possible determinations of its changes. A formula which encapsulates this view of human life, and its transformation centered around the notions of presence and present, was provided by Heidegger in the summer term course he held in 1931 (GA 33 1981): movement (change) = being in act = presence + absence.

Movement (*kinesis*), which corresponds to change (*metabole*), is an effective mode of being—the enduring presence of the substrate (*ousia*, *energeia*)—which represents the transition from something which is no longer present (*apousia*) to something which comes into presence (*parousia*) in a different and opposite way from the starting conditions. Just as the actuality of the work (i.e., the table) lies in its form, so man's being in act manifests itself through its appearance and coming into presence and, at the same time, through its going out of presence (*dikos*). Youth vanishes as old age appears, or a disorder vanishes as well-being emerges. If we think of this from a Greek perspective, the manifestation of well-being coincides with the emergence of the absence of the disorder. This opposition between form and the privation of form implies that every form of movement, every alteration, constitutes a transformation of what endures, which, being always there, can change in terms of presence and the lack of presence (privation).

The lack of presence corresponds to a positive position, in the sense that it indicates a living absence: as when we say that we miss someone. The absence of this person is rather a determined mode of his/her presence. This privation of presence is itself constitutive of being-there. In the case of a young person who grows old, for example, the growing old is already part of the human being as a non-presence which may or may not manifest itself when—and if—the person's being young is no longer present.

This becoming being is what characterizes the condition of instability of sublunary entities. An entity in movement is not yet what it will be: it is an unfinished present being. The crucial point, however, is that change, the transformation of human life, is bound to a permanent presence. Every possibility, in other words, is abolished and reduced to a present essence, which endures while changing in terms of presence and non-presence.[25] In this sense, every possibility is to be understood as an availability which may or may not become actual, depending on whether it comes into presence or not, but which in no way corresponds to sheer potentiality. That is to say that, when Aristotle shows the changeability of sublunary entities, he grasps it on the basis of the present by fixing and neutralizing what is not yet within the permanent presence. Every past and every possible future are therefore reduced to this presence, to a permanent present. Historicity is thus erased.

Heidegger's most thrilling discovery, based on an analysis of movedness (*Bewegtheit*), lies precisely in having identified this view of being as being present or being available (*ousia*) as the ontological orientation of the ancient world.[26] This meaning endured as the precondition for thinking of man and his possibility for change in Western thought. It is no coincidence that man came to be referred to as subject (*subjectum*), i.e., that which is self-subsistent. With Decartes the I of man then became the unquestionable foundation of knowledge: the very I for which objectivity comes into being.

Bibliography

Berti E (2009) Nuovi studi aristotelici, vol 4. Morcelliana, Brescia, p 1

Brague R (2009) Aristote et la question du monde: essai sur le contexte cosmologique et anthropologique de l'ontologie. Cerf, Paris

Chantraine P (1977) Dictionnaire etymologique grec. Klincksieck, Paris

Festugière AJ (1950) Contemplation Et Vie Contemplative Selon Platon. J. Vrin, Paris

Foucault M (2001) L'herméneutique du sujet. Cours au Collège de France 1981-1982. Gallimard-Seuil, Paris

Friedländer P. (1928–2004) Platon, 3 vols. Berlin: De Gruyter (Trad. It. Platone Bompiani Ed.)

Hadot P (1995). Qu'est-ce que la philosophie antique?. (eng. Trans.) P. Hadot, What is ancient philosophy?, transl. by Michael Chase, Belknap Press, Cambridge, 2002

Heidegger GA 18 2002

Heidegger GA 24 1975

[25] The argument that acts are antecedent to possibilities rests on a supreme entity, the unmoved mover, whose essence is the act of moving. This entity moves things just as desire for the loved one moves the lover.

[26] In the Greek world, the term *ousia* is used to describe both an entity and its being. The ontological perspective, whereby the being of an entity is grasped as the being available of any entity in general, coexists with the common sense (ontic) perspective, whereby *ousia* is used to describe a specific available entity—in Homer, fishing nets, for instance (see Marx 1961).

Heidegger, GA 27 (1996)
Heidegger GA 33 1981
Heidegger GA 62 2005
Heidegger GA 9 (1976)
Jaeger G (1944) Paideia. Oxford University Press, New York
Malingrey AM (1961) "Philosophia" Étude d'Un Groupe de Mots Dans la Littérture Grecque, des Présocratiques au Ive Siècle Après J.-C. Klincksieck, Paris
Marx W (1961) The meaning of Aristotle's ontology. Martinus Nijhoff, Hague
Nightingale AW (2004) Spectacles of truth in classical Greek philosophy: theoria in its cultural context. Cambridge University Press, Cambridge
Patočka J (1976) Le monde naturel comme problème philosophique. Martinus Nijhoff, La Haye
Patočka J (1997) Platone e l'Europa, vol 55, Milan. Vita e Pensiero
Patočka J (2003) Socrate: Lezioni di Filosofia Antica [Socrates: Lessons of the Ancient Philosophy]. Bompiani, Milan
Reale, G. (2004) Metafísica di Aritotele. Bompiani
Riedweg C (2005) Pythagoras. His life, teaching, and influence. Translated from the German by Steven Rendall. Cornell University Press, Ithaca/London
Volpi F (2008) Guida ad Heidegger. Laterza, Roma-Bari

Creatures, Technology, and Scientific Psychology

2

2.1 The Finitude of the Created Entity

The ontological horizon developed by the Greeks, and which provided the background to their interpretation of God, the soul, and nature, was gradually transformed and disintegrated across the various stages of its transmission. A crucial turning point was certainly the rediscovery of Aristotle in the thirteenth century, which led to the grafting of Christian theology onto Greek ontology. Christian theology thus came to interpret from the ontological point of view of production both God's act of creation, as described in Genesis, and the mode of being of the subject, which—from the same perspective—came to be understood as a self-subsisting created entity. In this respect, Patočka is quite right to note that "the Renaissance science of Copernicus, Kepler, Galileo is still recognizably continuous with medieval *theoria* as a dimension of the care of the soul" (p. 85, 1996). The medieval rediscovery of Aristotle was further developed by the Spanish philosophical schools and in particular by the Jesuit Suarez (Courtine 1990). And it was from the Jesuits of La Flèche that Descartes received his education.

A few decades ago, in an epoch-making book, Colin Morris (1972) argued that the encounter between the Classical heritage and Christianity—over the course of a period which began in 1050 and reached its climax in the twelfth century—is what shaped many of the personality traits of Western man, down to the modern age. Morris goes so far as to speak of the discovery of individuality. In support of this thesis, he adduces a wide range of material, which helps grasp the transformation of late-antique individual consciousness: confession, the first attempts at autobiographical writing, the transition from icon to portrait, love poetry, the birth of psychological introspection, the "humanization" of Christ, etc. Such elements are all cornerstones for the development of new orientations, which lie at the origin of the notion of the individual (Gurevič 1996).

We cannot delve into the themes which emerged at this remarkable stage of the development of the West. However, it is clear that this continuity—the traits of modern man, which, according to Morris, were foreshadowed by the twelfth-century

G. Arciero et al., *The Foundations of Phenomenological Psychotherapy*,
https://doi.org/10.1007/978-3-319-78087-0_2

discovery of the individual as an object of reflection—has to do with the way in which ancient ontology assimilated the new question of the self-affirmation of subjectivity. As we will see, Foucault traces the patristic imprint of this period as far back as the fourth to fifth century, identifying the emergence of the Western self (and of the technologies for governing it) in monks' interpretative analysis of the content of personal experience. This is the very same tendency, which later established itself with philosophies of the subject, to seek an objective knowledge of the human being starting from one's own subjective experience.[1] In any case, what would appear to underlie the continuity emphasized by Morris (and overlooked by Foucault)—and which spans over a millennium of Western history—is the same ontological horizon, marked by the paradigm of production.

On the one hand, the new problem of individuality, which was quite unknown in Antiquity, completely redirected the development of philosophy by steering it toward subjectivity. On the other hand, this problem was interpreted on the basis of the ontology which had been transmitted and in the light of its understanding of natural entities: the exemplary model of production. All the new questions about the subject, which from Aquinas to Descartes and Kant, down to the present day, have been at the center of the development of modern thought and have always been conceived within the framework of ancient ontology. The perspective of production, in other words, remained the ontological horizon within which the new interpretation of the subject became rooted. Descartes was the first to envisage the I as a finite substance (*res cogitans*), separated from the equally finite substance of the world (*res extensa*) by an abyss of meaning, and he based this distinction precisely on the ancient understanding of the world as substance (*ousia*).[2] Kant made this view his own, along with the accompanying method, and radicalized to the point of claiming that the genuine perception of an entity is only possible for its creator, for the one who has produced it.

Only the producer can know the essence of the entity, whereas all that human beings can know is what they themselves do, and not things as they subsist in their genuine being. In his *Reflexionen zur Kritik der reinen Vernuft*, Kant writes: "It is difficult to comprehend how an intuitive intelligence other than divine intelligence could exist. For this cognizes within itself the possibility of all things as primordial causes (and archetypes); but finite beings cannot by themselves know other things, if not the mere appearances which they can know a priori, for they are not the producers of those things. It is believed that every a priori knowledge is knowledge of things in themselves; actually, the very opposite is the case, since it is always only

[1] According to Foucault, the twelfth and thirteenth centuries marked the beginning of a judicial age characterized by the ambition to govern whole peoples through juridical structures.

[2] It is on this ontological basis that Descartes grasps the subject as the foundation of metaphysics, as the origin of methodological certainty. The subject of the method is the formal I which stands vis-à-vis objects. It is this "methodical" I, the same subject envisaged by ancient ontology, that possesses knowledge of... (GA 28 1997). The I is the source of judgment by virtue of its essence. ("Dieses Urteilssubjekt wird vom Urteilen nicht erst aufgesucht und vorgefunden, sondern es liegt im Urteilen selbst – schlechthin," p. 120).

knowledge of things as phenomena, i.e. as objects of experience. We can therefore only cognize things in themselves in God" (929 p. 261).[3]

Ultimately, what Kant did was develop the full implications of Descartes' thought. What separates the self-aware entity (*res cogitans*) from reality (*res extensa*) is not an abyss, but a constitutive impossibility of access. If the I knows itself (qua finite spiritual substance) as the foundation of its multiple determinations and attitudes, and thus possesses itself, and if all we can know is the effect which things produce on us and not the things themselves, since we are not their creators, then the knowledge of objects is bound to have to do with the alterations of the state of my I in relation to such objects—i.e. "affections." Indeed, the thinking substance for Kant is not merely intelligence (or, more generally, activity) but also receptivity.[4] A passage from Kant's 1782–1783 lectures reads: "Sensible representations are sensible representations according to the manner in which I'm affected by things; intellectual representations are ones independent of that. Sensible cognition that rests on receptivity is sensible, that which rests on spontaneity is intellectual. Through sensibility thus I cognize things as they are for us; but through understanding as they are in themselves" (p. 249 Kant Metaphysik Mrongovius). From this it follows that the only way to approach the knowledge of objects is by dealing with "the concepts and principles of the intellect only insofar as these refer to the objects that may present themselves to our senses" (1793–1967, p. 66).

This is a crucial point: the given as given can count as an element of knowledge for the knowing subject only to the extent that it is mediated by the subject's own intelligence and rationality. Hence Kant's famous observation that intuition without conceptual activity would be "blind ... less than a dream." The transcendental turn consists precisely in the establishment of a science capable of investigating our a priori possibilities to know objects.

Kantian transcendental philosophy, therefore, has to do with a limit: human finitude, which consists in the fact that, since we are not the creators of things, it is only possible to gain knowledge of an entity starting from the cognitive structure of the thinking subject.[5] This finitude is the original debt which does not allow us to forget that being for a person amounts to having being produced: to being an *ens creatum*.

---

[3] "Es ist schwerlich zu begreifen, wie ein anderer intuitiver Verstand stattfinden sollte als der göttliche. Denn der erkennt in sich als Urgründe (und archetypo) aller Dinge Möglichkeit; aber endliche Wesen können nicht aus sich selbst andere Dinge erkennen, weil sie nicht ihre Urheber sind, es sei denn die blossen Erscheinungen , die sie a priori erkennen können. Man meint aber, dass alle Erkenntnis a priori Erkenntnis der Dinge an sich selbst ist; sie ist aber gerade das Gegenteil, sondern allemal nur der Dinge als Erscheinungen, d. i. als Gegenstände der Erfahrung. Daher können wir die Dinge an sich selbst nur in Gott erkennen*)."

[4] Intelligence is not confined to the activity of knowledge acquisition, but extends to acting morally, as an end in itself. The "I act" provides a crucial link between the *personalitas trascendentalis* and the *personalitas moralis*.

[5] Persons, conceived as intelligences, are distinguished qua spiritual substances from natural entities, understood as corporeal substances. As intelligences, they are characterized by spontaneity—the capacity to bring about certain effects starting from themselves—and by receptivity, which is to say, the capacity to receive the "affections" engendered by another substance.

In this respect, Kant follows in the trail of ancient ontology. For Kant, the *ens creatum* corresponds to that I of pure self-awareness—the spiritual substance—which underpins every experience and which, as such, is destined to remain unknowable.[6] On the other hand, the interaction with finite (spiritual or corporeal) substances produces certain effects within us. As already noted, this generally corresponds to our way of knowing things. Hence, according to Kant, the being of an entity coincides with its being perceived. The "affections" roused within us by our dealings with others and with objects manifest themselves as states of the self to our inner sense—"that is to say, the faculty of making one's own representations the objects of one's thought" (Kant 1992, p. 104).

From this view of personality, a *positum* emerged which was destined to shape psychology for centuries to come: the notion of the states of the I as the objects of inner sense (as opposed to the objects of outer sense).[7] Psychology thus became the science of inner experience, ontologically conceived as something measurable and hence open to investigation through the methodology of the natural sciences.[8]

2.2 Action, Embodiment, and the Transcendental Perspective

To continue our account of the view of man on which the providing of care is based, it is necessary to show how the influence of Kant, and later of Kantianism, represents the fundamental turning point which enables us to grasp the origin of the main currents of contemporary psychology and psychotherapy.

The problem which the kind of psychology inaugurated by Kant bequeathed to later centuries concerns precisely the relation between the I of pure self-awareness and the psychological I, between pure I and empirical I, between unity and multiplicity, between the intellectual and the sensible, and between understanding and intuition. As is widely known, one century later, within the context of American

[6] The expression *personalitas trascendentalis* describes I-ness, understood in general ontological terms—that I-ness which makes empirical determinations possible. The experience of this I-ness is bound to underlie every factual experience. It is crucial to stress that the psychological subject that experiences inner sense must not be confused with the subject of the pure self-awareness on which it rests. In other words, the range of inner states where the I is the object of inner sense always presupposes an awareness of that substantial and immutable I (pure I), which is known as the ground of one's experience, in all of its variety. According to Kant, by its very essence, the I of pure self-awareness can only be a subject, whereas the psychological I can only be an object.

[7] "The faculty of inner intuition is inner sense. Inner sense is the consciousness of our representations themselves. It has the soul as its object" (p. 250. Kant's Lectures on Metaphysics).

[8] "The fact of being aware of myself is a thought that already entails a double I: the I as subject and the I as object. ... But what is intended by this is not a double personality, because only the I who think and intuit are the person, whereas the I of the object that is intuited by me is, like other objects outside me, the thing" (Kant, p. 249). Heidegger observes: "That the I of transcendental apperception is logical ... only means that the being of this I is problematic; according to Kant it is in general indeterminable, and in any case in principle not capable of determination by means of psychology. The *personalitas psychologica* presupposes the *personalitis trascendentalis*" (cit. G 24 p. 125).

pragmatism, James too was to articulate his own view of the self according to these two poles. However, in contrast to the Kantian approach, he did so without making any explicit distinction between psychological and philosophical inquiries in the study of subjectivity (Arciero and Bondolfi 2009; Pihlström 2009).

A new unitary vision of the subject was developed by Fichte, who lent new impetus to Kant's tripartition of the human person into *personalitas trascendentalis*, *psychologica*, and *moralis*. Regarding his own position as being not only consistent with Kant's philosophy but an essential part of it (at any rate in the positivist decade of his career), he brought about that crucial turn which enabled Kant's view to be assimilated by the natural sciences and hence by psychology. Fichte once again, and in more radical terms, posed the Kantian question "How are knowledge and experience possible?"

The crucial issue for Fichte concerns the search for unity between the subject of cognition and that of action: the search for that I-ness (*Ichheit*) on which they rest before consciousness separates them into knowing I and acting I. In this respect, Fichte brings the Kantian project to completion. However, his method consists in connecting the knowledge of truth to that of its sources (Hyppolite 1959). Before any distinction between subjective and objective, inner and outer, through a radical process of reduction which sets all forms of awareness aside, Fichte searches for a means to access a territory in which meaning emerges without the need to invoke subjectivity. This is a quest for the original experience. It constitutes the underlying theme of the Introduction to *The System of Ethics*—an introduction which is actually more of a succinct reformulation of Fichte's position. In what follows, we will therefore recall its salient points.

The Introduction to *The System of Ethics* says nothing about ethics or the work it is intended to introduce. Instead, it outlines the basic tenets of Fichte's theory of subjectivity.[9] Fichte sets out by establishing as the starting point of his inquiry the absolute, yet hidden, identity between the I as object and the I as subject. This unity become separation as soon as any real awareness emerges, be it only self-awareness. The most obvious consequence of this rift is the primacy assigned to the "I think" over the "I act." Within the framework of this approach, and in line with the Cartesian tradition, Fichte at the same time addresses the problem of methodology.

While modern thought has largely striven to understand how correspondences may exist between representations and things (mind-to-world direction of fit), it has found it is perfectly natural that acting upon the world may lead to expected results (world-to-mind direction of fit). It is starting from this second point that Fichte's practical philosophy takes shape, with a much broader approach compared to Kant's focus on the exclusively moral dimension of action.[10]

---

[9] Along with the first two books of *The System of Ethics* and the first book of *Foundations of Natural Right*, this Introduction contributes to a clearer understanding of *Foundations of the Entire Science of Knowledge* (1794) and *Foundations of Transcendental Philosophy (Wissenschaftslehre) nova methodo* (1798), which provide an overview of the issues which Fichte investigated in his Jena years.

[10] This issue is chiefly explored by Kant in *Foundations of the Metaphysics of Morals*.

Constantly engaged in the search for the unity between the subject-I and the object-I prior to the separation of the two through awareness, Fichte is interested in the experience of the I acting within an objective world and in all the conditions of possibility for the experience of effective agency in the world. Thus Section 4 opens by placing the very experience of effective agency at the center of everything: "I find myself to be acting efficaciously in the world of sense".[11] For example, in the case of the transformation of a log into a table, I may assume that the knowledge of the materials, of their properties, and of the processes of transformation which will lead me to craft the artifact are acquired from the outside; yet this assumption is not enough to account for the fact that I am the origin of the transformation of the log into a table.

In other words, discovering oneself as the foundation of the changes which the I can engender in the world is not an act of reflection on sensory information from the world; rather, it is something which emerges precisely by acting upon the world. Only the subject who acts with a given aim in mind can know the transformation he effects within things in the world, as he stands at its origin: knowing is doing.[12] This origin is immediately posited with myself as an active subject. This is the axis of Fichte's philosophy: the absolute position of the I as something active. "*Ich setze mich als tätig*" ("I posit myself as active"): Fichte repeats this like a mantra at the beginning of three successive paragraphs[13]!

As the foundation of all change, the subject has both knowledge of the world and of itself as a subject capable of acting upon the world. The sense of agency is therefore inextricably bound to the sense of ownership of one's own body as a source of changes within the world.[14] This awareness is devoid of content—it is nonrepresentational: it is simply an activity. The genesis of sense lies in practical action.

On the other hand, the sense of agency acquires its determinations in relation to concrete activity, to the actual performance of certain actions in a world that may prove an obstacle to the accomplishment of our aims. So whereas the I manifests—and knows—itself through given forms of activity, the world (the non-I) simultaneously reveals itself as resistance to the accomplishment of one's aims. "A pure activity—Fichte writes—cannot be intuited as such; it can be intuited only insofar as it encounters some resistance, and then it is called an 'action'. This is because an action has to be directed at some object, which our language correctly designates 'what stands in opposition', for this object is what resists activity" (p. 166 Fichte

[11] "Ich finde mich als wirkend in der Sinnenwelt."

[12] "The subject of consciousness and the principle of agency are one" (sec. 4). Herein lies the unity between theoretical and practical knowledge.

[13] Emphasizing how this vision of the I tallies with the Kantian perspective of the critique of immediacy (according to which no datum is "what" it is for a human knower by virtue of its giveness alone), Michael Baur (2003) writes: "Any suggestion that the self might be more adequately defined by reference to something other than its own activity implicitly involves the problematic claim that there is some datum or content, as merely given, that is necessarily determinative for the self's activity in knowing" (p. 106).

[14] One can hardly overlook the link between current neuroscientific studies on the sense of *ownership* and *agency* (Gallagher 2000) and the misleading characterization of schizophrenia as a disorder of ipseity (Nelson et al. 2014).

and Breazeale 1998). In this respect, for Fichte the self must be understood as the activity of being aware of oneself as radically free, undetermined by any given content, and its nature is to only exist in relation to an "other" (a nonself). The intertwining between the I and the world, the mutual determination between self-knowledge and the encounter with the "other," emerges from the founding character of practical action.[15]

To actualize itself, the I selects specific possibilities, starting from what Fichte acknowledges to be a "determinable" horizon with respect to the determination of the I: "The practical activity is not constrained in making its selection, for then it would cease to be freedom; it is constrained in this sense, however: i.e., in that it has to make its selection exclusively from what is determinable" (57–156).[16] The possibility of this mutual determination and the related concept of purpose are founded on a sensibility system which represents the unity connecting a multiplicity of sensations which are woven together as the background to a concrete situation.[17] "This system will eventually prove to be our body, understood as a system of affectability and spontaneity" (WNM p. 90). The body is thus understood as a whole that realizes itself through the interconnection of its elements but which in turn makes the existence of its parts possible by organizing them into a totality. In his 1855 lecture on Kant, Helmholtz saluted this view as being "in the most exact agreement with the conclusions which the physiology of the senses later drew from the facts of experience."

After having established the sense of agency as the a priori condition for experience, the Introduction takes its boldest step by introducing the famous notion of deduction of the body. Since it is unthinkable to act upon the material world if not by means of something which is itself matter, when I think of myself as an acting subject, I grasp myself as a material body and hence as a causal force within the world of bodies. "To represent myself as a causal force in that world—Martin (2003) summarizes—is to represent my agency as a part of it; this for Fichte, is the phenomenological structure of our embodiment: the body is the point of subject-object unity in agency" (p. 15).[18] Not only that, but the body also becomes the means by which to enter materially in contact with other subjectivities, similar to

[15] In order to account for this intertwining, Fichte develops concepts such as those of purpose, embodiment, will, and freedom, which were carried through the nineteenth century into modern cognitive sciences. These concepts inspired contemporary themes such as those of self-reference, self-organization, ownership, agency, etc.

[16] What we have here is the concept of task (*Zweckbegriff*), which was destined to play a crucial role in von Helmholtz's epistemology.

[17] This view has consequences for the concept of perception, which according to Fichte has no direct relation to the object: "I sense only myself and my own state, not the state of the object." It is this framework of sensations that constitutes the transcendental condition of the determination of "alterability" or "affectability."

[18] Angelica Nuzzo writes: "Fichte's account of the human body—first *Korper* and then *Leib*—is framed by three different but interconnected issues: (a) the issue of the self's or person's act of embodiment; (b) the issue of the person's relation to another individual as a similarly embodied rational being; (c) and the assumption (or the possibility for me of assuming) that the other's body is the same in kind as my own" (p. 73, 2006).

one's own. The encounter with the other's body (external reality) summons one to act and to engage in a reciprocal action of mutual recognition. In the Introduction, Fichte makes no mention of intersubjectivity or, better, interpersonality, a theme which was to provide inspiration for phenomenology from Husserl to Sartre (Nuzzo 2010; Lopez-Dominguez 2010). However, Fichte's argument may largely be traced back to sections 5 and 6 of part 1 of the *Foundations of Natural Right*, written a few years before his *System of Ethics*. Such, then, is the broad horizon related to Fichte's Jena years and, more generally, the "positivist" themes in his philosophy, which the Introduction discloses.

2.3 Human Agency and Natural Epistemology

It is only in recent times that, in line with this perspective, attention has been paid to the influence which Fichte's conception of man exerted on one of the key figures in nineteenth-century science: Hermann von Helmholtz. However, the extent of this influence is still far from clear, as are the consequences of Helmholtz's appropriation of Fichte's vision for physiology, neuroscience, psychology, epistemology, and contemporary culture more generally.

The main documents attesting to Fichte's influence come from Helmholtz's correspondence with his father Ferdinand, a staunch supporter of Fichtian philosophy and a close friend of Fichte's son Immanuel. However, the significance of the references to Fichte in Helmholtz's scientific works, such as *The Facts of Perception*, is highly ambiguous: according to De Kock, this may be the reason why "the Fichtean undercurrent in Helmholtz's work has remained somewhat neglected" (p. 59, 2011).

Be that as it may, leaving aside this ambiguity, which the German scientist apparently displayed in other areas of life too (see the biographical sketch in Cahan 1995), the emphasis on the constitutive nature of human agency is no doubt the most evident indicator of Fichte's influence on Helmholtz's work. This conception of man is the unthematized presupposition that underlies his epistemology and the question on which it rests, namely, how are we to account for the objective knowledge assumed by the natural sciences, if sense perceptions tell us nothing at all about the real nature of external objects (i.e., things in themselves), but rather reflect the effects produced by these objects on the various sensory apparatuses?

"The same aerial vibrations which the skin feels as whirring motions, the ear feels as sound" (p. 346), Helmholtz states, giving an example of how the perturbation produced by an external event may take different forms depending on the sensory apparatus affected by the impression.[19] In support of this approach to the problem, Helmholtz refers to the psychophysiological discoveries made by his master, Johannes Müller, according to whom sensation is not the conduction of a quality

[19]Turner rightly notes that this way of framing the problem opened the way to study the sense organs mathematically and mechanically by analogy with inorganic models (1977, p. 54).

or state of external bodies to consciousness, but the conduction of a quality or state of our nerves to consciousness, excited by an external cause. Helmholtz writes: "We are led to the very important conclusions that our sensations are, as regards their quality, only signs of external objects and in no sense images of any degree of resemblance … They are signs which we have learned to decipher, and a language given us with our organization by which external objects discourse to us—a language, however, like our mother tongue that we can only learn by practice and experience" (p. 222, 1995).

External reality is translated into nervous impulses, so to speak, but these provide no information about the nature of the external stimulus, since they are only a sign of it and they are not bound to have any type of similarity with what they are a sign of. In this respect, as our sensations are signs which depend on our biological makeup, and which point to something enduring or occurring in the real world, they can delineate the law of this enduring or occurring. For instance, if stone heats up in the sunlight, the perception of the heat of stone and that of sunlight will be always found together in our sensation. This joint existence of the perception of sunlight and the perception of stone will occur as a datum each time sunlight and stone are set in relation to one another. Sunlight and stone are connected through a perceptual judgment based on the succession of perceptions pertaining to them, which in each case is only subjectively ascertainable. This is intuition.

The impressions which an object produces on the observer's nervous system take different forms depending on what senses are elicited. In order for these impressions to acquire an objective value, however, an act of interpretation is required that is capable of universalizing the relationship, in such a way as to make it universally applicable to all subjects. Instead of a connection depending on the subject's perception, then, what we would have is a relation which binds sunlight and stone through the concepts of cause and effect: if the sun shines, stone heats up regardless of whether an observer is present or not. "Every conclusion by induction—Helmholtz states—is based upon trust that a previously observed law-like behavior will maintain itself in all cases which have yet to come under observation ... Law-likeness is, however, the condition of conceivability" (1995, p. 363). This condition of relational stability between things, understood as that a priori mode of representation which anticipates data through concepts (cause-effect), is the causal law.

Clearly, this implies that the (a priori) conceptual anticipation of empirical data is what ensures the natural-scientific experiencing of an object—i.e., its objectivity. An object is such precisely insofar as it is defined in a universally valid way on the basis of a causal principle. In other words, the causal perspective is the vantage point from which we relate to an object before any other determination—i.e., a priori. When we observe sunshine and stone from the perspective of objectivity, therefore, we consider the objects "stone" and "sunshine" starting from how our thought anticipates them: through an a priori synthetic judgment.

Without a priori synthetic judgments, there could be no science, as no object would exist in its objectivity. This was Kant's great discovery, which is pithily expressed in a key proposition of the *Critique of Pure Reason*: "the conditions of possibility of experience in general are also the conditions of possibility of the objects of experience" (A158, B197).

2.4 The Unexpected Debt and Experimentation

Every possible object of knowledge, just as knowledge itself, is therefore subject to a double determination: by intuition and by the concept. Whereas intuition determines the (sensory) content of knowledge, the concept is what lends intuition stability and universality (causal law).[20] Thus far the indebtedness to Kant is clear and has long been recognized by scholars. It is an epistemological debt, insofar as the a priori elements of intuition and the intellect are identified as the conditions for the determination of any object of knowledge. However, there is also an unexpected, ontological debt toward Kant, which has gone completely unnoticed.

Kant drew upon ancient ontology and its distinctive view of production to clarify the "question of the thing": he argued that since an entity can only really be perceived by its producer, finite beings such as us have no way of accessing things in themselves. In doing so, Kant provided a remarkable epistemological indication for the subsequent generations nourished by the Enlightenment—an indication that to this day remains dominant, unchallenged, and unchallengeable.

The essence of Kant's indication is enclosed in a sentence which Emil du Bois-Reymond wrote to his friend Helmholtz in 1852, in order to convey his excitement about the latter's latest invention, an apparatus for measuring electric current during the contraction of muscles in a frog: "It is a spectacle for gods, to see the muscles working like the cylinders of a steam engine" (ip. 123, quoted Brain et al. 1999). The new course which du Bois-Reymond enthusiastically endorses in this message to his friend is that of technological reproducibility. The productive approach

[20] It is important to bear in mind that, according to Kant, it is the pure intellect which provides the foundations—the principles—for the objectivity of objects and that the objects he has in mind are those of physics and mathematics. What this means is that things, as physical-mathematical objects, are understood starting from the fundamental principle according to which each body left to itself will move with uniform rectilinear motion. In this respect, what determines each thing is understood in advance (a priori), and, at the same time, the (invariably uniform) setting in which things manifest themselves is defined. Things thus become physico-mathematical objects, insofar as they are set and arranged within the framework of the uniform spatiotemporal connection between movements. The fact that things are all mobile in terms of space, time, and relations of motion is what makes numerical measuring possible. Kant's research is intended to establish the pure intellect as the foundation of the objectivity of phenomena. Principles are precisely what provide the basis for this. Kant divides principles into four groups: (1) axioms of intuition (quantity), (2) anticipations of perception (quality), (3) analogies of experience (relation), and (4) postulates of experience in general. Along much the same lines, when medicine later established itself as a natural science, it came to understand the functions of the living body in the light of the fundamental principle of the logic of the dead body and hence of the mechanization of the living body (see par. ch. 3).

becomes the key for understanding natural phenomena and the very heart of scientific research.

This notion is emphatically confirmed in a course of lectures which Helmholtz delivered after the completion of his volumes on *Physiological Optics* and which were brought together as a paper, "The Recent Progress of the Theory of Vision." In the first section of this paper, significantly entitled "The eye as an optical instrument," Helmholtz infers certain imperfections of the human eye by comparison with the perfection of a series of optical instruments. He writes:

> Now it is not too much to say that if an optician wanted to sell me an instrument which had all these defects, I should think myself quite justified in blaming his carelessness in the strongest terms, and giving him back his instrument. Of course, I shall not do this with my eyes, and shall be only too glad to keep them as long as I can—defects and all. Still, the fact that, however bad they may be, I can get not others, does not at all diminish their defects, so long as I maintain the narrow but indisputable position of a critic on purely optical grounds. (p. 142)

In order to understand natural phenomena, the scientist reproduces them with an even higher degree of perfection. This is the spectacle fit for gods! The key point is that this ontology stands at the origins of the epistemic connections between technology and the body—the dead body which Bichat placed (as a model) at the very center of physiology—that were destined to characterize, on the one hand, the models and metaphors in brain research (Brock 2011), down to the contemporary cognitive sciences, and, on the other hand, the development of a technology mirroring neural functions by means of mechanical or electronic devices.

The remarkable context that favored the development of this new path emerged, partly as a consequence of considerable socioeconomic pressure (Lenoir 1988), through the establishment of the Berlin Physical Society in May 1845. Certainly, the most interesting fact from today's perspective is that, for the first time, an interdisciplinary effort was undertaken of the sort that was destined to characterize the cybernetic revolution—roughly 100 years later, within the same tradition and in another moment of major crisis. In addition to Helmholtz, the Society included a small group of doctors who had trained in physiology in Muller's laboratory. Twelve of the 154 members were engineers: 6 of them bore the title of "Mechanikus," while the other 6 were lieutenants—probably in the artillery and engineering corps. Among these was Werner Siemens, the future founder of the Siemens corporation. The Society further included Privatdozenten seeking to reform the university and to redefine its disciplinary boundaries, the Emeritus Professor in Physics Gustav Magnus, and members of various scientific, technological, and industrial institutions: the School of Industry, the United Artillery and Engineering School, the War College, and the Academy of Architecture (Brain et al. 1999).

The most distinctive aspect of the new epistemological position that found its sharpest representative in Helmholtz was the centrality assigned to experiments. Experimental knowledge had also been cultivated in Antiquity and the Middle Ages, as the basis of craftsmanship, but had then been redefined by modern science. From Galileo onward, experimentation had come to be understood as a verification conducted within the framework of a fundamental model of nature that predetermined the essential conditions according to which nature could answer to the questions posed to it.

As noted by Heidelberger (1993), however, for Helmholtz "the function of testing hypotheses is only secondary. For him experiment is far more an art of invention (*ars inveniendi*) than an art of demonstration (*ars demostrandi*)" (p. 483). According to Helmholtz, the function of an experiment is to find the causal conditions for an event; and to do so, it is necessary to intervene in the world by changing its circumstances. Therefore, in order to determine the causes of phenomena, science intervenes in the course of natural events, by isolating objects and exposing them to selected influencing factors: "in the experiment the causal chain runs throughout our self-consciousness. We know one member of these causes—our will's impulse—from inner intuition, and know the motive by which it has occurred. The chain of physical causes which transpires in the course of the experiment has its initial effect from it, as from one initial member known to us and to one point in time known to us" (p. 358 *The Facts in Perception*). An experiment, in other words, enables us to grasp the initial conditions of a causal chain of events that are produced starting from the experimenter's voluntary, intervening activity. Only bringing our bodies "in various relations to the objects" can we be "sure as to our judgments of the causes of our sensations."

In the light of this perspective, it is easier to understand the strange combination of elements which Helmholtz brings into play in his research on sensory physiology: on the one hand, introspection, employed as a mode of access to the realm of pure sensations; on the other, the remarkable use of a variety of new media technologies, electric, photographic, and telegraphic inscription devices. In all likelihood, Helmholtz developed this approach to introspection—which acquires a systematic character in his *Optics*—through an engagement with the teachings of his master Johannes Muller.[21] What lies behind this unique combination is actually a specific mode of operating: introspection enables the distinction of the phenomenon to be studied. Indeed, in volume 3 of his *Optics*, Helmholtz explicitly states: "the first thing we have to learn is to pay heed to our individual sensations" (p. 7, vol. 3).[22] Helmholtz displayed his mastery not just through his ability to discern a set of fundamental or primary tones in physiological acoustics but also through his studies on the perception of color, which confirmed Young's hypotheses; his investigations into afterimages, simultaneous contrasts, irradiation, and retinal rivalry; and his observations on the perception of space relations.

[21] In a study of visual hallucinations, *Ueber die phantastischen Gesichtserscheinungen* ("On Fantasy Images"), Muller had shown that the visual system is an active rather than passive recorder of external events. In a series of rigorous self-experiments, having noticed that just before falling asleep he could sometimes see imaginary people and scenes, he tried to manipulate these figures (Otis 2007).

[22] These opening pages of volume 3 show a striking affinity with the notion of phenomenological reduction that is realized by putting natural aptitudes aside—what enables the constitution of objects according to Husserl's phenomenology.

On the other hand, in order to explain each distinct phenomenon, a mechanism was put forward, "materialized" by a device capable not just of reproducing the phenomenon itself through its functioning but also of enabling the manipulation, rearrangement, and comparing of the data thus obtained. As Lenoir insightfully emphasizes: "the new technologies were a resource for representing the scientific object, and ... in their material form they were not just 'representatives' of an object described by theory; rather they created the space within which the scientific objects, 'eye and ear', existed in a material form" (p. 205, 1994). From this perspective it is possible to understand why Helmholtz conceived of the nervous system as a telegraph and of its appendages—sensory organs—as a media apparatus: the eye as a photometer and the ear as a tuning-fork interrupter with attached resonators (Lenoir 1994).

2.5 The Physiological Basis of Technological Development

"Technological" reconstruction is based on the anatomical-physiological structure from which the behaviors observed originate. Thus, in the case of physiological acoustics, for example, the study of fundamental tones is based on Corti's rods, while in the case of color vision, it is based on the specific nerve endings in the rods and cones of the retina. But the key point is that the reconstruction is carried out by focusing on the conformity of the behavior reproduced to the one observed. From this perspective, the explanation of the behavior observed inevitably corresponds to the "materialized" mechanism capable of reproducing it. Hence the emphasis on experimentation and on the inadequacy of observation without active manipulation, not only in scientific research but for human knowledge more generally—as already Fichte himself had noted:

> This last point is, I believe, of great importance—Helmholtz says. The meaning we assign to our sensations depends upon experiment, and not upon mere observation of what takes place around us: we learn by experiment that the correspondence between two processes takes place at any moment that we choose, and under conditions which we can alter as we choose. (*The Theory of Vision*, p. 195)

The problem which Fichte had raised concerned precisely the origin of the divergence between the awareness of the external world and that of the self, which he had identified in the unconstrained activity of the knowing subject (self) that finds effective determination through the resistance (nonself) put up by the world.

Helmholtz drew upon this way of envisaging man. Indeed, the core of his physiological epistemology consists in the fact that self-certainty and the different awareness of self, compared to the nonself, derive precisely from the practical manipulation of a world that exists in its own right. It is only in such a way that while grasping the limits of his action (self)—continuously discovering himself in experience—the subject distinguishes the spatially determined external world (nonself). The self as pure activity finds its own determinations through the nonself.

Helmholtz makes Fichte's notion of the founding character of practical action his own. He draws upon the concept of purpose (*Zweckbegriff*), understood as the accomplishment of an action for the sake of a goal—starting from a "determinable" world—and extends it to the scientific domain by making experimentation the centerpiece of the process of formation of scientific knowledge. This lies at the origin of the peculiar juxtaposition between gnoseology and epistemology which, as we will see, was destined to characterize the cognitivist and constructivist psychology of the twentieth century and which is best exemplified by Kelly's notion of "man-the-scientist".[23]

Like the scientific experience of the external world, human experience springs from active intervention in this world: "Each of our voluntary motions by which we modify the manner of appearance of objects, is to be considered as an experiment by which we test whether we have correctly conceived the lawful behavior of the phenomenon in question, that is, its presumed existence in a definite spatial order" (p. 358, *The Facts of Perception*).

In other words, the voluntary motions of our body constantly alter our spatial relation with objects and, at the same time, change the impressions which these objects make on us. Helmholtz writes: "From the first step on, where we perceive the abiding objects distributed before us in space, this perception is the recognition of a law-like connection between our movements and the sensations appearing therefrom" (p. 360, *The Facts of Perception*). When we perceive an object, the sensations which emerge tell us nothing about the nature of the object itself, as what is directly perceived is a sensory-motor coordination which, once stabilized, automatically occurs each time we encounter the object in question. Helmholtz called this process "unconscious inferences".[24] On the other hand, this sensory-motor coordination may newly be generated by reiterating those motions required to produce the form of the objects, in such a way as to confirm the notion of the shape we have formed into reality: depending on its degree of correspondence with the original, we will become convinced of the accuracy of our conception.

Moreover, active experience, which is generated by a subject with a given aim in view,[25] also lies at the basis of the differential awareness of sensations from the world and sensations resulting from our action upon the world—and which we can produce or

---

[23] "Relatively few but well performed experiments are enough to allow me to see the original causal conditions of an event with greater certainty than a millionfold observations by which I could not arbitrarily vary the conditions. . . . We learn how to make reliable judgments of the causes of our sense perceptions only when we place, through our own will, our sense organs into different perspectives to the objects. Such experimenting happens from early youth onwards" (*Optics*, p. 451, quoted in Cahan 1993 p. 483).

[24] "The psychic activities that lead us to infer that there in front of us at a certain place there is a certain object of a certain character are generally not conscious activities, but unconscious ones ... it may be permissible to speak of the psychic acts of ordinary perception as unconscious conclusions, thereby making a distinction of some sort between them and the common so-called conscious conclusions" (p. 4, Vol. 3).

[25] "The chief reason, however, why the power of any experiment to convince is so much greater than that of observing a process going on without our assistance is that with the experiment the chain of causes runs through our own self-consciousness. We know one member of these causes—our will's impulse—from inner intuition and know the motive by which it has occurred" (p. 358).

revoke.[26] According to Helmholtz, this prereflective awareness of limits and possibilities emerges from the physiological mechanism associated with muscular feeling, understood as that condition which enables us to perceive changes of position of the parts of the body through muscular action and hence to grasp what the I can or cannot change. Helmholtz therefore anchors the body, which Fichte had envisaged as a system of affectability and spontaneity, in muscular feeling. According to Helmholtz this term includes:

1. The intensity of the effort of will, whereby we endeavor to bring the muscles into action
2. The tension of the muscles, that is, the force by which they try to act
3. The result of the effort, which, regardless of its being perceived by other organs of sense, such as sight and touch, makes itself felt in the muscle by a contraction which actually takes place and in which it may be possible to perceive after a fashion the change of tension of the skin over the parts affected (p. 243, Vol. 3 *Treatise on Physiological Optics*)

Helmholtz, who explores this theme within the context of a study of vision, notes that muscular feeling is not just the foundation of perceptual awareness but also its perceptual outcome. In the same section, he continues as follows: "Thus in the case of muscles that are much fatigued, I may be able to perceive that I have to make the utmost exertion of the will to produce tension in the muscles, but that their tension is no longer sufficient to obtain the result. On the other hand, in the case of powerful muscles, with a moderate effort of will, I can produce a distinctly perceptible tension in them, yet, owing to some external opposition, without obtaining the desired result." What emerges here is the key Fichtian idea—expressed in the *Foundations of Natural Right*—that the body, understood as a sensory-motor apparatus, constitutes the basis for grasping the distinction between the sensation of external objects and inner sensations in actual experience.

This way of conceiving corporeality, which is the cornerstone of von Helmholtz's view, was destined to exert considerable influence on Husserl's phenomenology—enabling us to interpret his cycle of lectures "*Ding und Raum*" from a new perspective (Feist and Sweet 2003; Sheets-Johnstone 2011)—and even more so on the attempt to develop a synthesis between the cognitive sciences and phenomenology which Varela has termed neurophenomenology.[27]

[26] If the experience in question does not lead to the desired results, or if our sensations change independently of our will, then we will posit the existence of external objects as the cause of the observed changes. "Those alterations which we can produce and revoke by conscious impulses of the will are distinct from those which are not consequences of such impulses and cannot be eliminated by them."

[27] It is interesting to note that although these authors, with different emphases, have all grasped the relation between Helmholtz and Husserl, they have completely overlooked—as indeed Husserl himself did in his 1903 seminars on Fichte (Hart 1995, *Husserl Studies*)—the importance of the sense of agency for Fichte and hence for Helmholtz's thought. For a critical overview of the relation between Husserl's philosophy as expressed in *Ding und Raum* and the naturalistic approach, see Costa's introduction to the Italian translation of this text.

Between the folds of the awareness of agency in perception—which was first grasped by Fichte, then served as a basis for Helmholtz's physiological psychology, and ultimately reached neurophenomenology—a completely new conception of the a priori emerged in the second half of the ninetenth century. As we shall see, it was to provide the starting point for Rudolf Magnus' research program.

2.6 Experimental Epistemology

In 1871 Helmholtz left the chair of Physiology at Heidelberg for that of Physics at Berlin University. His place was filled by Wilhelm Kühne, a student of du Bois-Reymond, who in turn was a friend of Helmholtz's and, like him, a pupil of Johannes Müller. In 1898 Rudolf Magnus prepared his dissertation under the aegis of Kühne: the research program he set himself makes him, in a way, the last great heir to Helmholtz. A glance at the lectures on Goethe, which Magnus delivered when he had just passed the age of 30 (1906), clearly reveals how deeply acquainted he was with the work of Helmholtz—who, like Magnus himself, had harbored a keen interest in Goethe throughout his life.

The central theme which Magnus explores in his research—in keeping with the neo-Kantian tradition—concerns the identification of those conditions which enable us to establish the body, grasped as a unity, as the precondition for knowledge. Helmoltz had stressed that the nature of our impressions is determined a priori by our sensory apparatuses anchored in the body's muscular feeling. Magnus broadens and further develops this perspective by addressing the problem of how to account for the conditions of possibility of action itself in physiological terms. He does so within a broader conception of the a priori, which leads him to conclude that the founding character of practical action central to the thought of Fichte and, later, Helmholtz, is inextricably bound to a priori determined postures that set and stabilize the body orientation against external perturbations. From this perspective, we can understand why Magnus' program of empirical research is directed toward the study of those reflexes which are present a priori and which, through coordination by subcortical centers, form "posture," understood as an active process of global organization of the musculature of the body with respect to gravity (Magnus 1926).[28] Magnus can therefore argue that "every movement starts from and ends in some posture" (1925). The action of reflexes thus helps to determine our sensory perceptions a priori, by de facto configuring the physiological a priori in terms of an ongoing sensory-motor unity.

[28] Magnus identifies the following starting conditions for the study of animal posture: (1) reflex standing, (2) normal distribution of tone, (3) attitude, intended as the harmonizing of the position of the different parts of the body with each other, and (4) righting function, that is, if an animal is brought out of the normal resting posture, then a series of reflexes are evoked to reach the normal position again (Magnus 1925).

In the light of this new idea of a priori and through sustained experimentation, Magnus draws upon the work on the "integrative action of the nervous system" conducted by Sherrington, whom he quotes with the utmost admiration in every paper. "And integration—Magnus writes—is especially necessary in the case of posture, because nervous excitations arising from very different sense organs are flowing towards the postural centers in the brain-stem, and must be combined so that a harmonizing effect will result" (p. 340, 1925).[29]

The strange fate of this man, who died only a few months before he was in all likelihood about to receive the Nobel Prize, is curiously reflected by the transmission of his theoretical testament: a lecture significantly entitled "The Physiological A Priori" (1930)—one of a series of lectures he was meant to deliver at Stanford University and which were published posthumously. What makes this an interesting text is not only the overview it offers of Magnus' understanding of man but also the fact that an imaginary dialogue may be read between its lines, a dialogue which had evidently shaped Magnus' research and has Helmholtz as its interlocutor.

This emerges in a remarkably clear way when we compare this lecture with "The Facts of Perceptions," the paper which Helmholtz had given 50 years before. Magnus draws upon its arguments, further explores its themes, develops its suggestions, and clarifies the various points made by Helmholtz, setting his own contribution against such background. Magnus writes: "We possess numerous mechanisms acting unconsciously and partly sub-cortically which prepare the work beforehand for our psyche, and the results of which are a priori present before sensory perception and psychological awareness come into play" (p. 103). The stance of our body, which is determined not just by our sensory structure but also by our orientation with respect to gravity, is the a priori foundation for how we intuit and think of the world, the reality of which would otherwise be ungraspable. Thus, if a person is color-blind (red-green "dichromatic"), their sense impressions of the outside world—say, a green apple tree with red fruits—will be incomprehensible to a normal-sighted ("trichromatic") person with the normal ability to distinguish colors. "The nature of our sensory impressions—Magnus states—is thus determined a priori, i.e. before any experience, by this physiological apparatus of our senses, sensory nerves and sensory nerve centers." These are the fixed mechanisms of our sensory apparatus which determine the nature of experiences. "But beside these, other 'active' processes (reflexes), acting through the central nervous system, also influence our sensory observations and help to determine them a priori" (p. 99). What Magnus is referring to is precisely a series of reflexes—constituting "a real physiological a priori "(i.e. the righting reflexes)—which enable the "normal" positioning and orientation in space typical of each species. This is the Kantian-Fichtian heart of Magnus' work. But if this is the case, it is reasonable to posit an interplay between the processes of knowledge-acquisition and their results: "Since all study, analysis, and understanding of the events in the outer world are conducted through

[29] Nervous impulses, which can influence posture, arise (1) from the labyrinths, (2) from the proprioceptive sense organs, (3) from exteroceptive nerve endings of the body surface, and (4) from teleceptors, reacting to distance stimuli, such as the eye, the ear, and the nose (Magnus 1925).

the senses, a scientific worker surely ought to know what are the fundamental mechanisms of his body and of his nervous system which determine the results of his work."[30] With this final statement, questioning the dependence of scientific observations upon the observer's biology, a lecture that sheds light on two centuries of natural science comes to a close. As we shall see, these pages resonate with the extraordinary view which Magnus had borrowed from von Uexkull (1903).

Upon Magnus' death, the chair of Physiology fell vacant. Dusser de Barenne, who had been a close associate of his, applied for the position, but for reasons that have nothing to do with science—as is all too often the case in academia—the chair was assigned to someone else. Dusser de Barenne quit Holland for Yale. Here, in order to delineate the entire central system subserving bodily sensation, he continued to carry on the same experimental procedure he had developed in his studies with Magnus: local strychninization. This technique consisted in the local application of strychnine to various areas of the cerebral cortex, combined with clinical observation of the animal's exaggerated responses to stimulation.

Once at Yale, Dusser de Barenne demonstrated in a series of experiments that this procedure, being a more precise tool, was preferable to the two most common methods at the time: the method of extirpation—employed by Magnus, for example—which consisted in removing areas of the brain in order to study residual functions and that of electrical stimulation, based on the electrical stimulation of different brain loci and peripheral nerves as a means to map sensory and motor functions. Both techniques presented some significant limits.

The crucial point that emerges from Dusser de Barenne's work, in line with the neo-Kantian physiological tradition to which he belonged, is once again an innovative search for the physiological foundations of the a priori. Dusser de Barenne believed these could be found in cortical integration: he constantly put his practical skills to the service of this principle, which would appear to have been the lodestar of his experimental intelligence. This explains the considerable attention he paid to the development of a wide range of techniques: from local strychninization to laminar thermocoagulation,[31] from the later coupling of the strychnine method with the recording of the electrical activity of the cortex (Dusser de Barenne and McCulloch 1938, 1939) down to chemical neuronography as a means to study communication between different regions of the brain. On the other hand, from this perspective we can also understand his constant attacks on the classic localization theory, "with its assumption of a sharp, point-to-point, geometric projection of the body on the cortex, with its centers for separate psychic functions" (Dusser de Barenne and McCulloch, 1939). Dusser de Barenne endeavored to demonstrate that the a priori was rather to be grasped in the organization and integration of the sensory-motor

[30] Foreshadowed here is the key theme of the second cybernetics.

[31] "The principle of the method is very simple. Local application of moderate heat to the exposed cortex for short periods of time results in destruction of the nervous tissue. By using different temperatures and different periods of application it is possible to 'dose' the depth of the destruction, and thus to produce laminar lesions" (p. 517) (1934).

functions in the cerebral cortex, which was seen to work as a whole through its system of neural connections.

This integration was studied by employing both electrical methods (electrical recording) and chemical ones (local strychninization), which represent "a powerful tool for delimiting the origin and ending of neurons in the central nervous system" (Abraham 2003). At the same time, these methods paved the way for the momentous transformation which brought about the development of the cognitive sciences and technologies, particularly with McCulloch.

This mode of research sheds light on one important aspect of the transformation highlighted by Sheets-Johnstone (p. 181), when she observes that the de-animation of perception and the rise of cognitivist science are two intimately related phenomena. The reason for this is clearly connected to experimental praxis, as McCulloch explicitly confirms: "When one is working on the physics and chemistry of the anesthetized brain, as I was, one is doing biophysics and biochemistry necessary for neurophysiology, but falling short of physiology because the nervous system is then deprived of its functions; but even if it were working properly it would still be only physics and chemistry and not physiology unless one were studying the function also" (McCulloch 1974). Indeed, the neurophysiologist who anesthetizes the brain to study its mechanisms, connections, and organization isolates the brain, as a coherent whole, from any context and therefore ignores its relations with the outside. The remarkable efforts made by cybernetics researchers largely go in this direction.[32]

2.7 The Cognitive Turn

> "It is Kant's conception of the synthetic a priori and his conviction that this depended upon the physical, the anatomy and physiology of the body, that, coming through Helmholtz became the Leitmotiv of Rudolf Magnus, perhaps most explicitly in his Linnaean Lecture on 'The Physiological A Priori.' Magnus' student, J.G. Dusser de Barenne, was my teacher" (p. 1, 1974).

McCulloch's research is rooted in this tradition and it is only in the light of it that it is possible to appreciate the scope of his project. McCulloch took the bold step of investigating the physiological foundations of pure intellect, the source of the principles of all truth according to Kant: a priori synthetic judgments. When faced with such a task, Helmholtz had stopped short a century before. "The causal law—he had written—is really an a priori given, a transcendental law. It is not possible to prove it by experience because, as we have seen, not even the first steps of experience are possible without the application of inductive conclusions, i.e., without the causal law ... Only one piece of advice is valid here: trust and act!" (p. 363).

[32] Clearly, it is also by ignoring these aspects that the various currents of psychotherapy have been able to casually adopt modelizations based on this approach, without realizing that they are in fact impossible to apply, precisely on account of the defining features of the kind of practice in which they are rooted.

McCulloch undertook this formidable challenge by adopting the same research strategy as Helmholtz, supported by the same ontology—and revealing the same unexpected indebtedness to Kant. The ontology in question is the unthematized one that stands in continuity with ancient ontology. It grasps the nature of things starting from the paradigmatic view of production, whereby things, which are created according to their creator's plan, are only fully accessible to him, while everyone else can only have a partial perception of them.[33]

Just to what extent this hidden ontology steered McCulloch's scientific praxis is revealed by a brief description provided by Lettvin, an associate as well as lifelong friend of his: "He devoted himself to discover how the brain works in the same way that an inventor knows exactly every single cog in the machine he has created. The key to such knowledge lies, not in observation, but in the construction of models which are then compared to the data available ... And McCulloch preferred to run the risk of failure in his attempt to create a brain, rather than succeed in furnishing an improved description of existing brains" (Dupuy 1985, p. 78).

This hidden ontology lay at the basis of a new way of conceiving experimentation, understood by Helmholtz as the actual reconstruction of the mechanism which underlies and generates the phenomenon to be explained; and what McCulloch sought to explain was the production of a priori synthetic judgments!

The focus on the brain in terms of connections and organization constituted the perspective from which de Barenne had developed his research, as well as the outcome of a whole tradition. By bringing forth this view, McCulloch—like Helmholtz—resorted not just to new technologies, but to all practical-conceptual resources available to reproduce the phenomena he had observed and identified in the laboratory.

In retrospect, McCulloch acknowledged that Leibniz's logical machine, just like Turing's idea of a "logical engine" for the process of mathematical computation, provided a significant source of inspiration (Abraham 2003). However, the most important role was arguably played by the mathematical biophysics of the Rashevsky school and, within this context, by McCulloch's encounter with a mathematical genius, the then 17-year-old Walter Pitts (Smalheiser 2000). Pitts allowed McCulloch to axiomatize an intuition which he had had in the 1930s by providing him with the logico-mathematical means to analyze excitatory and inhibitory activity in a simple neural circuit. Thus, through a binary logic whose principles were embodied in the brain and its neurons—logically defined neurons characterized by on-off (0–1) functioning—it was in theory possible to understand the properties of neural nets. "The all-or-none law of nervous activity—McCulloch and Pitts write in their famous 1943 paper—is sufficient to insure that the activity of any neuron may be represented as a proposition. Physiological relations existing among nervous activities correspond, of course, to relations among propositions" (p. 21).

[33] Wenn ich aber Körper sehe so sehe Ich keine Substanzen sondern Erscheinungen. Ich kann auch gar nicht di Substanzen vernehmen; denn kein Wesen, als der Schöpfer allein, kann die Substanzen eines andern Dinges vernehmen (p. 97 Kant 1972). "When I see bodies, I do not see any substances, but rather phenomena. I cannot perceive substances at all. No being except the creator can perceive the substance of another thing."

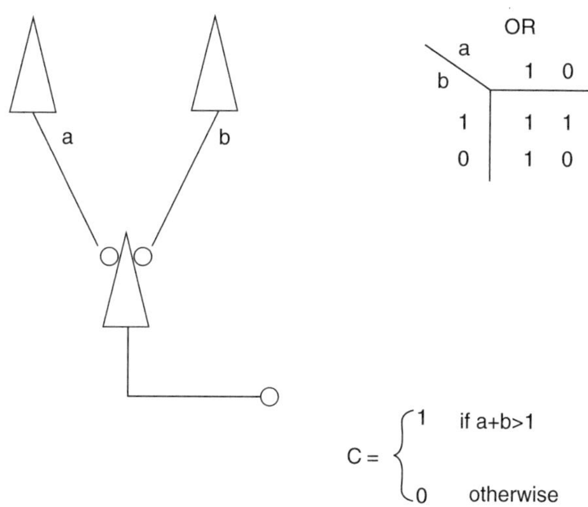

Fig. 2.1 McCulloch, Warren S. Introduction by Seymour A. Papert., Embodiments of the Mind, figure, © 1965 Massachusetts institute of Thechnology, by permission of the MIT Press

Hence, starting (a) from the states of activity and inactivity of every single neuron which correspond, respectively, to the logical values true and false (0 and 1), (b) from their connections—so that if two neurons tend to be active together, connecting up is facilitated, whereas the opposite state inhibits any connection (the rule governing the change is a [Boolean] function of two arguments [such as "and" "or"]) , we obtain (c) that the brain is comparable to a machine operating by deduction (Varela et al. 1991) (Fig. 2.1).

The spirit is embodied in the mechanism. Starting from basic operational rules, a machine would be capable of ordering concrete experience, that is, it would be capable of "thinking." In such a way, following the Kantian tradition inaugurated by Helmholtz, it would be possible to reconstruct the physiological foundation of a priori synthetic judgments by identifying it with a network of logically defined neurons embodied in the brain.[34] Unlike Helmholtz, who conceived experimentation as the reproduction of a chain of physical causes that could account for the phenomenon observed on the basis of the voluntary, intervening activity of the observer, McCulloch freed experiments from the concrete action of the researcher, so to speak. Like Maxwell in physics, McCulloch dematerialized neurophysiology by assigning a new role to mathematical modeling and ultimately mathematizing neural phenomena.[35] In the introduction to another famous paper on the perception of

[34]Among the many interpreters of McCulloch's work, only Arbib (2000) would appear to have acknowledged the important role which the Kantian perspective played in the development of his research.

[35]From this perspective it is possible to understand McCulloch's great interest in the work of Craik, who had been the first to develop a psychology in which mental models played a leading role. "Craik thought of human memory as a model of the world with us in it, which we update every tenth of a second for position, every two tenths for velocity, and every three tenths for acceleration as long as we are awake" (McCulloch p. 10, 1974). McCulloch was also responsible for the posthumous publication of Craik's works. This line of research spawned the psychology of mental models (Johnson-Laird 2005).

auditory and visual forms, Pitts and McCulloch write: "To demonstrate existential consequences of known characters of neurons, any theoretically conceivable net embodying the possibility will serve. It is equally legitimate to have every net accompanied by anatomical directions as to where to record the action of its supposed components ... But it is wise to construct even these nets so that their principal function is little perturbed by small perturbations in excitation, threshold, or detail of connection within the same neighborhood" (1947, p. 46).

Axiomatization—understood as a fundamental framework and support for any theoretically conceivable net—thus opens up a new space that is intrinsically hybrid, since it enables the reconstruction of a pre-existing reality which it helps confirm while at the same time resting on those very elements (actual neurons) from which it cuts itself off (Dahan and Pestre 2004). Moreover, if the activity of any neuron may be represented as a proposition, the most significant result is that this logico-mathematical way of conceiving facts pertaining to the brain provides the fundamental criterion and framework for the quantitative understanding of mental state as well. The mind which emerges through the regularity of neural interactions is a measurable, well-localized object. Hence, to the question, "Why is the mind in the head?", McCulloch can only answer: "Because there, and only there, are hosts of possible connections to be formed as time and circumstance demand" (p. 86, 1951). So if a correspondence is drawn between the operations of human reason and those of binary-logic neurons, the mind that carries out the axiomatization and the one that is its object must represent two related poles, although the actual relationship between them remains problematic.

The different orientations that shape the cognitive sciences reflect different ways of envisaging this dialectic. When cognition is viewed from the perspective of a computer, which is to say, that of the modeling of thought—as has been the case with cognitivism (linguistics, AI, cognitive psychology) since the late 1950s—it is bound to be defined as a form of information processing focused on symbolic representations: as the (explicit or implicit) rule-based processing of more or less appropriate representations of the real world. Mental representations are taken to be occurrences of a formal system, and the mind's activity is what gives these representations their attitudinal color—beliefs, desires, intentions, etc. We will see what influence this perspective was to exert on cognitive psychotherapy.

By contrast, if we view cognition once again in its mechanized version, as in the previous case, but attribute its organization to the cooperation between logical neurons whose rules of connection vary on the basis of experience, as opposed to a framework of explicitly designed logical circuits, cognition emerges as an overall property of the system. This trajectory is what led to the idea of a connectionist modeling framework and to self-organization theories.

Finally, a third position, which brings together the two poles, is to be found in the new course inaugurated by Lettvin and Maturana's famous 1959 paper "What the Frog's Eye Tells the Frog's Brain." The authors showed not just that the visual process is mediated by a topographically organized activity parallel distributed over layers but, most significantly, that the retina processes features of the visual input in a way related to the frog's ethology, instead of transmitting some more or less accurate copy of the distribution of light. In other words, the frog they studied generated and specified its own relation to the environment on the basis of a mode of organization of its visual system, whereby it visually identified what had species-specific significance in the context (i.e., insects or worms, any object of the same size, or any possible enemies to avoid).Therefore, the different perceptual experiences which the frog had of its own world did not so much represent external reality as reflect the animal's perceptual-motor structure. This is the starting point of second cybernetics, whose mantra is that reality is constructed rather than represented.

Hence the circularity predicted by Magnus—"those strange loops the cyberneticians were so crazy about" (Dupuy 2009)—which later enabled Maturana and Varela (1973) to grasp the mutual specification between a particular way of being and the appearance of the world, i.e., between the knowing subject and the objectivity adopted, that was only intuited in the 1959 paper. Therefore, if every organism negotiates its own survival by reacting to significant environmental stimuli through the generation of perceptual-motor structures determined by its own biological organization, the physiological a priori cannot be localized in the head, but must concern the relationship which all living systems have with the environment which they inhabit and bring forth.

Fifty years after McCulloch posed his famous question, Varela was to state that the mind "is in this non-place of the co-determination of inner and outer" (Varela and Shear 1999a, b).

Bibliography

Abraham TH (2003) Integrating mind and brain: Warren S. McCulloch, cerebral localization, and experimental epistemology. Endeavour 27(1):32–36

Arbib MA (2000) Warren McCulloch's search for the logic of the nervous system. Perspect Biol Med 43(2):193–216

Arciero G, Bondolfi G (2009) Selfhood, identity and personality styles. John Wiley & Sons, Chichester

Baur M (2003) Kant, Lonergan, and Fichte on the critique of immediacy and the epistemology of constraint in human knowing. Int Philos Q 43(1):91–112

Brain RM, Wise MN, Biagioli M (1999) The science studies reader. Psychology Press, Hove

Brock C (2011) Toys are us: models and metaphors in brain research. In: Choudhury S, Slaby J (eds) Critical neuroscience: a handbook of the social and cultural contexts of neuroscience. Wiley-Blackwell, Oxford, pp 113–134

Cahan D (ed) (1993) Hermann von Helmholtz and the foundations of nineteenth-century science. In: University of California Press, Berkeley, CA

Cahan D (1995) Science and culture: popular and philosophical essays. University of Chicago Press, Chicago

Courtine JF (1990) Suarez et le système de la métaphysique. Presses universitaires de France, Paris

Dahan A, Pestre D (2004) Transferring Formal and mathematical Tools from war management to Political, Technological, and social Intervention (1940-1960). In: Lucertini M, Gasca AM, Nicolo F (eds) Technological concepts and mathematical models in the evolution of modern engineering systems. Birkhäuser, Basel, pp 79–100

de Barenne JD (1934) The disturbances after laminar thermo-coagulation of the motor cerebral cortex. Brain 57(4):517–526

de Barenne JD, McCulloch WS (1938) The direct functional interrelation of sensory cortex and optic thalamus. J Neurophysiol 1:176–186
De Barenne JD, McCulloch WS (1939) Physiological delimitation of neurones in the central nervous system. Am J Physiol 127(4):620–628
De Kock L (2011) Some preliminary considerations on Helmholtz's Fichte: towards a naturalized epistemology of constraint? Revista de Estud(i)os sobre Fichte, 2 | 2011
Dupuy JP (1985) L'essor de la première cybernétique (1943-1953). Cahiers du CREA 7:7–140
Dupuy JP (2009) On the origins of cognitive science: the mechanization of the mind. MIT Press, Cambridge, MA
Feist R, Sweet W (2003) Husserl and Stein. Council for Research in Values and Philosophy, Washington
Fichte JG, Breazeale D (1998) Foundations of transcendental philosophy (Wissenschaftslehre) nova methodo (1796/99). Cornell University Press, Ithaca
Gallagher S (2000) Philosophical conceptions of the self: implications for cognitive science. Trends Cogn Sci 4:14–21
Gurevič AJ (1996) La nascita dell'individuo nell'Europa medievale. Laterza, Roma-Bari
Hart JG (1995) Husserl and Fichte: with special regard to Husserl's lectures on "Fichte's ideal of humanity". Husserl Stud 12(2):135–163
Heidegger GA 28 1997
Heidelberger M (1993) Force, law, and experiment: the evolution of Helmholtz's philosophy of science. In: Cahan D (ed) Hermann von Helmholtz and the foundations of nineteenth-century science. University of California Press, Berkeley
Hyppolite J (1959) L'idée fichtéenne de la doctrine de la science et le projet husserlien. Husserl et la Pensée Moderne–Husserl und das Denken der Neuzeit. Nijhoff, The Hague
Johnson-Laird PN (2005) Mental models and thought. In: Holyoak KJ, Morrison RG (eds) The Cambridge handbook of thinking and reasoning. Cambridge University Press, Cambridge, pp 185–208
Kant I (1972) Vorlesungen über Metaphysik und Rationaltheologie. (Metaphysik L1). Gerhard Lehmann
Kant E (1992) On the form and principles of the sensible and the intelligible world [Inaugural Dissertation]. In: Walford D, Meerbote R (eds) Theoretical philosophy, 1755–1770 (The Cambridge Edition of the Works of Immanuel Kant). Cambridge University Press, Cambridge
Lenoir T (1988) Social interests and the organic physics of 1847. In: Science in reflection. Springer, Netherlands, pp 169–191
Lenoir T (1994) Helmholtz and the materialities of communication. Osiris 9:184–207
Lopez-Dominguez V (2010) Political realism in idealism: Fichte versus Hegel and their different versions of the foundation of right. In: Breazeale D, Rockmore T (eds) Fichte, German idealism, and early romanticism, vol 24. Rodopi, Amsterdam
Magnus R (1906) Goethe als Naturforscher Vorlesungen. Barth, Leipzig
Magnus R (1925) Croonian lecture: animal posture. Proc R Soc Lond Ser B Biol Sci 98(690):339–353
Magnus R (1926) The physiology of posture: cameron lectures. Lancet 211:531–536
Martin WM (2003) Nothing more or less than logic: general logic, transcendental philosophy, and Kant's repudiation of Fichte's Wissenschaftslehre. Topoi 22(1):29–39
Maturana H, Varela F (1973) De Máquinas y Seres Vivos: una caracterización de la organización biológica. Editorial Universitaria, Santiago
McCulloch WS (1951) Why the mind is in die head. In: Jeffress LA (ed) Cerebral mechanisms in behavior. The Hixon symposium. John Wiley and Sons, New York, pp 42–111
McCulloch WS (1974) Recollections of the many sources of cybernetics. ASC Forum 6(2):5–16
Morris C (1972) The discovery of the individual, 1050-1200, vol 5. University of Toronto Press, Toronto
Nelson B, Parnas J, Sass LA (2014) Disturbance of minimal self (ipseity) in schizophrenia: clarification and current status. Schizophr Bull 40(3):479–482

Nuzzo A (2006) The role of the human body in Fichte's Grundlage des Naturrechts (1796–97). In: Breazeale D, Rockmore T (eds) Rights, bodies and recognition: new essays on Fichte's Foundations of natural right. Ashgate Publishing, Farnham

Nuzzo A (2010) Phenomenologies of intersubjectivity: Fichte between Hegel and Husserl. In: Waibel VL, Breazeale D, Rockmore T (eds) Fichte and the phenomenological tradition. De Gruyter, New York

Otis L (2007) Mullers Lab. Oxford University Press, Oxford

Patočka J (1996) Heretical essays in the philosophy of history. Open Court, Chicago

Pihlström S (2009) Pragmatism and naturalized transcendental subjectivity. Contemporary Pragmatism 6(1):1–13

Pitts W, McCulloch WS (1947) How we know universals the perception of auditory and visual forms. Bull Math Biophys 9(3):127–147

Sheets-Johnstone M (2011) The primacy of movement, vol 82. John Benjamins Publishing

Smalheiser NR (2000) Walter Pitts. Perspect Biol Med 43(2):217–226

Turner RS (1977) The Ohm-Seebeck dispute, Hermann von Helmholtz, and the origins of physiological acoustics. Br J Hist Sci 10(01):1–24

Uexkull JV (1903) Der biologische Bauplan des Sipunculus. Zeitschr f Biol 44:269–344

Varela F, Shear J (1999a) (eds) The view from within: first-person methodologies in the study of consciousness (Special Issue). J Consc Stud 6 (2–3)

Varela FJ, Shear J (1999b) First-person methodologies: what, why, how. J Consc Stud 6(2-3):1–14

Varela FJ, Thompson E, Rosch E (1991) The embodied mind: cognitive science and human experience. MIT press, Cambridge

"Nemo psychologus nisi physiologus" 3

3.1 The Transcendental Neurosciences

The identification of the processes of knowing with the processes of organization of life shifts the focus onto that which is common to all living beings. In other words, the need is felt to provide an operational, concrete description of the conditions for the emergence of those global properties characterizing every living unit. According to the two Chilean biologists, Maturana and Varela (1987), the key concept and general frame for all living systems is biological autonomy. The defining feature of the autonomous system is its closed, circular, self-referential organization, which through the mechanical generation of a stable operational coherence creates inner regularities and hence a distinction between interior and exterior. From the point of view of the system, the exterior is nonsensical, unless it engenders a change in one's inner condition—in which case it becomes a perturbation. An autonomous individuality thus preserves its identity by changing the dynamics of its operations in response to environmental modifications.

This path ultimately leads to neurophenomenology, whose research program revolves around a crucial problem: how can an objective account of biophysical processes explain the origin of subjective experience? On a different level, what is brought again into play is the dialectic between experience and its modelization introduced by the first cybernetics. Neurophenomenology lends this dialectic a new direction: "We need to advance a cognitive science where there is a true circulation between lived experience and the biological mechanisms in a seamless and mutually illuminating manner" (Varela 1996). It is precisely this path which we wish to follow, in order to investigate the way in which Kantism and Fichte's

No one can be a psychologist without being a physiologist.

G. Arciero et al., *The Foundations of Phenomenological Psychotherapy*,
https://doi.org/10.1007/978-3-319-78087-0_3

epistemology[1]—whose transformations we have traced, starting from its encounter with physiology through Helmoltz's work—have found their way into the neurosciences via neurophenomenology. It is necessary to further elucidate the two aspects thematized by neurophenomenology, aspects which continue to shape the contemporary debate on possible approaches to the study of experience.

On the one hand, by stressing the (operational) closure of each organism, conceived as an autonomous system, neurophenomenology focuses on the mechanical generation—mechanical, that is, in terms of its inner operations—of the world of saliences, which we bring forth throughout our lives. This world is inseparable from the constant sensory-motor coping between the system and its surroundings. "Perception—Varela writes—does not consist in the discovery of a predefined world; rather, it is based on the perceptual regulation of action in a world that is inseparable from our sensory-motor abilities" (p. 20, 1992).

On the other hand, an emphasis is being placed on first-person experience: on the conscious individual who experiences himself thorough his encounter with the world he inhabits. From the point of view of biological organization, this experience would appear to be the product of the operations performed by the system in order to maintain a stable identity as an organism. It is always combined and coupled with the domain of those significant environmental perturbations, which an embodied individual, situated within the world, distinguishes in terms of personal experience in his everyday life.

The two above perspectives require two methods of investigation which neurophenomenology seeks to combine: the first is a third-person approach that grasps the I from the point of view of the mechanisms and organizational dynamics of the system; the second method is a first-person approach that grasps the I from an absolute position.

As we shall soon see, in the late twentieth century, the question of how to reconcile these two perspectives fired a debate on several fronts and across different disciplines, within a context marked by the new issues raised by the impact of brain imaging technologies on the study of personal experience (Chalmers 1995; Arciero 2006). Within this new cultural climate, the problem emerged of how to bring together a methodology designed for grasping first-person events, which is to say, individual observation and experiences, and a third-person approach to natural phenomena, whose defining characteristics refer to the properties of world events.[2] The methodological suggestion put forward by neurophenomenology is an interesting one: it newly thematizes the kind of paradoxical combination of introspection and the objective gaze which was the hallmark of Helmholtz's research, connecting the two through the notion of reflective complementariness.

---

[1] "La traduction de *Wissenschaftslehere* par *Doctrine de la Science* est usuelle. Nous nous y confermerons donc; le terme d'epistemologie conviendrait mieux" (Hyppolite 1959, p. 36).

[2] "Our stance in regard to first-person methodologies is this: don't leave home without it, but do not forget to bring along third-person accounts as well" (Varela p. 2, 1999)—such is the neurophenomenologist's mode of operation.

The relation between experience, and the means to reproduce and explain it, first introduced by Helmholtz and later formalized by McCulloch, is redefined by Varela according to the notion of recursive dynamics , which is to say, a kind of reflectivity whose structure refers back to itself over the course of its implementation. Hence, instead of studying experience through a model, Varela controls the investigated experience through given practices—e.g., phenomenological or meditational practices—in order to then study brain responses in the lab. In turn, however, the dynamical neural signatures interpreted on the basis of the first-person account bring up new questions and new prospects of development in relation to such practices. Varela (Varela and Shear 1999b) can therefore claim to be both naturalizing phenomenology and "phenomenologizing" neuroscience: "My proposal implies that every good student of cognitive science who is also interested in issues at the level of mental experience, must inescapably attain a level of mastery in phenomenological examination in order to work seriously with first-person accounts" (pag. 2). What we have here is a new call to redefine the style and values of the research community, which is all too often prey to the idolatry of data. But what does this new position actually consist in?

From the natural scientist's perspective, this suggestion translates into various different stages:

(a) The distinction, within the framework of subjective experience, of a specific domain that makes a well-defined first-person methodology necessary (i.e., the phenomenological approach or the introspective one, meditation, etc.). Through phenomenological, meditational, or introspective practices (Varela and Shear 1999a, b), the subject learns to develop a disciplined methodology to distinguish the phenomena which emerge through the practice itself.
(b) An explicit account of this distinction amenable to intersubjective validation through the common ground established by sharing the practice.
(c) The creation of a mechanism capable of generating the phenomenon observed as the result of the internal dynamics of its operation, i.e., of validating the practice through the results.[3]

This explanatory reconstruction of the phenomenon, however, is in turn contingent upon the validation of the results on the part of the subject who had the actual experience. Hence, first-hand experience (controlled through practice) is de facto assigned value as a means of epistemological confirmation, thereby opening up a space of co-determination between experience and its modelization: the reciprocal constraints between phenomenological experience and brain dynamics.

According to Depraz, who has crucially contributed to shaping such project, "Varela assigns phenomenology an effective role in empirical analysis and,

[3] Varela stands in the tradition of Helmholtz—who, as we have seen, assigned a key role to experimentation—and shares his ontology. In an interview conducted shortly before his death, Varela—who was as seduced by Kantian ontology as Helmholtz and McCulloch—did not hesitate to suggest the possibility of bringing the inorganic to life in a laboratory.

conversely, assigns empirical analysis factual causative power with respect to the meaning of phenomenology. Beyond the mere irreducibility of first-person experience, which is to say of isomorphism alone, which brings into play an external correspondence and compatibility between the two kinds of analysis, it is a matter of bringing out their co-productivity, by searching for those generative passages revealing their mutually effective bonds" (p. 259, 2006). As regards practical applications, we will refer here to the studies conducted according to this perspective (Lachaux et al. 1999; Lutz et al. 2002), foregoing any further investigation that would lead us away from our research trajectory and the journey we have embarked on.[4] One aspect worth clarifying, on account of its considerable conceptual interest, is the relation between cybernetics and phenomenology. On what bases does the tradition, which finds in Varela its last major representative, intersect with Husserlian phenomenology?

3.2 The Working Hypothesis of Neurophenomenology and the Absolute Position of the I

We have seen how the most significant step taken by the second cybernetics has been the attempt to develop an epistemology taking relations with lived experience seriously into account. For decades, the observer-observed problem remained one of the cornerstones of the movement which had developed around the Chilean school from the 1970s onward (Maturana and Varela 1973). Actually, the same spirit also marked the early experimental work conducted by Helmholtz, who—as already repeatedly noted—combined empirical research with the practice of introspection. We have also seen how Helmholtz's physiological epistemology was significantly influenced by Fichte's philosophical project and conception of man: the idea that, by manipulating the world, man discovers—via experience—the limits of his agency (self), distinguishing them from those of the world (non-self) which he thus comes to know.

This imprint, which brings the relation between a science of knowledge founded on physiology and lived experience into play, was transmitted to all the key figures belonging to this tradition under the name of physiological a priori, ultimately configuring itself as the working hypothesis of neurophenomenology. Magnus, who introduced von Uexkull's great insights (1903) into the field of physiology, drew upon his emphasis on the need to thematize the reflective circularity between the researcher's scientific experience and his knowledge of the fundamental mechanisms of his body and nervous system. This circularity was only thematized with

[4] It is important to stress that, over the past 10 years, this epistemological opening has led to the development of a series of disciplines which combine the prefix "neuro" with a series of fields of knowledge, ranging from ethics to aesthetics, from psychoanalysis and economics to gastronomy. Unfortunately, these new disciplines lack the kind of methodological rigor which distinguishes neurophenomenology as an integral proposal for the codetermination of the two analysis perspectives: what they ultimately amount to is a kind of practice which correlates brain activity with the data of first-person experience elicited through tasks pertaining to the field chosen.

second-order cybernetics, without ever raising the problem of its conceptual origin: all traces of it were erased by the dominant technological and formal focus.

In the French-speaking world of the late 1980s, Livet explored the apparent similarities between Fichte's themes and those of Varela (1987). He intuited their shared aspects, yet without grasping the continuity between the two. Livet's suggestions were taken up a few decades later by Goddard (2003), who acknowledged the conceptual proximity between the two thinkers while voicing some strong reservations on the matter: "Nothing suggests that Francisco Varela was acquainted with German idealism" (p. 131).[5] Indeed, in order to really grasp Varela's association with Fichte, it is necessary to view it within a wider framework, one that follows—with remarkable continuity—precisely the trajectory we have been tracing and which finds its last major representative in Varela.

The interpretation which Varela came up with by grafting Husserlian phenomenology onto the biological egology developed by Maturana and himself is reminiscent of that put forward by one of the most authoritative French interpreters of German idealism: Jean Hyppolite. Hyppolite's reading of Fichte (1959) sets out from a core principle, through which he seeks to connect Fichte's philosophical intentions to the deep themes of Husserl's phenomenology. This core principle is constituted by the project of developing a science of sciences, an epistemology that in order to prove rigorous must have an absolute foundation that rests on nothing but itself. Hence, according to Hyppolite, Fichte proceeds through a methodology which, aside from revealing and articulating this science of sciences, aims to lay bare the original experience—i.e., lived experience—which stands at its basis and is formalized by that science. Hyppolite notes: "An absolute foundation can only be such if it is its own object and precondition" (p. 24).[6]

Fichte finds this crucial experience of the human spirit in practical action.[7] It is only by operating upon objects with a given end in view and by observing the results of his own free acting that the subject becomes aware of himself as self-determining.[8] But why should practical agency "with an end in view" be an absolute foundation? If it really were such, reflective action would only need to be directed toward oneself, instead of being centrifugal and directed toward an object.

Fichte's discovery, his pure I, corresponds to the recursive closing of reflection upon itself through an openness to encounter. Reflective agency is therefore

[5] "Rien n'indique que Francisco Varela ait fréquenté l'dealisme allemand."

[6] "Un fondement absolu ne peut etre tel que s'il est à lui meme son propre objet e sa propre garantie."

[7] Fichte characterizes the activity that the practical subject is supposed to ascribe to itself in order to be self-conscious as "the act *of forming* the concept of an intended efficacy outside us, or the concept of an *end*."

[8] Interaction among rational beings is possible because their free agency is mediated by bodies that inhabit the same sensible world. An important step in this proof is the argument of §5 that having a body is a necessary condition of self-consciousness, since the ability to carry out one's ends requires an immediate link between one's will and the sensible world in which the will's ends are to be achieved. Thus, human consciousness is necessarily embodied, and our bodies play an essential role in constituting us as rational beings (Neuhouser p. xviii, in Fichte et al. 2000).

centrifugal (or intentional, to quote Hyppolite) and hence open to encounter precisely because it is only through encounter that the I can reflect itself: the encounter with the other is nothing but the discovery of oneself. On the other hand, no encounter is possible, or meaningful, if not as reflection (p. 30).

This, then, is the crucial point which Hyppolite grasps in his interpretation of Fichte: "Absolute knowledge, knowledge within immanence, is not opposed to the infinite richness of experience. It shows how such richness is possible. The closure of absolute knowledge does not exclude the openness of experience" (p. 26).[9]

It is according to this perspective that Varela sets out in Fichte's footsteps. Like Fichte, he downplays the dualism between the subjective and the objective on which natural awareness is based, to discover a new domain of phenomena that, according to Fichte, correspond to the absolute position of the I (*Ichheit*) and, according to Varela, to biological autonomy. Before Husserl, through a radical phenomenological reduction, Fichte brought an original field to light, the experience of pure activity, and established it as the foundation of the production of sense. This is the absolute position of the I, which acquires its determinations only in relation to a world that it encounters and stands face to face with (*Anstoss*). According to Varela, this I corresponds to life, which cannot escape its organizational closure; the downplaying of the natural attitude, in other words, opens up the prospect of the organization of the living being—conceived as a system which constantly gives form to itself from within—in relation to everything which elicits transformations of the state or dynamics of the system (i.e., perturbations).[10]

This view of the living organism echoes Müller's teaching—taken up by Helmholtz and axiomatized by cybernetics—that nervous impulses do not provide any information on the nature of the external stimuli generating them but are only a sign of such stimuli. Varela radically increases this gap between the autonomy of the organism and the world, to the point of conceiving external stimuli as perturbations of inner dynamics: something nonsensical which only becomes meaningful through the alteration of the inner structure of the organism (structural change).[11]

---

[9] "Le savoir absolu, le savoir dans l'immanence, ne s'oppose pas à la richesse indéfinie de l'expérience, il montre comment cette richesse est possible; la fermature du savoir absolu n'exclut pas l'ouverture de lexperience."

[10] *Varela's* entire conception of mind, and ultimately of experience, is concerned with the constraints exerted by the specific phenomenology of our concrete coping upon our internal dynamics as autonomous systems, and reciprocally, the effects of the latter upon the former, in a circular framework (Rudrauf et al. 2003, p. 33).

[11] This I is constantly emerging and therefore constantly determinable: it represents a constant possibility of determination. It is the absolute subject of judgment. Certainly, in the light of this view, the fundamental question of the nature of the subject loses its meaning. Heidegger notes: "Sofern Subjekt als das unmittelbar und gewiß gegebene Selbst und als absolutes Urte

ilssubjekt fungiert, und diese Bestimmung einzig die Problematik des Subjekts umschreibt, heißt das: die Frage nach dem spezifischen Sinn des Subjekts als Ich wird überhaupt nicht gestellt" ("If the subject stands as the immediately and certainly given self and absolute subject of judgement, and this determination alone defines the problem of the subject, the question of the specific meaning of the subject as I does not arise at all") (GA 28 1997, p. 121).

It is evident, therefore, that one's relation to the environment cannot be understood in terms of information or contextualization. Goddard writes: "The autonomous self of Varela, like the I of Fichte, is without an outside or, to be more exact, it is from within that an environment, a world takes shape for it, as a possibility for its own dynamism, as that which is meaningful for it—which is to say, as a 'self'" (p. 133).[12]

Therefore, although perturbations come from the outside, they do not convey any sense from the exterior; rather, sense coincides with the reformation of inner "mechanics." "Cognition or behaviours are operational phenomena without final cause: they work in a particular way" (Varela 1986).

The most evident, and most problematic, consequence of this view is the fact that the recognition of alterity is always traced back to the properties of the system. Based on this integral immanence established as a foundation, the recognition of the other can only occur by depriving the latter of its condition of absolute alterity. What we have here is a theory of encounter with no absolute transcendence (Goddard 2003).

At the same time, however, this privation is what nourishes Fichte's I and Varela's autonomous system, insofar as it is precisely nonsense—conceived as an external perturbation—that, through the production of internal changes, fosters the closing of reflection upon itself, providing some content: the encounter with the nonself. By eliciting an inner change, this encounter contributes to the emergence of a new sense. This is the recursive circle which links encounter and cognition.

Through circularity—supported by the mechanism of the closure of the organization of the organismic operations, coupled with the possibility of modifying their inner dynamics—the I and the other are redefined and indissolubly bound in one sweep. They effectively turn into two perspectives on the same reality: self-referential identity or the absolute position of the I. And the latter can only take shape by relating to the nonself—to nonsensical perturbation—which, in turn, can only relate to the self-positing of the I.

In this respect, the You only exists for the I in relation to the dynamics of the inner operations establishing it as a system, as self-positing—and the same holds true of the I in relation to the You. The encounter (*Anstoss*) with the other, therefore, amounts to a call for self-determination, and the same applies to the encounter with the I for the other. Within this framework—that of intersubjectivity understood as the structural coupling of two organisms—the transformations of my self-concerned cognition represent perturbations for a You that, by making sense of them through the restructuring of its inner dynamics, generates perturbations for the I. Through this intertwining, one's own inner dynamics support, and in turn are supported by, those of the other. Hence, "The relation with the other is invariably a relation with oneself."[13]

[12] "Le soi autonome valerien, comme le moi pur fichtéen, est sans dehors ou, plus precisement, c'est de l'interieur que se definit et s'esquisse pour lui un environnement, un monde comme une possibilitè de son dynamisme propre, comme ce qui a sens pour lui, c'est-à-dire comme 'soi'."

[13] "le rapport à autrui n'est jamais que rapport à soi-même."

This reduction of the other to oneself entails a considerable danger: confusing one's own consciousness with the self-consciousness of the other. By preserving autonomy as its core foundation, neurophenomenology envisages—or, rather, is forced to establish—intersubjective validation as a means of mediation between the subjective and the objective perspective. It is here that the notion of empathy derived from Husserlian phenomenology comes into play, as both a conceptual requirement and a methodological guideline: the second-person position, which takes shape through a shared, disciplined experiential praxis. Nowhere is the kind of empathic resonance which Husserl described as "taking another's place in one's imagination" ("*sich Hineinphantasieren*") better realized than in the laboratory.[14] With regard to this "egoistic" position of Fichte's subject, on 28 October 1794 Schiller wrote to Goethe:

> According to some verbal communications of Fichte, for he had not yet come to the point in his book, the I is the creator of its own images, and all reality is in the I itself. The world is but a ball which the I has thrown up, and catches again by reflection. (quoted in GA 49 1991 pag. 161)

Practical action, the inalterable and underivable facticity envisaged as the extreme limit which cannot be questioned any further—the limit to (self-referential) reflection itself—is a fact of the I for Fichte and of biological organization for Varela. As Kiesel emphasizes, "The self-positing absolute I of Fichte still does not feel its thrownness, it rather posits and 'throws the world'" (p. 64, 2008). From the perspective of Varela's biology as well, we might argue that the organism preserves its autonomous stability by referring the perturbations coming from the world back to itself in meaningful terms (self-reference), thereby bringing forth the world.

3.3 Narrativist Therapies and the Ontology of Presence

How can the healer be oriented by this way of conceiving the nature of man, i.e., by the notion of a constantly emerging and determinable self, whose organization endures without change as existence undergoes transformation? The notion, that is, of a self in process which remains self-identical and hence absolute precisely through its determinability?

The first and second cybernetic revolution, the various neurophysiological currents, and the currents of psychology sprung from them have given rise to different forms of psychotherapy: cognitive-behavioral therapy, cognitive therapy, the various schools of constructivist therapy, systemic therapies, analytic therapies, and, more indirectly, humanistic therapies—all of which are marked by the same way of conceiving the existence of man.

[14] "I am here and I imagine I am going there to the place where you are just now; conversely, you are here (the there where I am going to) and you imagine you are going there, to the place where I am (my here)" (Depraz in Thompson 2001, p. 173).

One version of this conception of the self, which apparently has nothing to do with the tradition we have been outlining, but rather purports to be making a break with the ontology on which it rests, is the perspective developed within the field of language studies and discursive practices, and which has engendered a current known as social constructivism. This current serves as a point of reference for the various forms of psychotherapy that are brought together under the label of narrative therapies and which to different extents and with various nuances all roughly share the same mantra: "you are not the problem, the problem is the problem" (Boje D.M. 1999–2005). In other words, insofar as the problem which the patient is experiencing has been dialogically and relationally built with others within a society, and within a context of organizations and family structures, it may be deconstructed and re-authored by pulling down its language scaffolding (White and Epston 1990; Anderson 1997; Berg and de Shazer 1993). Harrè sums it up as follows: "An individual emerges through the processes of social interaction, not as a relatively fixed end product but as one who is constituted and reconstituted through the various discursive practices in which they participate. Accordingly, who one is, is always an open question with a shifting answer depending upon the positions made available within one's own and others' discursive practices and within those practices, the stories through which we make sense of our own and others' lives" (p. 45 Davies and Harré 1990).

Within this context, the constantly emerging, and in each instance determinable, self becomes a sentence element, a subject of assertion within the context of an ongoing dialogue. The ever-existing possibility of determination of the subject is realized through linguistic action, via participation in a form of dialogue, whose development is related to the constant social construction and reconstruction of the linguistic action of each of the participants. From this perspective, the linguistic subject can no longer be ontologically conceived as a substance which endures unchanged while its properties undergo transformation. The subject of the sentence is instead constantly determined through what is predicated of it in the sentence itself. It is evident, therefore, that a predicate must not necessarily have a substance as its foundation. This determines a significant extension of the concept of subject from the ontological sphere to the logical-grammatical one: a thing's properties become the subject's predicates.[15]

When something is asserted about something else, what endures is not necessarily an immutable substrate, understood as the ontological basis of all possible transformations; rather, a foundation is to be sought in the reference of what is being asserted (predication) to that something as such (the subject). The subject of an assertion may, for instance, be change itself, without the subject "change" thereby implying the endurance of something that remains the same in an ontological sense. The subject of the sentence—the foundation of every predicate—is grasped as that which is spoken of and which in each case is determined in a sentence, i.e., as a subject understood as a linguistic element that takes shape in a sentence in relation to its predicative determinations, which emerge through a dialogue between different speakers. This way of

[15] Fichte's subject has already undergone this extension, which began with Kant.

framing the problem of the self is the basic assumption underlying narrativist psychotherapies. Narrativists can therefore argue that people have multiple identities and that their identities are shaped and reshaped in social interaction.

In this case too, however, as in the case of the biological and philosophical-transcendental approaches, what is overlooked is precisely the ancient ontology governing these perspectives, in the belief of having resolved and dissolved the substantialistic formulation of the problem of the subject through an emphasis on the impermanent, constantly emerging and multiple aspect of the self.

What remains unquestioned is the principle governing the constant process of emergence of the self in its various articulations, namely, the very ontology which had prevented Fichte from grasping the fact that his absolute I throws the world like a ball, only to catch it again as it bounces back, so as to always perceive itself as the same. Differently from Fichte's case, as in a team match, the ball here reaches the player in ever-new game situations, and in each case, the player repositions himself in relation to the match he is engaged in, and which throws the ball back to him, so that he ultimately forgets that he himself—a person in flesh and blood—is actually on the playground!

Once again, then, the crucial question of the nature of the subject dwindles into obscurity. It slips under the radar precisely because of the failure to grasp the key point of ancient ontology, whereby every possibility of determination—every possible form of emergence, every linguistic position of the self in conversation—is understood in the light of time, conceived as present time: as a coming into presence and withdrawing from it, as the constant resolving and consuming of events under the observer's gaze.

What remains unthought—while being implicitly adopted—is Aristotle's way of conceiving the motility of life in the light of the paradigm of production, by tracing every "not yet" back to an enduring presence: that enduring present which makes an object available, manipulable, and theoretically graspable. Just as in the case of a table, whose actuality—in the process of production—always lies in the presence of its appearance, the motility of life is grasped from this perspective as perfect presence, insofar as it is accomplished in its predicative determination at every moment. Harlene Anderson (2003) writes: "The person and *self*, including development and human agency, are viewed as interdependent, communal, and dialogic entities and processes rather than as isolated autonomous interior ones" (pag. 184).

Movement is thus associated with dialogically and relationally constructed presence, while life itself—which supports and constantly regenerates it—is lost behind words, narratives, and meta-narratives. Certainly, against the view, one might conclude—to borrow a statement by a Chinese critic quoted by Hadot (p. 152, 2008)—that "anything that can be enunciated lacks importance." What remains concealed, and therefore eludes investigation, is the fact that the person and self, before constituting itself through linguistic practices, progressively takes form via the praxis of existence as a selfhood that renews its incompleteness through its always-to-be-established relation with the world, without which any linguistic configuration would remain up in the air.[16]

[16] On the other hand, understanding this ever-incomplete structure of selfhood as a self-reference means grasping it as an I-pole, which certainly has nothing to do with Heidegger's notion of experience as always being mine.

If the time of life is traced back to the present time, the condition of instability and incompleteness, which characterizes the being-there of man over the course of his existence, can only attain completion with death: for it is only with death that the presence becomes complete, as a whole which no longer lacks anything. Death is what enables one to account for life, which only appears truly complete with its end—like the table, which subsists in itself, in the full presence of itself, only once it has been fashioned. Human perfection, conceived on the basis of the ontological categories of practical experience, is realized according to the same movement which governs the process of production, i.e., as a presence constantly moving toward the completeness of the finished product. And when the presence attains this fullness, when it no longer lacks anything, at that very moment, perfection vanishes and ceases to exist.

3.4 The Transcendental Body of Medicine and Psychology

The reverse understanding of life just outlined, based on the notion of the completeness of its end, is responsible for the transformation of physiology into natural science and, through it, for the birth of anatomical-clinical medicine (Foucault 1963) and the development of the key idea of organism which governs physiological psychology. Bichat was the first to discover that knowledge about death provides a key to access the living being, insofar as it is death that brings to completion and hence manifests the mutual conditioning between the various organs. "Any kind of sudden death—Bichat writes—starts with the interruption of circulation, of breathing or of brain activity. One of these three functions ceases first. All the others end next; in order to illustrate the phenomenon of this kind of death, one must consider it from the point of view of these three essential relations" (Bichat, p. 156).[17]

Every possible death—every individual death—is reflected by the way in which the interruption of the functions of an organ, e.g., the heart, lungs, and brain, affects all other organs. The invariance of this concatenation, which reveals the mutual dependence between organs, provides the foundation for the establishment of a science of life. It is this simultaneous and invisible connection between the various organs that only comes to light when death fixes their invariable, (anatomopatho) logical concatenation. This network of relations is the "vital" object of experimental physiology (Huneman 2005).[18] It is death, therefore, which reveals life through its no longer being there, through its being complete. Existence means presence: an incomplete presence which remains unchanged with respect to all possible transformations and which reaches completion with its end.

---

[17] "Toute espèce de mort subite—Bichat writes—commence en effet par l'interruption de la circulation, de la respiration ou de l'action du cerveau.L'une de ces trois functions cesse d'abord.Toutes les autres finissent ensuite successivement;en sorte que pour exposer avec précision le phénomènes de ces genres de morts, il faut les considérer sous ces trois rapports essentiels."

[18] As Huneman (2008) emphasizes, it is Kant's theory of organisms in the third *Critique*—understood as an organized and self-organizing whole—that had delineated the metaphysical and epistemological context for the emergence of the new physiology.

Setting the theme in relation to a much broader historical perspective—which first emerged in the context of the Orphic religion and was then incorporated within Christian and Gnostic conceptions—Jonas (1967) brilliantly shows that one of the defining features of the modern age is submission "to the ontological dominance of death." Modern thought strives to reveal the laws of inert matter governing life as though the living organism were nothing but "a *ludibrium materiae*, a subtle hoax of matter" (p. 12). Thus death, which determines the perfection of existence, at the same time, annihilates it.

On the one hand, starting with Helmoltz, this way of envisaging organisms—which constitutes the basis of physiology—has provided psychology with a solid foundation for the development of its experimental project. On the other hand, psychology sets out from the same way of envisaging man in order to focus on the self by subordinating change to what remains unchanged, and this necessarily precludes the possibility of grasping the mobility, mutability, and historicity of the self. As psychology is oriented by the ontology of presence—conceived as the ontology of death in physiology—it cannot account for the human power of historical transformation, for the simultaneous coexistence of existence and nonexistence (within possibility): the very path which Parmenides of Elea had ruled out by excluding nonbeing from the problem of being.

These two approaches within psychology, which may be distinguished in terms of the objective observation (e.g., reaction times, the magic number seven, behaviors, etc.) and the subjective observation of inner states (thoughts, feelings, imaginings, etc.), both take shape through objectivization: the theoretical gaze tears experience away from actual life and, by isolating it, strips it of its context of expression. Experience thereby becomes an immobile object viewed under the light of categories. Hence, the famous statement by Müller—Helmholtz's master—that "*nemo psychologus nisi physiologus*" ("no one can be a psychologist without being a physiologist") remains the most compelling expression of the principle that there exists a fundamental relationship between psychology and physiology: an idea that with the rise of neuroscience has made psychology superfluous as a discipline.

3.5 First and Third Person: The Outcome of a Debate and the End of Psychology

On 17 July 1990, the then US President George Bush issued a public statement in which he described the decade just begun as the "decade of the brain." The most surprising aspect of this address to the nation is precisely its emphasis on a 10-year research program focusing on an organ: the brain. It was as though a new space mission was being launched!

Much the same message is conveyed in an editorial which Jones and Mendell (1999) published in "Science" 10 years later. The two authors—the former president of the American Society for Neuroscience and its current one—drew a remarkable balance of the decade which had just come to a close and which had made neuroscience a point of reference for many fields of knowledge, assigning it a role comparable to that played by mathematical physics for much of the twentieth century. The considerable visibility and public acknowledgement enjoyed by neuroscience was not only due to its evident successes, and to the new hope it had sparked in the fight against serious illnesses, but also the explosive growth of the number of people describing themselves as neuroscientists: researchers coming from a range of disciplines that apparently had nothing to do with neuroscience. Jones and Mendell mentioned around a thousand new memberships a year, pointing to the rapid spread of new technologies related to functional imaging or molecular genetics as a possible reason for the phenomenon. The year was 1999.

Along with the triumph of neuroscience, the same decade witnessed something completely unexpected, which Bush's speech had certainly not foreseen and which remained largely undetected even in the report delivered 10 years later. With the celebrated decade of the brain, the problem of consciousness returned to the center of the scientific stage. The novelty did not lie in the effort to thematize consciousness or the mind: the Anglo-Saxon philosophical tradition had continued to do so throughout the twentieth century, articulating the mind-body problem according to the various "-isms" of philosophy. Arguably the most intriguing point made within this tradition was formulated by Nagel in a famous 1974 article boldly entitled "What Is It Like to Be a Bat?." Here the philosopher noted that the use of an objective explanation to account for what it is like to be a particular organism leaves an irremovable trace and that is to say: what remains of what it is like to be a bat, if one removes the point of view of the bat[19]?

With the rise of neuroscience, the problem of consciousness emerged from a new angle. The novelty lays in the attempt to study consciousness using the tools and methods of the natural sciences. The title of the first of a series of interdisciplinary conferences that were held in Tucson each year from 1994 onward, "under the Clear Desert Skies of Arizona," is highly revealing in this respect: "Toward a Scientific Basis for Consciousness." It was in a paper delivered at this first Arizona meeting that David Chalmers outlined for the first time what has become a well-established distinction: that between the easy problem of consciousness, involving cognitive functions that can be explained by invoking neural or computational mechanisms, and the hard problem, which remains such even once the mechanisms and modes of execution of relevant functions have been explained.

Within the framework of this new context, the hard problem newly raises the question of what it is like to be a bat. While the challenge consists in attempting to scientifically explain the hard problem or—as was argued a few years after Chalmer's paper—to naturalize the study of the phenomena of consciousness, the very possibility of posing the question of first-person experience is due to the remarkable impact of the new brain imaging technologies (FMRI, SPECT, PET, etc.). A series of gradual advances in mathematics, physics, and computer and clinical imaging—followed by a series of legal battles over competing claims and disputes concerning patent priorities, down to the public dispute over the Nobel

[19] As we shall see in the second part of the present work, an attempt can be made to answer this question only by resorting to a privative zoology, i.e. one that understands animality by taking its lack of humanness as a starting point.

Prize—has spawned a new generation of machines that provide access to the inner recesses of the most sacred of all organs. The new machines enable the "profanation" of the living brain, the brain of a living body that feels, acts, and thinks. This is the dense organ which medical science had hitherto studied only on the dissection table or by grasping a few superficial expressions of it as a living organ through EEG. As with Galileo and the telescope, imaging technology has generated a new sphere of observation and elicits a renewed conceptuality, not only in the field of established medical culture but also across the various disciplines which are gradually starting to investigate the brain (Arciero 2006). Only from this perspective can a truly unique aspect be understood: consciousness has returned to being a focus of inquiry for the very reason that it could no longer be such roughly a century earlier. This, then, is the field in which the problem of the first and the third person, and of the relation between the two, emerged and gradually became dominant.

The interdisciplinary debate on these themes—which for a few decades continued to be framed according to the trajectories traced by the annual Tucson conferences and essentially developed through the "Journal of Consciousness Studies"—is also marked by another peculiarity. Psychology, which up until the early 1990s still played a prominent role, gradually vanished from the scene. After a whole century had passed, just when the study of experience required a science finally capable of resuming the research which James had made his own, and which logical positivism had broken off, psychology dissolved. Little by little it was swallowed up by neuroscience, which assimilated it and even absorbed all its different branches: these became the cognitive, affective, social, behavioral, evolutionary, clinical, etc. neurosciences. So much so that, from the very university in which one of the authors of the present volume studied the fluxes and variations of individual consciousness together with Mahoney in the 1980s, Gazzaniga announced the death of psychology: "Psychology itself is dead" (1998).

Indeed, as first-person experience became an object of investigation for the natural sciences, it ceased being of interest to psychology, which had neither the tools nor methods—if not those of the natural sciences—to undertake its study. This raises a series of questions. What is the relation between the end of psychology, the development of brain imaging technologies, and the first- and third-person problem? Why has psychology failed to live up to the task set by the latest research developments? On what basis has it been absorbed by neuroscience? And how does neuroscience approach the first- and third-person question?

As already noted, all research on man is shaped by the way in which man's nature or essence is conceived. This perspective, which is often adopted unquestionably, determines the framing of the questions and hence the very course of the research. The formulation of a problem already suggests the possible answers and is inevitably associated with what we may term a—more or less conscious—ontological decision that concerns the way in which the preliminary question of the nature of man is tackled.

Framing the first-person problem in terms of how we can scientifically account for what it is like to be a self implies a factual view of the nature of that self. In other words, the problem of the first-person position is only addressed after having already established what it means to be a self.

Besides, the answer which Chalmers provides in his famous 1995 paper betrays precisely this ontological approach, which was to become the common point of reference for all researchers seriously considering the problem of consciousness in the wake of the Tucson conferences. Chalmers writes: "I argue that if we move to a new kind of nonreductive explanation, a naturalistic account of consciousness can be given. I put forward my own candidate for such an account: a nonreductive theory based on principles of structural coherence and organizational invariance and a double-aspect view of information" (pag. 200). What stands in the background is a well-defined view of who man is: man is "the self," the invariant, that which, by remaining the same over the course of its existence, is conscious of itself and at the same time of its manifold experiences.

It makes little difference that, in the light of this way of conceiving man's being, some researchers have endeavored to grasp phenomenal regularities by interpreting them as an expression of the subject's beliefs or desires. This is the case with Dennett (2004), for example, who based his conclusions on data from Damasio (1994, 1999, 2003). Other researchers have instead sought to identify, by means of eidetic reduction, the experiential invariants that emerge in variations, in order to study the emergence of conscious experience in relation to the neural activity that accompanies it. This is the case with neurophenomenologists who have chosen to follow Varela's lead (Petitot et al. 1999).

In 2004, 10 years after the first Tucson conference, which had marked the comeback of first-person experience as one of the hot topics in natural science, C. Whitehead was invited by the "Journal of Consciousness Studies" to review the themes of the conference. Whitehead provided the following picture of the different degrees of influence of the two most representative approaches: "By 2002 the triumph of cognocentrism was virtually complete, at least at the plenary level. Of 40 plenary papers, 32 came from cognitive neuroscientists and artificial intelligence investigators. There were only five plenary papers on phenomenology, and three on emergent phenomena and downward causation" (p. 71). This disparity may be due to the greater operational simplicity of the cognoparadigm compared to the transcendental approach, although the ontological assumptions are always the same.[20]

Thus even in this new context, chiefly dominated by neuroscience, the unquestioned underlying perspective has engendered the same kind of epistemological tension between different currents as had emerged over the course of the cognitive revolution. Just as back then, the actors involved—all striving to defend their arguments and research programs, and to manage their spheres of influence–never consider the problem of the ontological perspective shaping their shared view. All disputes aside, these researchers share the same conception of the nature of man, which they unquestionably assume: what guides them is the same ancient view, marked by the predetermination of the essence of man according to the categories of production.

[20] C. Whitehead remarks: "I have nothing against the cognitive sciences, but the cognoparadigm is not only disembodied and individualistic, it is profoundly impoverished."

According to this view, natural entities are present in the world not as entities produced by someone but as emerging, self-organizing selves. A natural entity, the self, emerges and comes into presence and thus remains unchanged over time and immutably stands at the basis of all changeable qualities.

According to this perspective, therefore, what it means to be a self has already been established; and what is also preestablished, on the basis of this determination, is the trajectory of sense orienting the research. A shared underlying foundation is provided by the conception of the motility of life in terms of the paradigm of production, whereby every "not yet" is traced back to enduring presence: that enduring present which makes objects available, manipulable, and theoretically graspable.

Such is the idea of man adopted by all those researchers who, from the first Tucson conference to the present day, have directed their efforts toward the development of a natural science of consciousness. It is hardly surprising, therefore, that the new wave created over the last 20 years by the impact of brain imaging technology on the study of first-person experience has exhausted itself, without even touching upon the theme which it sought to clarify.

If ancient ontology is what has shaped the cognitive sciences, and subsequently the natural-scientific research on consciousness, with its conception of the living organism as a technological-computational artefact, it is evident that the study of first- and third-hand experience can only proceed according to this perspective. As already noted, the way in which a problem is formulated already hints at the possible answers. In the light of this, we can understand why the triumph of neuroscience goes hand in hand with the terminal crisis of psychology. The latter came about when the guiding notion of organism—which stands at the basis of physiological psychology—in conjunction with the new technologies enabled neuroscience to grasp experience in its nascent state by examining its source, the brain, in the present.

Remarkable evidence of these developments is to be found in the way in which the social neurosciences approach and claim to grasp the problem of the understanding of meaning by tracing its origin back to the neural dynamics at work, thereby effectively devitalizing actual experience and dehistoricizing life. On these conceptual and methodological bases, the prefix "neuro" is attached to any discipline whose practice may be correlated with the brain activity of the people exercising it. The result is a wide range of "neuro-" disciplines, from neuroeconomics to neurogastronomy. Neuroscience is therefore acquiring the role which mathematical physics played in the years of logical positivism. Regrettably, it is guided by the same principle: the idea that its own level of theorization is enough to prove the incompleteness, or insufficiency, of the level of theorization characterizing other sciences! This clearly seems to be the case in the relation between psychology and neuroscience. A banal example would be the study of emotional experience.

From the perspective of the psychologist, who never loses sight of actual experience, a particular emotion, e.g., sadness, corresponds to the way in which a person positions himself with respect to the existing circumstances: to the meaning of the ongoing situation and, at the same time, to a specific way of perceiving oneself within it. For the neuroscientist, sadness—understood as lived experience—cannot be regarded as a scientific datum, unless it is traced back to stable and objectifiable criteria, such as facial expressions or the main cerebral structures activated by it. The essentially nonsensical passage which occurs in the laboratory—in keeping with the distinctive perspective of the neurosciences—is the reduction and transposition of the sphere of individual experience, which is what psychology ought to study, into that of objectiveness, which is completely foreign to the former sphere, with the alleged purpose of gaining a more in-depth understanding of it. The evident result of this operation is that sadness as a neuroscientific datum usually has very little to do with sadness as a psychological datum. As a consequence of this mode of operating, psychology, which shares the same ontological foundation as neuroscience, becomes redundant: "it's a dead topic," as Gazzaniga preached years ago (1998).

The primacy of presence, which served as the basis for the assimilation of psychology to natural science and still constitutes the positum for contemporary psychology, thus calls for a new debate on the founding principles of the discipline and its relation to the natural sciences: for it is precisely this assumption which remains unchallenged when it comes to the conception of man which psychology has continued to uphold up until its recent dissolution.

It is in the light of this disturbing condition of the discipline that we now embark on a journey to investigate the foundations of psychotherapy and possible ways to reclaim them. The inadequacy of certain concepts with respect to the thematic field, whose essence they purport to express, and the need to elucidate the bases of our practice are what lead us to raise some questions concerning the origins of the discipline. Finally, the requirements directly imposed by therapeutic practice in terms of the concepts used to express it, and which by now have made many of such terms outdated, lead us to follow in the footsteps of phenomenology, by providing an ontological redefinition of the thematic domain itself.

Bibliography

Anderson H (1997) Conversation, language and possibilities: a postmodern approach to practice. Basic Books, New York

Anderson H (2003) Postmodern social construction therapies. In: Sexton T, Weeks G, Robbins M (eds) Handbook of family therapy. Brunner-Routledge, New York, pp 125–146

Arciero G (2006) Sulle tracce di sé. Bollati Boringhieri, Torino

Berg IK, de Shazer S (1993) Making numbers talk: language in therapy. In: Friedman S (ed) The new language of change: constructive collaboration in psychotherapy. Guilford Press, New York, pp 5–24

Chalmers DJ (1995) Facing up to the problem of consciousness. J Consc Stud 2(3):200–219

Damasio AR (1994) Descartes' error: emotion, rationality and the human brain. Putnam, New York

Damasio AR (1999) The feeling of what happens: body and emotion in the making of consciousness. Mifflin Harcourt, Houghton

Damasio AR (2003) Looking for Spinoza: joy, sorrow, and the feeling brain. Mifflin Harcourt, Houghton

Davies B, Harré R (1990) Positioning: the discursive production of selves. J Theory Soc Behav 20(1):44–63

Dennett DC (2004) Freedom evolves. Penguin, London
Depraz N (2006) Mettere al lavoro il metodo fenomenologico nei protocolli sperimentali."Passaggi generativi" tra l'empirico e il trascendentale. In: Cappuccio M (ed) Neurofenomenologia. Le scienze della mente e la sfida dell'esperienza cosciente. Bruno Mondadori, Milano
Fichte JG, Neuhouser F, Baur M (2000) Foundations of natural right. Cambridge University Press, Cambridge
Foucault M (1963) Naissance de la clinique. Une archeologie du regard medical. Quadrige, PUF
Gazzaniga MS (1998) The mind's past. University of California Press, Berkeley
Goddard JC (2003) Autonomie, réduction et réflexivité: la philosophie naturelle de Francisco. J Varela et le projet transcendental Intellectica 36(37):205–225
Hadot P (2008) La filosofia come modo di vivere. Conversazioni con Jeannie Carlier e Arnold I. Davidson, 3-71. Einaudi
Heidegger GA 28 1997
Heidegger GA 49 1991
Huneman P (2005) Espèce et adaptation chez Kant et Buffon. In: Theis R, Ferrari J, Ruffin M (eds) Kant et la France-Kant und Frankreich. Olms, Hildesheim
Huneman P (2008) Métaphysique et biologie: Kant et la constitution du concept d'organisme. Kimé, Paris
Hyppolite J (1959) L'idée fichtéenne de la doctrine de la science et le projet husserlien. Husserl et la Pensée Moderne–Husserl und das Denken der Neuzeit. Nijhoff, The Hague
Jonas H (1967) The phenomenon of life: toward a philosophical biology. Harper & Row, New York
Jones EG, Mendell LM (1999) Assessing the decade of the brain. Science 284(5415):739–739
Kisiel T (2008) On the genesis of Heidegger's formally indicative hermeneutics of facticity. In: Raffoul F, Nelson ES (eds) Rethinking Facticity. State University of New York Press, New York
Lachaux JP, Rodriguez E, Martinerie J, Varela FJ (1999) Measuring phase synchrony in brain signals. Hum Brain Mapp 8(4):194–208
Lutz A, Lachaux JP, Martinerie J, Varela FJ (2002) Guiding the study of brain dynamics by using first-person data: synchrony patterns correlate with ongoing conscious states during a simple visual task. Proc Natl Acad Sci 99(3):1586–1591
Maturana H, Varela F (1973) De Máquinas y Seres Vivos: una caracterización de la organización biológica. Editorial Universitaria, Santiago
Maturana HR, Varela FJ (1987) The tree of knowledge: the biological roots of human understanding. New Science Library/Shambhala Publications, Boston
Petitot J et al (1999) Naturalizing phenomenology: issues in contemporary phenomenology and cognitive science. Stanford University Press, Stanford
Rudrauf D, Lutz A, Cosmelli D, Lachaux JP, Le Van Quyen M (2003) From autopoiesis to neurophenomenology: Francisco Varela's exploration of the biophysics of being. Biol Res 36(1):27–65
Thompson E (2001) Between ourselves: second-person issues in the study of consciousness. Imprint Academic, Charlottesville
Uexkull JV (1903) Der biologische Bauplan des Sipunculus. Zeitschr f Biol 44:269–344
Varela F (1986) Laying down a path in walking: a biologist's look at a new biology. Cybernetic 2:6–15
Varela F (1987) The science and technology of cognition: emergent directions. In: Roos JL (ed) Economics and artificial intelligence, 1st IFAC international symposium. Pergamon Press, Oxford/New York, pp 1–9
Varela F (1992) Un know-how per l'etica. Laterza, Bari-Roma
Varela F (1996) Neurophenomenology: a methodological remedy to the hard problem. J Consc Stud 3:330–350
Varela FJ (1999) The specious present: a neurophenomenology of time consciousness. In: Petitot J et al (eds) Naturalizing phenomenology: issues in contemporary phenomenology and cognitive science, vol 64. Stanford University Press, Stanford, pp 266–329
Varela F, Shear J (1999a) (eds) The view from within: first-person methodologies in the study of consciousness (Special Issue). J Consc Stud 6 (2–3)
Varela FJ, Shear J (1999b) First-person methodologies: what, why, how. J Consc Stud 6(2-3):1–14
White M, Epston D (1990) Narrative means to therapeutic ends. WW Norton & Company, New York

Part II

A New Beginning: Formal Indication, Non-rationalist Psychology, and Phenomenological Psychotherapy

The Accesses to Oneself

4

The terminal crisis of psychology, triggered by the comeback of first-person experience, would appear to be the final outcome of a way of addressing the problem of the self according to the direction imparted to the analysis by the theoretical approach. This perspective, which rests on ancient thought and defines psychology as a natural science, has shaped the methodological development of the discipline, and its research themes up to the present wreckage. It seems evident that by setting out from this mode of access and following its trajectory, psychology has given up on a radical question which is now proving of crucial importance to ensure the very survival of psychology and its capacity for transformation. The problem concerns the possibility of studying the world of experience, of scientifically grasping that self which every one of us is, without the need to search for what endures unchanged in the life of this or that person—i.e., invariant patterns—in the light of theories, presuppositions, or a priori categories. The key point, therefore, concerns the clarification and definition of the thematic field of psychology, its positum, but also the issue of how the world of human experience may be scientifically grasped in its general and individual determinations without immobilizing it into an object, organization, personality, system, or process. In other words, what is required, along with a demarcation of the field of study, is the development of a method capable of conceptually expressing the essence which distinguishes the very thematic area of this science—the incompleteness of man's being-there—without betraying its nature.

If the self is no longer posited as a natural entity—as a thing among other things under the eyes of the observer—it is necessary first of all to redefine the positive feature which distinguishes psychology as a science. But in order to take this step, it is necessary to call into question the very idea of man which, as has been broadly illustrated, predetermines the course of the research. As we have seen, the mode of access to the thematic field of a science—of any science—is always shaped by a preestablished understanding of man that determines its development (and frequently remains unquestioned).

G. Arciero et al., *The Foundations of Phenomenological Psychotherapy*,
https://doi.org/10.1007/978-3-319-78087-0_4

What is at stake, then, is not just the reconfiguration of the object of study of psychology and the possibilities to access this object and conceptualize the knowledge of it while remaining faithful to its mode of being and nature: what is primarily at stake is the search for a new approach capable of setting psychology on the right course to develop and maintain its distinctive research dimension. Psychology must engage with phenomenology precisely because phenomenology alone—understood as the science investigating the very emergence of meaning—can provide a formal indication for developing the kind of concrete, positive content capable of reorienting psychological research and hence the view of therapeutic care and possible courses of treatment. It is necessary, therefore, to clarify what the positum—or positive feature—is that allows psychology to be defined as a science and, based on this distinguishing feature, explain in what respects phenomenology co-determines its content.

4.1 Phenomenology and the Positive Feature of Psychology

Psychology is a form of conceptual knowledge pertaining to psychic phenomena. Over the course of its short history and across its various currents, scientific psychology has investigated psychic phenomena with the aim of establishing them as the ultimate elements of an objective domain whose functioning could be explained by laws, in accordance with the principle of causality. The epistemological assumption which, from Müller's time onward, has served as the scientific foundation of this approach is the separation between a subsistent subject, responsible for the generation of sense, and a world in which actual events take place. The gulf between sense and facts, between lived experience and events in the world, and between subject and object is the underlying epistemological question which has governed the development of psychology. The discipline has become caught up in this question, spawning discussions which have been dragging on for years, with no real results. The problem has been framed according to two perspectives, from which two different conceptions of the psychic sphere derive. As we have seen when discussing physiological psychology in the previous chapter, these two conceptions are actually two forms taken by the same unquestioned ontology, which continue to shape the contemporary debate.

On the one hand, we have the transcendental perspective, according to which the impact of events from the external world (perturbations) on the subject engenders brute sense data that only acquire meaning through the order ensured by categories—the laws of logical thought. From this point of view, it is evident that the events of the world, and its objects, are only real if the subject is aware of them. Neo-Kantian psychology (from Natorp to radical constructivism), natural-scientific psychology, Husserlian phenomenological psychology, and analytical psychology all adopt this perspective. On the other hand, we have the critical realist view, according to which the events of the world that strike the subject, while causing sensations, cannot be traced back to a subjective psychic sphere. In other words, sense data differ from the events and objects eliciting them: the world exists independently of the subjective sphere. This second perspective is shared by empiricist psychology in all of its various versions, as well as by behavioral and cognitive-behavioral psychology, and most kinds of cognitive psychology.

According to both these perspectives, the immediate experience of a situation or entity which takes place through perception consists in the grasping of an epistemological object in the light of a theoretical attitude—as though factual life developed according to epistemological scenarios!

But what do we mean by immediate experience? From a transcendental perspective, this corresponds to a sensation that occurs in the organism in response to an environmental situation of which it constitutes a brute sign but which, as a psychic datum, will become the object of an ordering consciousness. A sensation, therefore, is present and perceivable only if, through a theoretical attitude, I destroy, set aside, or turn into something incomprehensible (into a nonsensical perturbation) the environmental context providing a framework for the sensation in question. The centrality of sensation therefore stems from this theoretical stance (the epistemological gaze), which in its actual application implies the destruction of the environmental experience. Such analytical praxis inevitably also does away with the self which encounters the environmental context of its historicity.

From the point of view of critical realism, which takes the evidence from reality as its starting point, independently of the subject observing it, sense data are not pure products of the subjective sphere. Their validity, like that of knowledge in general, is to be sought in the objective causes of sensations. Thus in the name of objectivity, realists bring about the same destruction of the environmental experience as the neo-Kantians and do so through the same theoretical attitude. By reducing what environmentally surrounds a sensation to a given meaning, established on the basis of the particular objective qualities of the environmental context, realists ultimately erase the meaningfulness of the environmental experience.

Taking the existence of sense data as their starting point, the two perspectives just outlined address the problem of how to scientifically account for the functioning of the subjective sphere. The transcendental approach seeks to attain objective knowledge by remaining within the subjective sphere, whereas the realist approach seeks to move beyond the subjective sphere by invoking the objectivity of the data of sense perception and hence conceives the self according to the same categories it uses to conceive the world. The efforts made in the twentieth-century psychology were largely aimed at grasping those natural-scientific laws through which to explain the functioning of the mind according to the two opposite perspectives. Constructivist psychology and cognitive-behavioral psychology are two evident examples of these efforts. Besides, the latest effort that has been made—namely, to establish a natural science of the first person—is once again structured along these two lines: it represents the last stage of a journey which over the last few years has led psychology to dissolve into neuroscience.

Now, over the course of this complex journey, psychology has been systematically avoiding one radical question. The psychic sphere, or rather the domain of lived experience, is not an object and therefore cannot be grasped as a datum to be statistically processed—as a coherent system of experiences, an I that is the center of intentional acts, an organizer of recurrent patterns, a system of beliefs and desires, a system for the processing of information, etc. In other words, it cannot be grasped as a self that is always self-identical and determinable once and for all, which lies behind factual life and remains untouched by the unfolding of events.

What all this means is that the conceptual system which psychology shares with the natural sciences, and whose defining feature is the genesis of forms of objectivization, does not allow us to access the sphere of the experience of self, if not by denaturing, betraying, and devitalizing the spontaneous motility that represents its distinguishing feature. The categoriality of the theoretical perspective—according to which it is possible to observe and explain one's inner states (or those of others) in the same way an observer might gaze at the stars—is unsuited to conceptualizing the original dynamics and essential incompleteness of feeling alive.

4.2 The Unfulfilled *Positum*

Because life manifests itself as constant motility—understood as what happens or may happen to someone at every moment—it is necessary to outline a new perspective for the scientific investigation of the experience of living. Precisely because individuals relate to their own living on the basis of the current experience they have of themselves, and because it is only starting from such a condition that they can access the full domain of their existence, this relation cannot be one of observation but must rather be a relation of practical, actualizing co-determination. In other words, I can observe my own sense of anxiety—or that of another person—from the perspective of a theoretical relation with my own existence, or that of the other person, as though I were observing the histological structure of my liver. I will thus grasp anxiety as an emotion, for example, characterized by certain physiological parameters, neuronal correlates, and a certain propensity to act and perceive things. Lived experience in this case will emerge through its reflective objectification: it will be grasped from the outside, so to speak, and from this perspective, the unfolding of lived experience will be viewed as a process, a theoretical object, to be disentangled into its fundamental elements (i.e., arousal, valence, etc.).

In doing so, however—in transforming my emotional experience or that of another person into an object of classification—I fail to access the sense of the experience. This sense only comes to light when my experience is grasped not as psychic event that unfolds under the gaze of an observer but rather in terms of its actualization, of its taking shape in relation to this or that circumstance, as is indeed the case in everyday life. What we have here is the pre-theorical domain of life: a domain that can only be grasped through a methodology capable of understanding the experience of living by foregoing all a priori points of view and directions, so as to enable phenomena to manifest themselves for what they are. As we shall see, a methodology of this sort can only be non-theoretical and hence nonreflective, for else the domain itself would cease to be.

From this vantage point, my anxiety—like any other lived experience—emerges as a position with respect to the circumstances at hand. Therefore, starting from this situation, it surfaces as the living act of taking charge of oneself and of one's own history within a given context: as the actual interpreting of myself in the condition in which *I find myself*, in the light of my immediate future. What lies within my condition are the motives and tendencies from which it draws its "raison d'être" qua actualization and which an objectifying gaze is bound to leave out. Thus the self—i.e., ipseity—emerges in particular situations and is one not just with the more general domain of my experience, of which it is the manifestation, but also with the particular circumstances in which it realizes itself. In each particular instance, the self *finds itself* through that which it encounters, and this discovery varies from one factual situation to the next.

The crucial point, then, is that ipseity—i.e., the perception of oneself in relation to ongoing events—is actualized prior to any possible knowledge of it and is immediately grasped without any need for reflection. To understand the sense of experience without betraying the motility of life, then, is to move along with its taking shape as the experience of this or that and at the same time as the experience of oneself. Therefore, plunging into the experiential flow and swimming along with it means putting oneself in a position to interpret such flow according to its sense of enactment.

Those who, in the footsteps of James, seek to grasp factual experience in relation to a theoretical I or a narrative I—as do many psychotherapy acrobats who operate a conceptual travesty of this theoretical I through narrativist approaches—completely miss the mark. On the other hand, one cannot grasp ipseity and the whole horizon of phenomena it discloses by seeking to understand it as a first-person stance. If the experience of the self occurs on each successive occasion and if the current situation is part of this occurrence, the nature of the experience is that of an actual involvement which cannot be reduced to self-referential experience. The boundaries between experiencing and what is experienced, between feeling and what is felt, between living and what is lived become blurred. The experience of the self resonates with the way in which the world allows itself to be grasped, but the one thing loses itself in the other, precisely because the one thing *is* the other. In this respect, the experience of the self too is given at every moment.

It is evident that this impersonal occurring of life is exactly the opposite of that of an I which observes experience in order to lend meaning to it. When we live through a situation, the experience we have of it certainly does not correspond to a psychic process unfolding within us or before the eyes of our mind: we find no I within it or no reference to an I, with its individual peculiarities, but only the experience of something. We are there, completely, in every situation which offers itself to us, who experience it as being immediately meaningful: experience manifests itself not as a thing but as an event! And this event consists in the manifestation of a phenomenal dimension that lends determination to ipseity as much as to all that appears.

The way in which I perceive myself in any given situation (i.e., the motility of life) can be understood as a "may be," as a possibility, as something I can or cannot do—and, in doing it or not, what is at stake is always myself. Thus ipseity is given at every moment without presuppositions or theories; ipseity emerges at every moment as a "how" filled with content that each person experiences differently.[1]

[1]The young Heidegger wrote: "both the life-content of non-theoretical life and the what-contents of the material consideration always present themselves in a particular way in a '*how*' and factically an intensifying-concentration (*Zugespitztheit*) upon the factical self-life of particular, many, whole generations is found in life, just as factically as the what-diversity of the subject-areas. This intensifying-concentration is not a what-content, but rather a how-content, in which each still so varied what- content can stand" (p. 67–68 , (84–85) Heidegger GA 58).

4.3 Between the Psyche and Facts

The unavoidable problem is how to grasp this actual involvement of life in its originality and fullness. If psychology is the science of psychic phenomena and if psychic phenomena are an area of actual life, the question which emerges concerns the way in which psychologists encounter experience before they develop a composite discipline, articulated into a range of scientific determinations of psychic life.

Like all sciences, psychology defines its specific field of knowledge—its positum—by distinguishing and isolating within factual life situations, which it can use to construct its fields of knowledge and inquiry, while at the same time developing suitable analytical and technical means for their study, along with a conceptual framework of reference. How do psychologists interpret the actual foundation on which their science rests and which it seeks to make manifest?

We have emphasized the fact that natural-scientific psychology grasps experience as an epistemological object structured in accordance with the logical laws of thought. But in actual life, in the experience of living, situations as such are simply experienced as being more or less meaningful. Let us stress this once more: there is no need to tune in to the laws of knowledge in order to live our everyday life. But if psychology is to relinquish the ancient ontology which supports natural-scientific conceptuality, on what basis can it interpret the factual, original experience on which it seeks to establish its foundations as a science?

This is where the need emerges for an ontology that never loses sight of what happens or may happen to someone at every moment and strives to conceptualize that motility of life on which all positive sciences are founded, without ever crystallizing it into a theory: an ontology, that is, seeking to grasp the meaningfulness of life in its originality and fullness and capable of grasping the experience of living in its nascent state. An ontology of this sort—which we have described as a phenomenological ontology by virtue of its mode of operating, thereby emphasizing its method of investigation—presents psychology with a new and unavoidable task: to redefine its positive domain.[2]

The ontological corrective provided by the phenomenological gaze forces psychology to rethink its conception of the psyche and understood as that independently existing internal realm governed by laws to which all psychic activities may be traced back. If factual life actualizes itself within the framework of meaningful contexts, so much so that in each instance the experience of self appears to be integrated within the contexts that accompany it and are part of it, phenomenological ontology leads us to rethink one of the cornerstones of scientific psychology: the separating of sense and facts, of psychic processes and events.

While the phenomenological highlighting of an original determination of existence formally indicates the ontological features of the sphere of actual life on

[2]The phenomenological method—which is a specific method and hence one devoid and independent of any predefined thematic area—has enabled ontology to bring to light the structures of being in general. This method, then—which is not a particular philosophical discipline—allows us to reveal what is concealed and what withdraws itself from manifestation (Sheehan 2014).

which psychology is based and within which it must necessarily keep, it also enables psychology to redefine its thematic field and conceptual framework. The formal indication which psychology receives from phenomenological ontology does not merely translate into the thematizing of those concepts and themes that are regarded as fundamental: by setting out from the research horizon which distinguishes it as a discipline, psychology is forced to reconsider its encounter with psychic experience and hence its thematic object. This is required by the very nature of psychology as a discipline focusing on the individual generation of sense.

Only in this way can psychology open up to a process of renewal that calls into question its very foundations as a science. It is worth quoting Heidegger here: "It is true that someone engaged in research can master, in addition to his own positive science, phenomenology as well, or at least follow its steps and investigations. But philosophical knowledge can become genuinely relevant and fertile for his own positive science *only when*, within the problematic that stems from such positive deliberation on the ontic correlations in his field, he comes upon the basic concepts of his science and, furthermore, questions the suitability of traditional fundamental concepts with respect to those beings that are the theme of his science" (GA 9, p. 53).

In this respect, we envisage phenomenology as a possible ontological corrective which, in a moment of crisis for the very foundations of psychology, may help to find a new direction within the framework of psychology as a positive science by revealing its ontic content.

Freed from the theoretical outlook which invoked the categories of the intellect, on the one hand, and, on the other, the objectiveness and validity of knowledge, psychology finds itself having to carve out a domain for itself starting from a way of conceiving experience that overturns its underlying epistemological position, building a bridge across the abyss separating the psychic sphere from reality and sense from facts.

4.4 Psychology as a Positive Science and Its Method

If the psychic sphere no longer coincides with an isolated self that remains changeless with respect to life's situations, if lived experience can no longer be grasped as an ordinary object, since being oneself amounts to the full actualization of the experience of living as an involvement with one's own world and with others within the surrounding environment, what is called radically into question is the very positum of psychology. The theme of psychology, the psychic, emerges not as a sphere structured according to the laws of logical thought but rather as a dimension which manifests itself and constantly changes over the course of individual existence: an existence led together with others in the life-contexture of the world. This kind of feeling alive constantly acquires sense within and through those meaningful contexts that make up the situations *in which it occurs*. The meaningfulness of things—understood as a categorial determination of the world—approaches life by manifesting itself in a certain way, and in meeting it, life comes to expression;

hence, there is no dividing line between the act of experiencing and what is experienced.

Therefore, since the experience of self—i.e., the way the self reveals itself but also its self-interpretation—which comes to light over the course of a single life is bound to be a unique experience, psychology may be regarded as the science of personal experience. It investigates the origin of the configurations of sense determined by a particular person's understanding of the history of their production (Arciero 2006). In this respect, psychology is also a historical science, insofar as it conceives the sense of experience on the basis of the contextual and temporal actualization of life: existence unfolds in time, and only those who are, who exist, can have access to their sense of being alive. Finally, the fundamental feature of any experience of living is its corporeal aspect: psychology is therefore also the science of being embodied, and each of its branches must necessarily take this basic dimension into account.

The points just made shed new light not just on the aforementioned aphorism by Müller but—as we shall see—also on the relationship with physiology and neuroscience. It is immediately evident how, starting from the calling into question of the nature of the entity constituting the theme of psychology, the articulation of such a theme may lead to the redefinition of the range of disciplines that make up psychology: social psychology, self psychology, personality psychology, and the psychology of personal identity, where what is at stake is the grasping of the uniqueness of one's personal meaningful trajectory in a world shared with others; the psychology of language and cognition, which investigates the origin, construction, and role of conceptuality and knowledge; the psychology of emotions, perception, and attention, which opens up the field to the investigation of personal meanings, down to the very connections rooted in the flesh; and, finally, physiological and medical psychology, which sets the problem of sense at the very heart of the natural science of life, opening it up again to a kind of engagement which had been forestalled by the development of modern medicine.

The overturning of the relation between the theoretical position and the pre-theoretical one ultimately raises a question of method: a theme which proves crucial for any positive science. How can the experience of self be grasped if we forego the reflective approach? How can we scientifically grasp the self and the world of the self without any presuppositions or theories?

Adopting this perspective and addressing these questions may possibly lead us toward the kind of fallacy which Popper imputed to the empiricists: "Traditional epistemology (...) fails to take account of the fact that (...) there can be no pure perception, no pure datum; exactly as there can be no pure observational language, since all languages are impregnated with theories and myths" (1972, pp. 145–46). Even without being the most persuasive of all rationalists, is it not evident that each one of us in our daily lives grasps certain facts in relation to prejudices, beliefs, moods, expectations, or at any rate presuppositions of some sort? How, then, can we seek to establish a science—the science of the psyche, no less!—capable of redefining the relation between meaning and facts by doing away with those theories and presuppositions that shape our gaze on experience?

The answer to these questions is to be sought in things themselves. In order to scientifically grasp the experience of the self before its transformation into a theoretical object, it is crucial to understand it on the basis of factual life. In other words, it is a matter of grasping that the meaning of a phenomenon only becomes accessible in and through the situation in which it occurs. The self—my own self and that of every other person—manifests itself in this or that situation; this unique mode of manifestation, this "how," is embedded within the context in which it reveals itself and within the world of the self different distinctive aspects of which will emerge in different circumstances and different moments in the history of one's life. Ipseity only finds expression through its enactment. It coincides with the fact of turning, on each successive occasion, toward that which shows itself as it meets one and which in doing so discloses oneself. But what this also implies is that the meaning of the expression of something cannot be distinguished from the relation established with such thing and from its mode of enacting it.[3] Expression, in all its modes, also includes the meaningfulness of that which it encounters: by virtue of this implication, sense emerges, taking concrete shape through expression (and remaining implicit within it as a "being related to").

What all this means is that experience cannot be reduced to an object of consciousness and that consciousness is not the site of the experience of the self and the world. The factual realization of experience is itself the source of sense; therefore, the nature of this enactment and the issue of how it is possible to access "the self-experiencing itself" become the problem. The key question, in other words, is how to conceptualize the generativity of life, this endless emergence of sense, without betraying its nature. The study of that ipseity, which is experienced in each case through its engagement with this or that thing, enables us to redefine from the perspective of living actuality the underlying epistemological position of natural-scientific psychology: the relation between sense and facts and hence the mentalism of meaning. We regard the study of ipseity both in its general outline and in its individual determinations as the only possible starting point for psychology, conceived as the science of personal experience.

4.5 The Structure of Experience

If it is only within factual occurrence that the self opens up to itself as it encounters the world, every factual emergence of experience, every accomplishment of ipseity, corresponds to a way in which the self experiences itself vis-à-vis the world and the possibilities offered by the situation at hand. This means that in everyday occurrences, people relate to themselves by determining their own horizon of sense through the choice of given possibilities over others, which is to say through the kind of practical, actualizing self-determination that in each case enables me to take

[3] From this perspective, the various categories can be regarded as modes of being of life. A similar view was adopted by Peirce (Gethmann 1993 p. 272 ff.).

charge of my immediate future by projecting myself into what lies before me and which thereby reveals the world in all of its meaningfulness.

Each person unconcernedly lives in an ever-changing context of anticipation of sense, which takes shape again and again starting from factual experience. Within this constant shifting from context to context, within this flow of expectations and situations, each person is completely absorbed by what meaningfully meets him/her. The experience of self, which is absolutely personal, thus points to a shared world, conceived as a meaningful complex that is immediately grasped and on the basis of which something is brought to understanding, manifesting itself.

In the experience of this transient and ever-renewed opening to meaningfulness, the contexts which meet us cannot be reduced to their sensory perception. Indeed, in each case ipseity takes the concrete form of a mode of relating to, behaving toward, and becoming involved with that which lets itself be grasped as meaningful insofar as it concerns us. Ipseity, therefore, means both moving and making things appear.

If ipseity is primarily conceived as the relational sense for that toward which it is orientated, the content of such reference must be the world, understood as that to which it pertains and within which it abides: the world is grasped not as the sum of the all the objects of experience but as the totality of what is meaningful and which is presupposed by every object of experience. In each case, as it gives itself in its relation to things, this world simultaneously withdraws into the background. Ipseity, therefore, does not construct the world but rather makes it appear; and since it is part of it, it can always make it (the world) appear according to a given perspective, so that at the same time it (ipseity) also relates to itself. The experience of self emerges, then, insofar as the world becomes the world: in the choice of my immediate future, something gives itself in each encounter starting from a weave of references that "becomes" the world within occurrences, as ipseity becomes manifest, the boundaries of the self and those of one's world literally coincide. Hence the peculiar relation of self-permeating of the environing-word, with-word, and self-word, starting from which the unstable circumstantiality of life, determines itself over the course of its existence (GA 58).

If the world becomes such within events, we can argue that ipseity simultaneously occurs, comes to itself, is received, and is made its own. Through this occurrence, the self encounters and experiences itself on the basis of a given situation, taking charge of it in the light of this or that possibility. This appearance of the proper in the event and of what is extraneous in the proper (Greisch 2000) reveals the very fabric of the world of the self: the experience of the self is made in specific situations and becomes one's own within occurrences.

Lived experience is an event which in each case involves me (*er-eigen*) and brings me effectively into play, engendering with every occurrence a condition that is my own, an apparition of the self. The things I have experienced—expressions of myself—are available as part of a single complex of experiences, ready to be actively or spontaneously remobilized in relation to ongoing circumstances or to become sedimented as availability, motivational nodes or habits. Hence, the world of the self—the domain of one's experiences—cannot be mistaken for a self with an inner world.

The relation of the self to itself over the course of one's existence does not have the acquisition of knowledge as its aim and certainly does not manifest itself as a cognitive operation; rather, in each case it manifests itself as the taking up of a position, a stance with respect to possible weaves of meaningfulness. Within occurrences, ipseity in each case comes to itself, opening up to the dimension of possibility, not as the relation between an I and an object but through its practical self-determination, its very happening within and through such an encounter. As Quesne succinctly puts it: "Meaningfulness is included, resumed by the expression (of the self) and this resumption may even simply coincide with indifference to what is meaningful" (p. 165, Quesne 2004).

4.6 The Mystery of Sense

If self-recognition takes shape not as a reflective mirroring within the psyche but rather in the light of what we do or do not do, through the way in which the world reveals itself to us, how can we account for the mystery of sense? If ipseity also encompasses the world, it occurs as an opening up to the world, since the world approaches it and manifests itself to it. But why does the world let itself be approached? The remarkable discovery made by the young Heidegger—building upon Lask's suggestions (see Kisiel 2008)—is that the very movement of life, in its original opening up as a solicitous involvement with things, other people, and hence one's own self, discovers the world as a content of sense (*manifestatio sui*). In other words, sense is itself part of the world toward which I behave, within which I live, and in relation to which I *find myself*—my immediate future being that which is in each case at stake within this relation. The world here manifests itself in its meaningfulness. According to this perspective, unlike the transcendental subject, ipseity does not create or build the world but rather contributes to its appearance by conditioning its manifestation.

Through this way of understanding reality, which redefines the meaning of objects independently of any judgment characterizing it by newly situating form within matter itself, the world is grasped pre-theoretically, as the meaningful contexts including these objects. In these contexts, we live as though within categories that allow us to orient ourselves by deciphering the sense of things directly within the things themselves.[4] The attunement (*Stimmung*) between the contexts we encounter and life in its immediacy is in each case brought about by the encounter itself. In its original motility—understood as the fact of being directed toward

[4] In discussing Lask's influence on the young Heidegger, Kiesel notes "The relation of form to its matter is thus one of 'environment' (*Umgebung*). Matter is encompassed, embraced (*umgriffen*) surrounded or environed (*umgeben*), horizoned or bordered (*verbrämt*) by the form; it is enveloped (*humhullt*) enclosed (*umschlessen*) in the form" (2008, p. 53). As evidence of this radical departure from the neo-Kantian tradition to which Lask belonged, Kiesel mentions in the letter in which Lask informed Husserl that his master Rickert was accusing him of a "reactionary regression to antiquity." Rickert was here referring to the proximity of Lask's thought to that of Aristotle, who was seen as having abandoned the idealism of Plato in order to situate "forms" in "matter itself."

something—life distinguishes that toward which it is directed by making it appear and thereby finding both a sense and itself. From this perspective, the object is not separate from the subject that knows it but is rather part of the relation as the direction of one's behavior and turning toward something: an orienting oneself according to given tendencies with a given aim in view and, at the same time, a being oriented according to given motives. Ipseity reaches itself through that which it reveals.

The factual occurrence of life in its pure motility as the going toward something generates meaning and, with it, a unique time of its own that evolves with its history. This uniqueness, whereby I and I alone am involved in every act, at the same time is something which occurs; hence, it is not something which I can fully govern. In this kind of endless pursuit of a path leading the self toward itself, what emerges is the ambiguous condition of being the self, which is always mine yet at the same time does not belong to me: for it would be more accurate to say that it happens (Gadamer 1975).

This unique relation, constantly in the process of being actualized, between being oneself and the situation at hand is ipseity. How can we grasp this imminence? How can we grasp the famous "not yet" of life which, by taking shape on each subsequent occasion, has the power to distinguish the world and define the self? How can we conceptualize the possibility of the self, the potential for meaning to establish itself, without betraying its essence? How can we grasp the facticity of life?

Whereas according to Fichte, the genesis of sense lies in the practical action exercised by a subject who acts upon the world before him with an aim in view—an absolute I that acquires both knowledge of the world and self-knowledge through action (*"Ich setze mich als tätig"*)—according to the young Heidegger on each subsequent occasion, the world becomes the world through its encounter with the life which in each case reveals itself through the act of discovering it. Unlike Fichte—and neuroscientists and neurophilosophers, who have more or less consciously adopted his perspective—Heidegger grasps immediate experience not as blind factuality, as the brute fact which according to neo-Kantians cannot be called into question and which remains mute without the determinations of the intellect; rather, he grasps it as facticity connected to a given sense, within a categorially structured context.

The immediacy of experience, therefore, is not irrational or ineffable: it is the living experience of an event which reveals itself as such to each person to the extent that he or she makes it his or her own (*Es er-eignet sich*). Evidently, facticity thus conceived can only be pointed at formal indication; on the contrary, the a priori deduction of factualities, which tears the concrete individual away from the condition in which he is embedded, inevitably strips the experience of them of its vitality, meaning, and historical quality.[5] In order to grasp the dynamics of being out toward something as they take shape, Heidegger offers a new conceptuality, whose function is only to provide a formal determination.

[5] Greisch notes "It is easy to see how this de-historicization makes accessing the concept of objective history possible" (p. 45, 2000).

4.7 Formal Indication and the Prereflective Sphere

The methodological answer lies in grasping the imminent act—the "not yet," the simple motility of being directed toward something that is typical of intentionality—as a formal indication. In order to account for the incompleteness of each person's being-there, the indication in question, while concerning intentional movement, does not specify any content.[6] Heidegger identifies the schematism of the intentional relation as the original structure of experience—the *verbum internum* of the phenomenon—that accounts for the factual way in which life relates to the world in each situation, with a meaningfulness that varies from person to person. Heidegger articulates the intentional structure according to three directions of sense which effectively enable us to grasp the "preobjective incipience" of the motility of life in its "motivated tendencies and tending motivations," without freezing it under the reflective gaze: (1) the sense content of the world (*Gehaltssinn*), understood as that which life regards as important and which it aims for—that by which life is drawn, captivated, and absorbed. This is the world, conceived as that dynamic meaningfulness which is constantly disclosed through our behavior and our relating to things and others in a meaningful way; (2) the relational sense (*Bezugssinn*), the imminent act, being directed toward something—that historically charged opening up through which the world reveals itself in all of its meaningfulness, while at the same time providing life with a means to access the articulation and cohesion of one's own world; (3) the sense of enactment (*Vollzugssinn*) that corresponds to the concrete determination in each case of one's encounter with the world with a given aim in view and therefore with the effective realization of life's access to itself in a given situation. On each occasion, life acquires meaning through that which occurs and at the same time opens itself up to, or shuts itself off from, the domain of the proper and its historicity, according to the direction which emerges at the very heart of the event.

Through the actualization of facticity as an event, life becomes a self. The sense of actualization coincides not just with an appropriating articulation of the situation (*er-eignen*),but with an intuitive and direct apprehension of the cohesion of the sense of the self, of the immanent historicity of life (*Zeitigungssinn*). In other words, the sense of self reveals itself in concrete life, in effectively taking care of this or that thing, but at the same time, it emerges as a historical sense, as the immediate apprehension of the overall coherence of one's self. An event, therefore, is always a historical actualization of the self, and it is starting from the way in which it takes shape that the original meaning of experience is factually constructed.[7]

[6] Heidegger will therefore seek to develop a conceptuality indicative of facticity through impersonal phrases: "It's worlding, It's contextualizing, It's properizing, It's destining (indicating unique directions of sense and sending us on our way), It's temporalizing, It's tensing the stretch and reach of my temporal playing field, It's propriating my proper being and proper time" (p. 50 Kisiel 2008).

[7] "The self, in the present actualization of the experience of life, the self in the experiencing of itself, is the original reality" (GA 59 p. 173).

If at the heart of human experience there lies an intentional moment which on every occasion articulates the historical sense of the motility of life in its being directed/directing itself toward something, it is precisely this moment in its multiple articulations that must be the focus of the fundamental concepts of a pre-theoretical science of experience.

The prereflective access to meaningfulness, which—as we shall see—also grounds all discourse possibilities, completely overturns the epistemological basis of natural-scientific epistemology. The most radical consequence of this—and the most incomprehensible one for psychology, which instead ought to be dazzled by it—is the fact that the immediate apprehension of meaning is not merely sensorial. What makes immediate experience sensical is neither its reordering according to the categories of the intellect nor the objectiveness of the elements constituting it but rather something that is not visible: the meaningful complex which anchors the sensory datum connected to the object within a given context with respect to a discovering ipseity, whose opening up is entrusted to emotional states. Hence, the encounter with this or that thing is actually a coming across weaves of meaningfulness that appear in the background of the world and within which something approaches us as relevant and comes to manifestation.

Heidegger's famous example of the lectern (*Katheder*), from his early lessons at Freiburg, has taught us that there is more to the fact of seeing a table than the brute datum, the sensible datum striking the retina: in being seen, the table points to a web of references that situates it within a context whereby the sensory datum—the table I see—immediately acquires a certain meaning. The table is seen not as an object of perception but rather in terms of its meaningfulness in relation to the surrounding world. Experiencing things on the basis of horizons of sense is part of the essential constitution of human life.

I will thus "see" the table in my study in the light of my worry of not finding the annotation I am looking for, i.e., as a messy surface where I keep my papers and notes. My seeing the table is a seeing it as something. On the other hand, the cleaning lady about to tidy up my studio will see the same table as an area that is off-limits, an area which she must not lay her hands on. The postman who comes round now and then to deliver some mail will see the table as a place for the more or less organized arrangement of papers and books, i.e., as my work station. The electrician who is repairing the ceiling lamp that lights the table will view the latter as a foothold for reaching the lamp. The student who is sitting opposite me to discuss his thesis project will cast a curious gaze on the table, trying to catch some clues about the topics I am currently working on. But even those who have no familiarity with the horizon of sense within which the table is embedded will encounter the latter not as a perceptible thing but rather as something unfamiliar. In all these cases, in the very act of being seen, the table—which theoretical knowledge grasps as a thing and which assertion fixes through language—points to a mode of understanding, to a seeing of it as something, which "impregnates" the sensory datum with sense. This sense coincides with the fact of being directed toward something and at the same time with the "how" of explicating understanding. In this respect, it is evident that the first person is not a supreme authority that theorizes and imposes a name on things.

4.8 Ipseity and Historicity

Sense does not point to the a priori of a constitutive subjectivity—less still to a normative objectivity; rather, it appears together with ipseity. The latter factually takes shape on each occasion, temporally, as concrete history, as the fact of being on a journey. In this respect, the self is historical: not because it is an object of historical science but because the nature of its actualization is radically historical. History is therefore a field of possibilities that extends far beyond what is present. This is the great insight of Count York which the young Heidegger made his own.[8] In this fundamental sense, on every occasion, the fact of being oneself is fully realized as the historical actualization of the self "toward," "in," and "against" the world; in this fundamental sense, the concrete experience of being alive is the history of this having oneself in each case in relation to the circumstances at hand (GA 9). As Qoheleth states "To every thing there is a season, and a time for every purpose under the heaven" (Eccl. 3.1). Time, grasped in the simultaneity of its three dimensions, must here be understood as constituting the motility of life, which can have this time in different ways and thus experience itself in different ways in relation to existing events.

In each situation, we lend shape to a transient structure of time by recombining our expectations, memories, and being present. Moreover, in different periods and in different conditions of existence, one or another of these dimensions may gain preeminence over the others, changing the mode of opening of existence. In this respect, the time we have—the time which in each case is our own—is always a time for something.

> There is a time to be born, and a time to die; a time to plant, and a time to pluck up that which is planted;
> a time to kill, and a time to heal; a time to break down, and a time to build up;
> a time to weep, and a time to laugh; a time to mourn, and a time to dance;
> a time to cast away stones, and a time to gather stones together; a time to embrace, and a time to refrain from embracing;
> a time to get, and a time to lose; a time to keep, and a time to cast away;
> a time to rend, and a time to sew; a time to keep silence, and a time to speak;
> a time to love, and a time to hate; a time of war, and a time of peace. (Eccl. 3.2–8)

In this having of time and bringing forth of possible ways of being, an existence develops and matures.

What emerges, then, are multiple articulations of factual experience. Personal experience takes the form of a being oriented toward something which, by making the world appear, gives the individual a sense of relation and a way to relate to something. This something is the direction which is followed by such orientation and which at the same time absorbs the person relating to it.

It is in relating to this or that thing, however, that experience comes to be concretely defined by the way in which it occurs at a given moment in one's life—in the

---

[8] See § 77 Sein und Zeit (GA 2).

light of the imminent future—while at the same time disclosing the possibility of direct, intuitive access to the past, of historical appropriation and conservation of the event through the cohesion of sense. Temporality, therefore, does not emerge as a succession of indifferent "nows" but rather takes concrete form according to the "how" of the enactment of experience; and it through this "how" that one's personal history acquires meaning. So not only does time prove to be the source of sense, but its performative character accounts for the fact that there is a time for every matter under the sun. Meaning is not something static but is rather inherent to the facticity of life and must be grasped in terms of life's phenomenological relation to time (Bambach 1993).

It is in these terms that Heidegger drew upon a famous passage from book 11 of the *Confessions*, providing a modern take on Augustine's text: "In you, my spirit, I measure times; you I measure, thus I measure time ... In you, I say again and again, I measure time; the things encountered transitorily bring you into my self-perception that remains, while those things disappear. My self-perception I measure in present existence, and not the things that pass by in order that it may first arise. My *finding myself* is what I measure when I measure time.[9]"

If time is the "how," the being-there which on each successive occasion encounters the world in one way or another, and thereby gains itself, it is itself time. In other words, who each of us is corresponds to the time of his living, the times of his life.

A psychology that is devoted to the study of the uniqueness of existence, and is therefore searching in a life history for the succession of horizons of sense, grasps a crucial indication in certain time lapses, proper to each person, which are marked by an overall reconfiguration of one's feeling alive. These are the turning points of one's life: those elusive moments in which transformation occurs because one season of life reaches maturation and disappears. Each of these instants is an end, a completion, but also a beginning that proves disturbing and confusing, a losing oneself in the dark before a new horizon opens up. In those decisive instants in which the indifferent flow of chronological time is broken by the now, the origin of all the transformations of life is enshrined: an origin which the passing time will cover and conceal. But this is also the origin that stands at the basis of the interpretative categories which articulate how life understands itself starting from certain existential turning points; and the

[9] In dir, mein Geist, messe ich die Zeiten; dich messe ich, so ich die Zeit messe... In dir,sage ich immer wieder, messe ich die Zeit; die vorübergehend begegnenden Dinge bringen dich in eine Befindlichkeit, die bleibt, wahrend jene verschwinden. Die Befindlichkeit messe ich in dem gegenwartigen Dasein, nicht die Dinge, welche vorübergehen, dass sie erst entstünde. Mein Michbefinden selbst, ich wiederhole es, messe ich, wenn ich die Zeit messe (GA 64, p. 111). In te, anime meus, tempora metior...In te, inquam, tempora metior;affectionem quam res praetereuntes in te faciunt, et cum illae praeterierint manet, ipsam metior presentem,non eas quae praeterierunt ut fieret:ipsam metior, cum tempora metior" (Book 11, ch. 27.36).

retrieval of such origin through the repetition and remobilizing of the conditions which engendered it enables one to reach the prereflective source of meaning.

4.9 The Natural-Scientific Access to the World of the Self

Academic psychology and all forms of psychotherapy adopt an unequivocal approach to lived experience. The therapist who bases his orientation on natural-scientific psychology and takes this as his starting point for engaging with life stories in his everyday practice is not interested in lived experiences as such but only in those lived experiences that can be turned into knowledge—knowledge, that is, governed by a theory of knowledge.

It makes little difference, then, whether the theory of knowledge guiding the therapist's analysis is oriented by criteria that lead him to distinguish between rational and irrational thoughts; whether his guiding principles are motivational systems such as attachment, sexuality, and intersubjectivity; or whether what he is seeking to discover in the daily lives of his patients are inadaptive beliefs and thoughts and inadequate narratives. Certainly, this list could further be expanded; but in all cases, the mode of access to phenomena and their predetermination is always ensured by the kind of theoretical attitude that enables the therapist to conceive concrete experience in terms of objects of knowledge. In other words, in all cases, experience—uprooted from its original contexts—is assigned meaning on the basis of a priori categories, in line with the ideals of knowledge guiding one's investigation.

This mode of access to lived experience and life stories on the one hand removes experience from that background of living actualization within which it is embedded, dehistoricizing it, and stripping it of its meaningfulness; on the other hand, it recomposes experience within an impoverished dimension by presenting the patient with a filtered and resignified view of it. Thus one's relation with the experience under investigation is mediated and marked out by the theoretical orientation imposed upon the patient in the "therapeutic" relationship. In other words, what has happened in the life of a patient is grasped by the therapist objectively, according to a given theoretical determination.

The patient's way of relating to his own experience is therefore made to coincide with the actualization of a theoretical context (relative to the tendencies of the therapist's school), whereby what is understood as personal experience is actually found to be the correlate of the therapist's theoretical idealization. Consequently, even though the knowing subject—the patient focusing on his own history—is completely absorbed by the relation with his own lived experience, this occurs in such a way that while he epistemologically grasps himself in an objective way, he rules out any possible renewal within his own actual life. The patient knows himself very well, and precisely for this reason nothing changes, as may be inferred from the following notes taken by Mr Smith, a 70-year-old man who underwent a CBT therapy for 25 years.

Jun 2, walking 2 min in hot 36 deg, weightlessness/lost equilibrium-would have fallen if my son was not next to me to hold onto-triggered Panic Attack for several hrs.

Since then living in constant fear of falling off somewhere in street or alone at home.

2nd type but similar: for several months have had and deteriorating the following.

From end of most meals till 2 h afterwards—sudden feeling of fainting (not weightlessness), mouth watery, ext fatigue, nausea trouble focusing—Occasionally with hi twisting pain—feeling the pain as it goes down from stomach to abdomen—must lie down immediately.

Happens also after snacks in between meals.

Need to have snack between meals—hunger craving: must eat even in the street.

Cardiologist rules out the heart—Vestibular specialist does not think it is vestibular now –

They think it is a digestive problem—Vegal parasympathic nerve—or dumping syndrome—

For 25 years world renowned professors, vestibular specialists, psychiatrists, psychotherapists, using the latest meds with my limitations, not only have not produced meaningful results, actually I feel worse.

To be fair part of my problems are due to advancing age with major diseases.

Professors and medical professionals agree and have come to the conclusion that all of my problems are serious diseases or chronic, with no remedy.

Since none of the attempts to cure, alleviate, the symptoms have worked, clearly need to look at my deteriorating problems from a different angle.

A more fruitful endeavor would be not to try to cure the diseases, but to accept/embrace and live as well as possible while having the symptoms. It is ext difficult.

Do you agree?

*Have come up with this mental argument to tell myself whenever I feel dizzy with feeling of passing out, several hours daily now. Repeating this several times per day.

There is no cure for your dizziness, vertigo, GAD, IBS, MRV, incontinence, etc.

"You are not unconscious, so accept, Get used to and embrace your dizziness with other ailments that will be with you the rest of your life with advancing age.

Learn to live and function without having control"

This way (if you can really accept your most uncomfortable hrs of daily symptoms).

The solution is to start to become/to practice, in being without anx when with yourself.

Difficult to stay alone without anx because, only way you know yourself is being anx.

To try to accept the state of relaxation that is unsettling to me, particularly when alone.

The more I try, the more anx shoots up. The result is feeling even worse than before trying your method. Am becoming conditioned now to feel worse when trying your system.

Is there a way to do this without being conditioned for a worse experience the next time? Should we still try graded exposure therapy of CBT for the above?

Thu 23 Oct at 6 PM.

New developments—while you say I have no OCD, the following thoughts have come up.

Impulse Phobia

3 wks ago, at the end of dinner, wife had to turn her body and head back very close to me.

Suddenly got the frightening thought of hitting her on the head –

Becoming terrified of the thought, triggered immediate hot flash and mini panic.

Thinking, what if I go crazy and act irrationally.

Asking myself, why am I thinking like this especially regarding someone so dear to me?

Remembered/repeated mental arguments given to me by Mme. XXXXXX 20 yrs ago when I had the same impulse several times. The arguments were:

Phobia of impulse is never acted upon.

You have never tried to physically act on anyone in your life except to defend yourself.

This is just a weird thought.

These arguments comforted me immediately and the thought passed in a few min.

Discussing it with Prof. XXX. She said

- You call it impulse phobia.
- In fact that is an absolutely classical obsessive thought.
- What if I act on an impulse and harm someone. Particularly a loved one.
- Which would be the last thing you would want to do –is harm your wife.
- Also called a taboo thought—being the thought of causing harm.
- And your response was the absolutely correct one –
- Is to challenge it—to reality test it—and to dismiss it.

This passed more or less.

Now during meditation or wake-up in bed, I imagine a dragon is coming through the windows.

While I realize and am fully aware of the absurdity of this thought, it triggers hot flashes and panic symptoms that stick and get worse and worse, until I force myself to get up.

Even this is difficult as I feel I am frozen to the bed or recliner.

Quite worried about these new phobias that are sticking as soon as I am not occupied trying to rest in bed or recliner. Making it impossible for me to relax during the day.

Need positive mental argument response to calm me down.To prove this is not so .

Unfortunately when physical symptoms are being managed, then as if I am looking for something, I start imagining unpleasant things.

Scared of what ifs: am I going mad? Is this beginning of schizo?

Why do I always look at the remote possibility of bad neg catastrophic things happening.

How do I change this to a more balanced thought of positive things happening.

In accordance with this attitude, factual experiences are not conceived in terms of what they actually are in people's lives, but in the light of those determinations imposed by the theoretical perspective through which they accessed. The view which is presented to the patient interprets lived experience as though it were the particular instance of a general law, and in doing so inevitably discards or flattens all effective occurrences. In the light of an a priori systematics, the therapist identifies broken mechanisms and distills other people's experiences as though they were the constitutive elements of an unbroken process: the self in process. This unfolding of this process in time is guided by fundamental meaningful trajectories that the therapist alone is capable of recognizing.

4.10 Theoretically Oriented Psychotherapies

This approach to history, which subjects temporality to supratemporal principles and reflects a geometry of reason that demeans the motility of life, represents the basic feature shared by all contemporary forms of psychotherapy. Paradoxically, the products of this cognitive attitude are considered real precisely because they are based on natural-scientific evidence, which constitutes a guarantee for those psychotherapies capable of producing it.

During a course of treatment, the "therapeutic" reformulation of the sense of one's experience, burdened by the weight of the therapist's authority, is assimilated by the patients, who by accepting the therapist's interpretation end up applying his theories to their own lives. Consequently, in line with the philosophical tradition of healing based on the principles of reason, the therapeutic relation always takes the form of a (pedagogical) relationship that requires the patient to identify and learn a series of practices for managing present experience in the light of those supratemporal principles which the therapist alone can access by virtue of his knowledge.

A therapist, for example, may pass from an ongoing critical situation to an analysis of the patient's childhood behaviors and experiences, connecting them to the sentimental relationships of his adolescence and youth, and his present relationship with his spouse and parents. He will then interpret the therapeutic relationship itself according to the same interpretative key, identifying in each of these passages the same form of attachment as the invariant which, by enduring from the cradle to the grave, governs the construction and dissolution of the patient's ties of affection and emotional organization.

Alternatively, the ongoing situation of crisis may be understood as the consequence of the poor functioning of a system of information processing which through the therapist's a priori template is isolated, analyzed in its effective articulations, and corrected in the light of rational parameters for the implementation of efficacy or adjusted in such a way as to restore the patient's mentalizing capacities.

All this is even more evident in the metacognitive approach, which may be regarded as the most extreme form of this mode of operating, and hence the poorest one from the point of view of the understanding of lived experience. According to this perspective, it is assumed that every relation (with one's own lived experience) as such belongs to a particular class of objects—the class of relations—which entails a kind of access that is even more radical from a theoretical perspective: one based on an original distinction that is seen to characterize every relation in general. According to this mode of operating, the patient is required by his therapist to grasp and modify the way in which he relates to his own relation with lived experience, i.e., with the metacognitive beliefs that are said to be the beliefs underlying the individual's experiences, thoughts, and procedures. The implementation of this "therapeutically" determined cognitive behavior transforms the patient into a reclassifier of his own cognitive processes, creating an enormous distance between him and actual experience.

Therapeutic interventions of this sort become both a theory of knowledge and a destiny for the patient: a theory of knowledge through which to reformulate one's lived experience by resignifying it in the light of the principles learned from psychotherapy, but also a destiny, insofar as in each case concrete experience is newly understood precisely in the light of those invariant principles guiding one's personal development.

This process is also clearly reflected by the "historical" method shared by a large number of apparently different "schools." The most simple case is that of the techno approaches which regard each patient's ongoing problem as a kind of malfunction (corrupted file analysis, information processing, the substitution of narrative scenarios, the improvement of metacognitive skills, rational-irrational computation, pattern and systems analysis, etc.) and treatment as a way of repairing this malfunction by fixing the broken mechanism. Leaving this case aside, all those therapeutic practices that analyze one's personal history do so guided by the genealogical reflex. In other words, the problematic quality of the present situation is dealt with through an archaeological reconstruction of those invariants that, by remaining constant throughout the various stages of the patient's life, provide the coordinates for understanding why his current existential dimension is what it is.

The genealogical reflex, which owes its consolidated and unprecedented success to psychoanalysis, is the crudest expression of that fundamental orientation which psychotherapies have drawn from the ontology of presence and according to which what changes can only be explained in the light of what remains unchanged.

4.11 The Primacy of Theory in Constructivist Psychotherapy

Also proceeding along the same trajectory, but with a greater methodological emphasis, are those therapeutic practices—mostly constructivist ones—that promote a reflective focus graded according to different degrees of magnification and distinction, with the aim of reconstructing lived experience so as to gain access to its immediate dimension. In different ways they all unwittingly follow Natorp's (1912) attempt to regain the subjective sphere through the painstaking reconstruction of experiences: a reconstruction that purports to provide access to the immediacy of the sense of lost experience.

As is widely known, the reconstructive method was Natorp's answer to the limits of the intuitive-descriptive approach of Husserlian phenomenology. Natorp's criticism—the only kind of criticism to have put forward scientifically relevant objections to phenomenology, as Heidegger was to note—was chiefly directed toward the immobilization of lived experience brought about by phenomenological analysis. According to Natorp, phenomenological observation entails the objectivization of subjectivity and is therefore incompatible with that principle above all principles, the return to things themselves, by which it claims to be guided. The core of Natorp's criticism of Husserlian phenomenology lies in its attempt to restore the foundation and mobility of individual lived experiences, which the reflective-descriptive attitude of the phenomenologist inevitably immobilizes.

Natorp's argument may be summed up in two points: (1) if experience coincides with the relation between subject and object, in order to experience himself, the subject must reflectively grasp his own experiences as an object and (2) if, however, the individual grasps himself as an object, he cannot experience himself as a subject, since through reflection all that he grasps is a frozen version of the flowing of subjectivity. Natorp's solution is to circumvent the immobilizing and distorting aspect of reflection by grasping lived experiences as an objectification of subjective experience. Lived experience is the object on the basis of which it is possible to restore—through the reconstructive method—the immediacy of the very subjectivity that engendered it. The method, then, consists in retracing the process of objectivization of knowledge "down to what constituted the living element in consciousness prior to any objectivization" (p. 38, 1912).

Leaving aside the evident problem of how we may return to the sphere of immediacy through the mediation of reflection (GA 56/57), the point which links the reconstructive approach—shared by psychologists ranging from Kelly (2003) to von Glasersfeld (1995) and Guidano (1991)—to the other perspectives based on the primacy of theory is the notion that reconstruction only becomes possible on the basis of the objectivization of experiences accomplished according to an a priori order, understood as that to which history is subordinated.[10]

In other words, the breakdown of experience into discrete, punctuated elements, with the aim of retrieving the original life of one's consciousness, is achieved on the basis of objective criteria, which constructivists have striven to epistemologically justify from the very beginning. For example, over the course of the analysis of a relationship, the identification and detailed definition of different episodes related to the development of such emotional bond and of the experiences related to it are carried out in the light of a series of natural-scientific principles—e.g., cybernetic or ethological-evolutionary ones—that make it possible to lend shape to constellations of categories (constructs, meanings, organizational or motivational invariants, etc.) through which to articulate the patient's subjectivity. Unsurprisingly, the various currents of the twentieth-century constructivism have widely drawn upon evolutionary biology, general system theory, and the epistemology springing from it to find a justification and foundation for their a priori criteria.

The next step lies in actual reconstruction, which restores the continuity and unity of the selected experiences by reintegrating them as a more articulate and abstract sum of points, thereby enabling the restoration of the subject who has generated these lived experiences: the more the "immediate" elements of consciousness have been distinguished, the more effective is this step. As emphasized by Arrien (2009), for Natorp, it is a matter of tracing the various discrete moments (concrete lived experiences) back to a function expressing their possible continuity (consciousness); hence, his reconstructive psychology may be seen to rest on the mathematical model of the resolution of an integral.[11]

The determination of the subjective sphere being the purpose of psychology, neo-Kantism and the popular constructivist approach (stemming from it) reach this sphere starting from the logical determination of the multiplicity of experience, of immediacy, on the basis of which they then reconstruct the subjective domain. The reconstruction of lived experience therefore amounts to the grasping of immediate experience through the mediation of theory. As the determination of the subject follows upon the distinction of its immediate experiences—according to (scientific)

[10] The latest example of this mode of operating in psychotherapy is Guidano's techno-constructivism, which translates Natorp's reflections into a technical instrument, the "moviola" method. In the light of the orienting principles adopted in order to access the past, the "application" of this method to the flowing of memories slows down the sequences of images, making it possible to study them and bring them into focus through individual frames, as though lived experience were present and available to one's consciousness.

[11] In mathematical analysis, the integral is an operator which, in the case of a single-variable function, associates the function with the area bounded by its graph within a given interval (a, b) in the domain.

laws—the reconstruction itself is subject to the supremacy of theory. But what attests to the fact that the reconstructed consciousness matches that of the lived experience prior to the analysis? The only certain thing is that the theoretical access structurally rules out all possibilities of grasping experience if not through its objectivization, de facto allowing the reconstruction to access exclusively the logical, objectifying sphere. The reconstruction thus determines the modelization of the knowing subject—for instance, according to a hierarchy of constructs, as in Kelly; a system of knowledge, as in von Glasersfeld; or an organization of personal meaning, as in Guidano. The method of reconstruction leads from the unity of the multiplicity of facts, and hence the objective domain (e.g., the facts pertaining to the relationship analyzed during therapy), to the unity of consciousness (the system of knowledge which has generated it). The logic of knowledge is thus what grounds psychology. On these bases, all the various currents of constructivism, bar none, have made epistemology the foundation of psychology, as Kelly's metaphor of man as scientist emphasizes.

Consequently, attaining a greater, more abstract, or indeed inexhaustible knowledge of one's own cognitive processes has become the chief aim of treatment.

Both empiricist and transcendental psychology, but also descriptive and analytical psychology and phenomenological-eidetic psychology, while being based on distinctive modes of grasping the immediate data of experience and hence on different ways of approaching the study of the subject, ultimately all grasp the latter as a nobody, to use Ricoeur's wonderful expression: for they all grasp the subject according to a theoretical configuration which inevitably rules out the prereflective domain of life. So the forms of therapy deriving from these psychologies also adopt the underlying view of subjectivity that guides them and which justifies and orients their practice. In other words, the therapist's modelization of the patient's mode of being according to the guidelines of a general subjectivity—articulated in different ways according to one's epistemological perspective—shapes the practice of healing and hence the delineation of the patient's experiences.

From the patient's point of view, then, to undergo treatment is to reorder one's experience according to impersonal guiding principles and to upgrade one's system of functioning according to the more effective template provided by the therapist.

This is how lived experience becomes theoretically warped in psychotherapy.

References

Arciero G (2006) Sulle tracce di sé. Bollati Boringhieri, Torino

Arrien, Sophie-Jan. 2009. Natorp et Heidegger: Une science originaire est-elle possible? In Jollivet, S., & Romano, C. Heidegger en dialogue (1912-1930). Rencontres, affinités, confrontations, Paris, Vrin.

Bambach CR (1993) Phenomenological research as destruktion the early Heidegger's reading of Dilthey. Philos Today 37(2):115–132

Gadamer HG (1975) Truth and method (1960). Crossroad, New York

Gethmann CF (1993) Dasein: Erkennen und Handeln: Heidegger im phänomenologischen Kontext, vol 3. Walter de Gruyter, Berlin

Greisch J (2000) L'Arbre de vie et l'Arbre du savoir.: Le chemin phénoménologique de l'herméneutique heideggérienne (1919-1923). Cerf, Paris
Guidano VF (1991) The self in process: toward a post-rationalist cognitive therapy. Guilford Press, New York
Kelly G (2003) The psychology of personal constructs: volume two: clinical diagnosis and psychotherapy. Routledge, London
Kisiel T (2008) On the genesis of Heidegger's formally indicative hermeneutics of facticity. In: Raffoul F, Nelson ES (eds) Rethinking Facticity. State University of New York Press, New York
Natorp P (1912) Allgemeine Psychologie nach kritischer Methode, vol 1. JCB Mohr (P. Siebeck), Tubingen
Popper KR (1972) Objective knowledge: an evolutionary approach. Oxford University Press, Oxford
Quesne P (2004) Les recherches philosophiques du jeune Heidegger, vol 171. Springer Science & Business Media, Dordrecht
Sheehan T (2014) Making sense of Heidegger: a paradigm shift. Rowman & Littlefield International, London
Von Glasersfeld E (1995) Radical constructivism: a way of knowing and learning, Studies in mathematics education series, vol 6. Falmer Press, Taylor & Francis, Bristol

Self-Intimacy and Individuation

5

5.1 The Domain of the Proper and Situations

One's own world has nothing to do with the emergence of interiority through the observational and/or reconstructive retreat of self-consciousness within an inner space, a receptacle of occurrences that is ordered—and can be reordered—according to the laws of causality. As Dilthey discovered, we cannot be the spectators of our own lives because life is not given at a distance, but only through that life which we ourselves experience. "Life addresses itself in its own language" (GA 58 1992, p. 31), the young Heidegger observed, adopting this perspective.

I am concretely myself in particular, actual situations and in each situation I am, or am not, present to myself. Here presence to oneself is not to be understood as the presence of an object that is at hand, nor as a representing in one's mind, but rather as a becoming present, a presenting oneself, a letting oneself encounter what appears. In other words, while ipseity takes shape in an encounter, the underlying experience that precedes this encounter—ipseity in each case representing a kind of accentuation of the experience—opens up in and through the encounter. Thus the domain of the proper manifests itself in factual situations and given moments, and the situation is "its place," the field in which life addresses itself, in which I perceive the world I live in as being more or less familiar to me.

The category "world" simultaneously brings the content of experience and the sense of reference into play. As has been repeatedly noted, while on each occasion ipseity refers to that which manifests itself—i.e., the world—at the same time, it is within the world that ipseity finds what in each case presents itself to it: the content of lived experience, which is therefore the content of life itself.

If living may therefore be interpreted, according to the sense of reference, as the taking care of this or that, the world may be seen as the fundamental category of the sense of content, as it coincides with the multiplicity of references in life and hence with its very horizon.[1] The flowers, neurons, things, and stars encountered in terms

[1] The categories are rooted in actualization.

G. Arciero et al., *The Foundations of Phenomenological Psychotherapy*,
https://doi.org/10.1007/978-3-319-78087-0_5

of a "taking care of" are experienced in their meaningfulness, which is relative to the way in which an object subsists in the world and relates to it (the totality of references), to the way in which an object belongs to a given context. Conversely, the way in which experience gives itself, the way in which I am involved in experience, also expresses the manner in which I am attuned to the world in terms of meaningful spheres and directions. "Appearing—Barbaras writes (2007, p. 36)—entails in and of itself a reference to the subject of appearance, a subject which does not accomplish appearance but is founded upon it as one of its constitutive moments."

The world, understood as the fundamental category of the sense of content, constantly varies in relation to the way in which I enact my taking care of it. The different ways in which the sense of reference is realized thus correspond to different interpretative categories through which the world appears and, at the same time, to primary modes of life's access to itself.

It is evident that if we regard the self and the world as two objects facing each other, according to the perspective adopted by the natural sciences and, more generally, the theoretical approach, we lose the possibility of grasping the origin of meaning: the act of relating as something which is experienced and prereflectively intended. Life cannot be grasped and possessed by freezing it under the theoretical gaze, because life only understands itself—in each and every instance—through the world in which it finds itself. We are interested, seduced, captivated, stimulated, touched or angry in relation to the events which enable life to "notice" itself. Within this unstable set of circumstances, my own life always simultaneously coincides with everything which surrounds it and with the others it relates with. Yet life is synonymous, not just with the perception of the moment, but with the horizon of my expectations and the field of resonance of my memories. On each occasion, ipseity becomes manifest through that which it comes across and which takes shape within a phenomenal sphere, which is invariably historico-practical.

Not only do I always encounter the surrounding world according to a certain perspective—ensured by the nonrelevant role played by the domain of the proper and corresponding to the sphere of intelligibility which envelops factical life like an aura—but the situation in which *I find myself* is, at the same time, always a peculiar mode of access to myself: a way of being familiar, or unfamiliar, with how the cohesion of my life takes shape. This means that, on the basis of what I experience, by relying on particular motives and horizons, I have the possibility of becoming more or less intimate with that context of intelligibility which constitutes my own world and within which I exist (presence to oneself). A certain way of grasping the world (by finding oneself within it or not) will correspond to the accentuation of a series of aspects of the sphere of the proper, understood as that sphere of opening for and in a context of tendencies and motivations which takes shape ever anew in factual life.

This intimate, full, and obscure articulation of sense that life unfolds through living is profoundly different from the process of self-observation of the I. A psychology focusing on the individual person in his or her uniqueness is required to grasp the person's story by taking life as its starting point, rather than by destroying the vitality of life by tracing it back to the "simplest" elements identified according to the a priori determined by the theoretical approach. In other words, the possession of the self does not consist in isolating and immobilizing this or that experience, but rather in the living, critical conservation of one's past, stimulated by the present tendencies in one's life; in each case it comes into appearance, and in each case it is a phenomenon.

In the domain of the proper, then, the manifestations of the world and the various forms of expression of the self intertwine, according to the peculiar rhythm of each person's mode of experiencing.[2] The encounter between factical life and the sphere of meaningfulness offered by a particular occurrence co-determines the way in which the underlying experience that precedes and orients effective experience takes shape and with it the cohesion of life. In this respect, every situation is an event that becomes immediately present to me (and not others): an occurrence that becomes part of my own life and enters into resonance with my own story.[3]

5.2 Everyday Life

Let us focus on a banal sequence of experiences on an ordinary day in my life. I am expecting a couple of friends for dinner tonight. I go out to buy some wine, walk down the stairs, and do so with a sense that soon I will be going back up them for three floors. I hate these stairs. I have almost reached the bottom when a vague sense carries me back to 15 years ago, slipping in between my thoughts without even letting itself be touched: I was living in Rome at the time, on the last floor of a building with no elevator. All this is not even an image or memory: it is more like an atmosphere that I grasp for a fleeting instant. I am in the street. I head toward the wine store. As I cross the street, I am struck by the square: it looks like a square in Le Marais, in Paris. I had always wanted to move back there, but then I found a place here in Geneva with a Parisian feel to it. I am leaving for Paris tomorrow but will only be staying in the city until evening. I need to drop by my favorite bookshop to see if among the secondhand books I can find a certain volume by X that is out of print. I wonder if I will have enough time. Meanwhile, I have reached the Russian church. A blonde woman in her 50s makes her way down the steps of the church, shrouded in a gossamer veil. I am gripped by a vague feeling, like the trace of the presence of a Polish friend of mine, from my youth. Oh, youth! It is as though it were still here and not decades away… As I am about to cross a bridge, I imagine the view of the lake concealed by the grassy hillock. On the sidewalk three clerks in short sleeves are rushing off to have lunch. They look like bank clerks. They are having a listless conversation—perhaps they are discussing how to invest in the

[2] The domain of the proper is grasped as a relation that is always underway starting from the ongoing situation.

[3] From this perspective, the attempt to reduce the understanding of others to the exterior perception of acts or expressions of the human body—by tracing these back to psychic connections and ultimately completing the constitution of the flesh viewed from the outside by *Einfühlung*—seems contrived at best.

stock market... This is a sequence of experiences like many others from my everyday life. I go out to buy some wine, walk down the stairs, find myself in the street, prepare to cross a bridge, etc. As I experience these and other things, I experience myself – and now and then encounter myself.[4]

All the possible ways of reaching myself, and of experiencing the environmental contexts which I come across, emerge through my position in actual life. In other words, even in encountering the most banal circumstances, I am oriented by that sphere of possible understanding, by that more or less stable sphere of the possibility of sense, in which I largely unreflectively live according to certain tendencies and a network of motivations, which actual life sets in motion. At the same time, these also constitute an anchoring to the saliencies of the world and my relations with others, as well as a center and direction of gravity.

On the other hand, my position in the world, my familiarity with others, and the sense I have of myself are in each case elicited by the occurrences that, in meeting me, mark out as many specific "places" of self-understanding. The experience of novelty can only be understood by taking into account the inhibiting of the familiarity I have with myself in my mode of existing. In actual experience, it takes shape as that which falls out of the temporally particular context of expectations carried out by factical life.

What occurs is always experienced within the framework of a particular meaningful context and hence in the light of given motivations and expectations that lend life a direction, stability, rhythm, and sphere of intelligibility which every situation mobilizes anew. This is the always-to-be-established domain of the proper in the light of which one lives.

As already noted, the domain of the proper gives itself in an intelligible way in concrete situations, and experience represents the expression of it. Even the content of the most banal experiences presents itself in everyday existence according to a certain "how," which is to say, within a factual context of manifestation and hence in the light of possible motivations and tendencies.

Experiences, then, are not accessible as isolated entities, as things; rather, they appear to be the expression of actual life situations. Ipseity takes shape in each case according to a unique weave of occurrences and to a sphere of intelligibility which occurrences remobilize and recall on each occasion, de facto regenerating the access to what has been.

It is therefore according to a given access key that sense is elicited in terms of an indication, manifestation, or more or less acute accentuation of the domain of the proper: it is through what occurs that I have—or do not have—myself. And it is in this attempt to see myself on the basis of the unfolding of my life that the latter shows or conceals itself, since there is no need for life to step outside itself in order to grasp itself: life addresses itself in its own "language," to quote Dilthey. It traces, searches for, and grasps its own limits; it establishes its own objectives, perspectives, burdens, and questions; in other words, in each case, it brings its tendencies to

[4] Greisch is particularly fascinated by this patchwork aspect of the "carpet of life" and identifies it in Stefan George's poems (2000).

fulfillment in the context of its own spheres. I only grasp myself through this condition, which cannot be transcended and which determines all the directions of my inquiry.

Heidegger describes this basic aspect of life as self-sufficiency. This self-sufficiency is to be understood in phenomenological terms, and that is to say, as the generation of sense in relation to the meaningfulness we encounter. It has nothing to do with the biologization of life or with the positing of impulses or motivational principles determining its articulation. With its self-sufficiency, life always addresses itself in its own language, coping with its incompleteness through the creation of meaning: the original motility of existence makes the world appear as it penetrates it, becoming visible and traceable. Quesne (2004) observes: "The self-sufficient expression is ambivalent: at times it indicates a positive character of factical life, namely its movedness (*Bewegtheit*); at other times, it indicates the dangerous autonomy which this factical life can possess in relation to meaningfulness (meaningful contexts), insofar as it only possesses this ontological autonomy within this relationship of retrieval of meaningfulness" (p. 166).

The actual course of one's existence, then, appears to be marked by the possible opening or closing of the dimension of sense, by the loss or achievement of the presence to oneself in the various situations which punctuate the different periods of one's life according to a certain ever-renewed "how." Personality is this individual rhythm articulated over the course of one's existence: the emergence and loss of self-intimacy characterizing one's expressive intensity but also the disclosure or concealment of the past, connected to the mode of manifestation of the world one experiences. What we have here is a sphere of accessibility, a range of possible modes, a personal style with which we find ourselves among things and with other people in different moments of our life, a rhapsody of attitudes with respect to which the contents of life can be delineated in all of their diversity.[5]

If ipseity constitutes the starting point of psychology, one closely related question concerns the fact that its situational actualization corresponds to the disclosure of a coherent sense, of a style of experiencing, within each given experience. By adopting this perspective, in the subsequent chapters, we shall see how the movement of therapeutic care unfolds precisely within the experience of encountering and relating to our own lived experience, which is opened up and made possible by anything significant which addresses us. Therapeutic care consists precisely in the restless reclaiming of my own personal story in the light of the way in which the things I experience lead me to question myself and hence through the thematizing of the perception of self-familiarity distinguishing these experiences. In such a way, it is possible to bring to light the tendencies and motives supporting the factual articulation of existence on each occasion.

As will become evident, the therapeutic method has nothing to do with technical knowledge and its application to individual stories, since it is rather based on the very unfolding of life. It will be shown, therefore, that access to the self-intimacy of

[5] Personality, in other words, does not correspond to a theoretical category, but is rather a formal indication.

individual life begins with the articulation of sense through which it is factually realized. Psychotherapy sets out from life itself: for life itself is forced to choose the care of self as a means to achieve its fullness by its very unfolding and hence by its very motility, which constantly erodes the meanings it engenders. Only in such terms is it possible to understand why one of the crucial elements for the reconstruction-appropriation of a patient's personal story lies in the therapist's sensitivity to discover and indicate those events whose meaningfulness has shaped the actual course of the person's story; this sensitivity represents the cornerstone of every therapeutic relationship.

5.3 Ipseity and Individuation

The cohesion of my own life is mobilized on each occasion according to a given meaningful direction, determined by the ongoing event. The tendencies and motivations that are generated by the flow of existence—and which are never permanent—provide, on the one hand, a context of manifestation for the development of the mutability of everyday facts and, on the other, the kind of background of intelligibility and of possibilities for immediate access that accompanies every human being and by which he or she is relationally inclined in a unique way.

One is led to wonder, then, how prominence in the sphere of the proper—which is related to a certain way of perceiving the life revealing it—is generated, realized, and lifted out of a concrete and indistinct background of available experiences as one particular experience.

I walk down the stairs, cross the street, reach a bridge, and go and buy some wine: my everyday experience. I live within an actual experiential weave that coincides with my being absorbed by the immediacy of things, by the sounds of the world, by the fullness of meaning. Such is our banal everyday life. Ultimately, we can only grasp the act of experiencing as it produces itself: we can only intuit experience by apprehending it as it actually comes into being. No matter how banal it may be, every instant in this flow is significant, and anything that becomes part of my experience is contaminated by it, so to speak, as it engages and moves me according to ever-new contents.

Heidegger writes: "I live in the factical as in wholly particular contexts of meaningfulness, which are continually permeating one another, i.e. every meaningfulness is meaningfulness for and in a context of tendency and a context of expectation, which constructs itself ever anew in factical life" (GA 56/57 1999, p. 84).

As we have seen, being constantly enmeshed within a web of meaningfulness does not correspond to first-person experience—it has nothing to do with any self-referential system. As Richardson notes, it is something "pre-subjective" (1988, p. 188).

This is a crucial point with respect to that Kantian tradition which was made even more acute by Fichte: life is not a self-enclosed phenomenon. What is given is not a world, not objects or other individuals, and then, subsequently, the meaning they have for me, within me; nor is it a subject that constructs a world, generating within his interiority. Rather, what is given is a fundamental lack of distinction between what absorbs us and our being absorbed. Patočka calls this the sphere of appearance—of which the world and ipseity merely constitute two traits—the phenomenal field and outlines its composition as follows: "We consider this universal totality of what manifests itself, the great whole, as belonging to the structure of appearance as such, as well as to that to which what manifests itself appears, namely subjectivity (which has an empty pronominal structure, not to be identified with a single fixed subject) and the 'how' of appearance" (1995, p. 177).

Therefore, in each case, what appears coincides both with an appearance of lived experience and an appearance of the world. In every instant, feeling alive corresponds to this experience which on each occasion finds completion and renewal in the meaningful connections that we come across. We bear traces of these events, as a living text that is to be reinterpreted again and again in the light of the world.

So as the world becomes viable and visible, the experience I have of it becomes passively and actively "available," so to speak, in the form of elements that can be mobilized anew in relation to the course of my actual life—as conditions determining the production of subsequent experiences or as the entrenchment of an ability: a habit.

The world which life touches upon takes shape as the world of that life: a life which, without any first person, receives itself and historically appropriates itself in relation to the complexes of meaning which it comes across time after time through its own factical flow. The meaning of living is therefore generated in each case through the way in which life—my life—de facto understands, expects, imagines, desires, behaves, and remembers within the context of the fullness of the experience of how it perceives *itself*. This generation of meaning is always suspended upon an expectation, a "not yet," the structural deficiency of life, its need of enjoyment (Levinas 1979), its desire—which in every instant corresponds to a new context of expectation that enables the world to appear and life to actualize and manifest itself.

Clearly, all this cannot be described as a judgment on the part of the body, understood as a "mode of cognitive grasping of the world," which is how cognitivists have viewed it, surfing the recent trend of embodiment (Solomon 2004, p. 77). Less still can embodiment be conceived as the perception of oneself elicited by the immediate experience of the world and ensured by a body, understood as a material entity—a fragment of organized biological matter—which is how it is conceived by neurophenomenology, in keeping with cybernetics, thereby revealing the idealistic undercurrent of this discipline. Once again, the two perspectives overlap when it comes to this theme, since they share the same ontology and therefore the same view of embodiment, which they interpret by taking consciousness as their starting point, ultimately regarding embodiment as one of its modes. Whether we are speaking of embodied cognition, embodied simulation, the embodied theory of mind abilities, or the embodied affects, the body is conceived as a mode, a condition that enables the living consciousness to make sense of the world. This even applies to Merleau-Ponty's phenomenology, since—as Barbaras (2007) notes—it designates the body as an entity within the world whose property is to enable its perception, instead of thematizing the body as what emerges as such through its being-in-the-world. In other words, it does not grasp the body in the light of an ipseity which, in bringing the world into manifestation ever anew, reaches itself through that which it causes to appear, thereby establishing its substrate.

I walk down the stairs, cross the street, reach a bridge, and go and buy some wine: the concrete actualization of life corresponds to the direction it takes on every successive occasion, starting from the still undifferentiated motility that constitutes it and which in each case plunges, so to speak, into meanings, into lived experience. This tension toward something, this restless possibility of being something, finds its determinations in terms of content in the concrete enactment of directions of sense, in the establishment of living, actual relations. The way in which I encounter myself, then, the familiarity or lack of familiarity with my own self, the fact of possessing or not possessing myself, is not given theoretically, but in each case corresponds to that meaning which is generated—in a fully concrete unity—in a concrete vital relation, in the factical realization of life.

Seeking to confine life to stable forms, to a clear-cut conceptuality, as a psychology resorting to the method of natural science has done, means betraying the fundamental essence of life: the transitoriness of one's being on a journey, the motility of life (*kinesis tou biou)*. This constitutive transitoriness of life is such that in each case what is at stake, on a day-to-day basis, is oneself. Hence the rigor of psychology is understood as the science of personal meaning, which can only differ from that of the natural sciences: it has to do with the capacity to grasp the original production of sense in actual life.

Even in the most banal everyday experiences, when we attempt to grasp ourselves starting from our life, which is what we are, we once again come across those structural aspects which Aristotle regarded as the basic features of sublunar entities: the motility of life and its incompleteness—that incompleteness which characterizes every entity in motion.

Yet, at this stage, something new appears. The motility of life now manifests itself, not as the change from a condition of incompleteness to one of completeness (as selfsameness endures in its being), but rather as an inclination toward a "not yet," a still indeterminate tension, a being turned toward something (*Bezugsinn*) that in each case concretely emerges (*Vollzugsinn*) in the actual encounter with an occurrence (*Gehaltsinn*) and, crucially, manifests itself as self-intimacy. Movement, therefore, is not a passing from non-presence to presence, but rather a conducting and leading oneself toward what is experienced in its meaningfulness, a taking care of something which may be enacted in different ways. In other words, movement amounts to a mode of self-presence which, in renewing itself on every occasion with respect to that which thereby manifests itself, spells out the way in which this enactment becomes time and leaves a trace for each person, marking the path of individuation.

"Time—Heidegger states—is to be understood here not as the framework or dimension in which facts are ordered, or the (specifically formal) character of the connections among historical events, but rather as the *mode of motility*, meaning a character that not only makes motility possible by releasing it from within, but constitutes it as itself moving in an autonomously factical way" (GA 61 1985, p. 169).

As we have seen, time is one's concrete mode of comportment in this or that situation (i.e., the "how"). Thus the process of individuation takes shape and unfolds as a life trajectory, from cradle to grave, through the enactment of sense in relation to the events we encounter as our life journey unfolds. For each one of us, becoming a story in every instance amounts to the actual co-experiencing of our self-familiarity in grasping the meaningfulness of that which meets us. In this respect, becoming a story is the mode in which our relation to our own past is preserved according to the enactment of our being-there in the light of various fields of possibility. This relation is renewed in actual occurrences and points to a being-there that concretely exists precisely by virtue of its relation to that which it regards as its own story: being-there gains a story in which to live through its mode of updating its past—a past that is therefore a vital part of ipseity.

5.4 The Phenomenon

If ipseity is experienced through its relating to this or that thing, life may be seen to move toward itself in relation to the way in which things become present in one's encounter with the surrounding world. In this respect, ipseity does not belong to the world like all other entities, but represents the very condition for the appearance of the world, so to speak: by encompassing it, the world enables its enactment *qua* existence. The motility of life, which lends shape to itself through events, thus becomes rooted in the world, which becomes its own and by which it is in turn constantly summoned.

The occurrences which emerge time after time bring motility into play and inevitably absorb it, imposing a given sense upon it. From this perspective, the meaningful directions, which become manifest and accessible through the very enactment of motility, correspond to specific ways in which concrete life understands the events which it encounters (categories): these directions of sense come alive within life itself. Meaning, therefore, is not an entity characterized by a mental existence (*ens rationis*), but rather consists in the enactment of directions of existence, in the factual relation between life and the circumstances it finds itself in. This relation is not to be envisaged in terms of a self-standing biological whole referring back to itself a world that it observes or knows (e.g., as in the case of the various forms of embodiment mentioned above). Rather, it is to be understood as a mode of life, whose determination is perfectly integrated with the sense of the content (the meaningfulness of the world) that is brought into play each time according to a particular sense of reference. Experience, therefore, emerges as (1) a mode of interpreting the present situation by directing ourselves and being directed toward that which meets us (reference, *Bezugsinn*), (2) a possibility of gaining an understanding of our own life by concretely enacting it in the light of the directions offered by a present occurrence (enactment, *Vollzugsinn*), and (3) a content with respect to which we feel situated (content, *Gehaltsinn*). According to Heidegger, the sum of the sense of reference, its actual realization at a given moment (the sense of enactment) and the content of the experience (sense of content), is what constitutes full sense: the phenomenon (*Voller Sinn = Phänomen* GA 61 1985, p. 53).

Meaning in its fullness corresponds to the way in which, in all instances of everyday experience, life moves (yet not in motor terms!) toward itself, in the direction of the present occurrence, according to certain inclinations. As we shall see, phenomenalization, understood as motility, as the possibility of enactment, corresponds to the mode of being of corporeality. It combines the belonging to the world, conceived as the fact of relating to the world as a whole, with the possibility of making the world appear; and this relation is what characterizes the human movement of existence: it is what distinguishes man from animals and things.

Corporeal existence can only be understood, therefore, on the basis of this belonging—and not the other way round, as is (unquestionably) assumed by many researchers who have made embodiment their forte. Through its constant enactment, motility enables life to gain a direction to suit the circumstances it encounters and in accordance with its own tendencies and motives, since motility is at the same time the precondition for the phenomenalization of the world. Thus ipseity, which is situated in a world that contains it, by constantly realizing itself makes the world appear as it manifests itself.[6]

At every instant we live according to a direction that completely engulfs us, even when we are disoriented and lost. Within this constant process of situating ourselves, together with a world—and a related sphere of contents that are more or less stable over time—we carry a horizon that is more or less thematized. This horizon guides contexts of expectation that develop or dwindle (with distinctive rhythms and styles) and which are actualized through new passions and spheres of activity, conflicts and relations, desires and surprises, ideas and beliefs, friendships and loves, and births and losses. This is the world which we struggle to relate to, for example, when a loss tears it away from us or when a phase of our life comes to close.

This applies to each one of us, including those who are only consistent in their inconsistency, who feel themselves to be continually puzzled and piecemeal, who live their life in a picaresque fashion, who have no early ambition or later sense of vocation, no interest in climbing a career ladder, who go from one thing to another, aimlessly, and who are "somewhere down to the episodic end of the spectrum" (Strawson 1999, p. 15).[7]

---

[6] In its attempt to explain the generation of meaning according to a motor theory of mind founded upon the neo-Darwinist axiom of the survival of the fittest, evolutionary epistemology (Hayek 1952; Popper and Eccles 1986; Weimer 1977) turned to biology in search of the objectiveness required to ground such a theory—as already Pierce had done. In doing so, however, it turned away from actual life and from the possibility of finding the source of meaning in life itself, as it is actually experienced. On the other hand, this issue may be regarded as a point of contact with the tradition of American pragmatism, in the style of Ricoeur's hermeneutics.

[7] Heidegger notes: "Even the life which distracts itself meets the world as 'distraction', as something dispersive, multiple, which satisfies, engages, absorbs, empties and bores. This means that the inclination shows itself as something moving toward itself" (GA 61, pp. 119–150 it).

The structural need of life, the need of enjoyment (Levinas 1979), coincides with its possibility of movement. In our various modes of being constantly turned toward the future, we give ourselves a direction according to what, on each occasion, meets that trajectory in such a way as to engender it—situations, places, people, stories, encounters, objects, landscapes, scents, music, and images—and make ourselves arise with it. Everything we meet and which, as a situation, also corresponds to a kind of manifestation of ourselves, not only gives our life a direction, but becomes part of that sphere of the proper which accompanies the unfolding of existence at every moment.[8] What is experienced, and hence made one's own, becomes available to be newly mobilized in different ways in relation to those factual contexts which in each case take shape along with individual existence. On the other hand, what occurs, what we meet, always manifests itself within the framework of a situational condition of oneself. This context of intelligibility within which I live—which I may become more or less familiar with, which I may or may not appropriate, which I may or may not free of its opacity—constitutes a shifting background (that constructs itself ever anew in factical life) in the light of which the world becomes directly accessible. As has already been emphasized, the way in which ipseity is brought into play in the present situation constitutes the real *principium individuationis*.

5.5 Vantage Points, Passages, and Crossroads

With the temporal development of one's own world, the tendencies and motivations that have been enacted according to the concrete situations in one's life also emerge, become stabilized, or vanish. In this respect, we are always living according to a direction, attuned in some way or other, each according to his or her own distinctive vantage point and position within the world. This position, which differs from person to person, in each case corresponds to a sphere of possibilities of signification, which is always mobilized as it factually takes shape and which constantly affects the domain of its own intelligibility.[9] This direction corresponds to a turning toward a certain meaningful context, and hence to an inclination of one's sense of reference, which in its enactment determines the ever-restless and different margins of the world of one's life. It is evident that each person grasps meaningfulness in a

[8] Heidegger writes: "In factical life experiences run through each other in multiple ways, at one moment coming out of this life-world, at the next moment out of that one, entwining and reshaping themselves, growing over themselves—among the availabilities back and forth, over and away to other running motivations" (GA 58, p. 55).

[9] Boss (1977) grasps this point by grafting it upon the spatial harmonics which has always been dear to phenomenology: "Once human existence is seen as bearing the spatio-temporal worldwide realm of openness as, itself, a perceptive and actively responsive openness, the problem of location takes on a wholly different aspect. It is then self-evident that the possibilities for relatedness belong fundamentally and immediately to the whole realm of perceptive openness. They are, in fact, what constitute human being-in-the-world. This openness is the primordial spatiality or prespatiality in which the possibility of a phenomenon's being located at any point in space is grounded" (p. 132).

different manner from others and in a distinct way in each phase of his or her life. The phases of one's life are different periods characterized by the stabilization of a range of different horizons that intertwine to form a more or less stable unity of sense, a perspective: the separate motivations and tendencies constituting the familiarity which each person has with his or her life constantly coalesce and recombine with respect to present occurrences.

Perspective is a phenomenal field that draws us in and guides us. As we find ourselves concretely absorbed by the world, perspective lends immediate intelligibility and accessibility to occurrences. Each one of us can mindfully experience this network of tendencies and horizons of expectations, although for the most part it remains undetected in everyday life. In our ordinary experience, a given way of grasping the world will correspond to the more or less variable accentuation of a series of aspects of the sphere of the proper which, in the present situation, will intertwine with our encounter with things and other people, so as to shape in each case the way in which we feel emotionally situated.

The framework of perspective is an emotional one, in the sense that it is our emotional condition that concretely expresses, not just the way in which we actually feel in our relation with other people and in the world but also the space of memory and the horizon of our expectations (Arciero and Bondolfi 2009). Being emotionally situated therefore corresponds to a way of *finding oneself* in a given situation and with a way of facing the future; hence, it is always connected to a personal story which guides it and which in turn is guided by it (Arciero and Bondolfi 2009; Caputo 2015).

What becomes clearer within this framework is the prospect theory developed by Kahneman (2011), since we can now better understand the foundation of what he describes as one's "reference point," i.e., the state relative to which gains and losses are evaluated; "For an example, take the following scenarios: Today Jack and Jill each have a wealth of 5 million. Yesterday, Jack had 1 million and Jill had 9 million. Are they equally happy? (Do they have the same utility?) … The happiness that Jack and Jill experience is determined by the recent *change* in their wealth, relative to the different states of wealth that define their reference points (1 million for Jack, 9 million for Jill). This reference dependence—Kahneman observes—is ubiquitous in sensation and perception" (p. 274).

To different degrees, we all take charge of our past in the light of a complex of expectations, guided by a project suited to our tendencies, through which we determine our conduct and bearing in everyday life. In this grasping (or failing to grasp) the weight of our story by assigning ourselves a future as a task, we experience that through which we determine ourselves; and this determination occurs in our daily behavior and way of operating, guided by big or small passions. Day after day, the task in question—or its absence—takes place through the actual implementation of the tendencies of our historical being, supported by emotional dispositions which make us open up and become attuned to the world by situating us within it with and for others.[10] It is thanks to this taking charge of one's own history as a task, as labor and work, that we encounter the world and other people through given configurations and modes of relation. So as we mature, as embodied time, by following our life trajectories, supported, exposed, and directed by our emotional tones, we anchor ourselves by relating to things and other people.

Over the course of our life, these tendencies to situate ourselves in given ways may become stabilized, giving rise to different phases of our personal existence. These will be characterized by enduring (yet not permanent) configurations of emotional engagements which will determine the directions of our taking care of things: a given style and perspective of involvement—and hence given projects, relations, commitments, social contacts, objectives, and careers. This is what Patočka (1976) has in mind when he grasps one's being-in-the-world in terms of the overall movements pertaining to one's life journey and which are defined by their orientation: "an overall movement distinct from the partial and individual acts within which the total movement withdraws and concentrates at a given moment" (p. 106).

On each occasion, this range of life trajectories attains unity through the enactment of experience, through one's expressive situation. As it becomes visible, this unity of sense points to a range of different horizons, which are not brought together according to any associative logic or ordered through a theoretical approach that grasps each object as a basic element that congeals multiple appearances according to a unity of expression; rather, the opposite is the case: a range of different phenomena here attain unity through an expressive situation. In this respect, unity of expression does not have only one reference, but rather points to a heterogeneity that is to be traced, disentangled, and interpreted in relation to a particular period in one's life.

On the one hand, the stabilization of one's own situatedness, and hence of certain forms of manifestation which distinguish each phase of one's life, corresponds to the enduring delimitation of given fields of meaningfulness and thus too of stimuli and attractions; on the other hand, it corresponds to a persistent configuration of modes of reference through the actualization of forms of engagement that become structured in emotional terms. In this respect, we speak of "emotional style" (*Wiegehalt*) to describe the way in which life becomes stabilized over time according to certain tendencies but also the way in which it stably reaches itself in particular circumstances and in different periods, which clearly may vary from person to person and over the course of one's existence (Arciero and Bondolfi 2009). Although life is fully present in every situation, the notion of situatedness, as it is used here, refers to a stable perspective as a condition that is engendered over the course of an individual's existence by those key occurrences that constitute turning points and

[10] Heidegger grasps this relation with the other as the threat of subjection (*Botmassigkeit*) to the dictatorship of the anonymous "yes" (GA 2, p. 126). Radically reversing its meaning in ethical terms, Levinas instead grasps the same phenomenon as the command (*Gebot*) issued to me by the face of the other, an order to be obeyed (see Greisch 1994a).

hence mark some transformations and new life trajectories. These key occurrences trigger a re-attunement of the global sense of oneself, by reconfiguring the conditions of our openness to perceiving and dealing with what we encounter and, at the same time, of the way in which what we encounter attains manifestation. These crucial experiences, which can constitute a challenge for one's life as a whole, are closely reminiscent of the limit experiences described by Jaspers (1919) and engender the various contexts that characterize the stages of maturation of a human being; in this respect, they mark the passages and crossroads of existence.

Ever since Classical Antiquity, the changes just described have been ascribed to the body. Human life, in other words, has been divided into periods of maturation according to the development and transformation of the body—periods reflecting those differences which the various human groups have acknowledged in relation to the possibility of accessing specific shared domains. It is usually within the framework of these stages of evolution that the modes of participation in society have been determined, the specific roles and duties of citizens—which is to say their participation in public life—being age-dependent. This subdivision into different stages, which establishes a kind of socially acknowledged norm for emotional development—usually divided into three or four ages (childhood, youth, maturity, and old age)—runs through the iconography of the Renaissance (e.g., Giorgione, Titian) and informs an extensive literature that, in the modern age, extends from Kierkegaard (2013) to lifespan developmental psychology. The underlying assumption is always the same: the stability of invariants relative to the stages of bodily development. This stability is also emotional, as emphasized by Erickson (1982), who associates each passage with a crisis.[11]

This primacy assigned to the development of the body as an entity within the world, whose transformations also mark the stages of maturation of the self, is the underlying approach embedded in the genetic code of evolutionary psychology. It conceives development in relation to the embodied being as an entity that belongs to the world, instead of grasping the taking shape of a body—my, your, or their body—in the light of individual experience.

From the perspective we are endorsing, although the various phases of life may coincide with stages of maturation that are clearly articulated in ways that differ from person to person, the turning points in one's life history cannot be reduced to these transformations. In the various phases of our existence, the temporary stability of being emotionally situated determines, in different ways, the mode of enactment of our sense of reference with respect to that which acquires meaningfulness as we encounter it. This temporary stability manifests itself as an orientation toward a world that gives life directionality and anchorage. What also emerges as the world is brought into manifestation is a mode of existing in the light of which the body becomes such, in each case becoming established in relation to the bonds and sensibilities that characterize it at a given stage of its evolutionary journey, thereby determining its possibilities in terms of development. In other words, through enactment, at each stage of evolution over the course of one's life, the sense of reference given within—and with—life itself delineates a world possessing a force of attraction that at the same time lends life a gravitational direction.

[11] For a review, see Baltes et al. (2006).

5.6 The Prominent and the Unprominent: Life's Relation to Itself in the Light of This or That

To understand the motility of life and, with it, meaning in the light of our turning toward something, our concerning ourselves with something is not only to grasp the articulation of sense in occurrences but also to reveal how we gain possession of ourselves with respect to the existing circumstances: how through occurrences we redefine the relation of coherence and cohesion between our lived experiences (the domain of the proper), whose sense and possession is renewed through the way in which situations present themselves to us. Therefore, if on the one hand life finds sense as it comes to itself, being reflected by the things in the world, on the other hand, the way in which it experiences itself in such enactment is not only hazy but is usually also far from evident from the point of view of life. In experiencing things, we live unreflectingly; life essentially manifests itself as though it were absorbed by the things that occur, while the tendencies and motivations supporting this accentuation and providing the (ever-renewed) sphere of their enactment remain indistinct.

I may be disgusted by a certain thing, alarmed by a situation, enchanted by some music, absorbed in some work, and irritated or fascinated by a given person, but my mode of encountering myself with respect to such a person or situation will remain obscure and unprominent. The fact of being disposed in different ways in relation to different experiences is something which I only experience in relation to some content. Finding myself engaged in a trivial conversation with an acquaintance of mine has a different effect on me than listening to a cello sonata by Bach.

What remain invisible are precisely those prefiguring motivating tendencies that make up a context of intelligibility, a horizon of expectations, and thus enable the advancing development of the weave of actual life to be in each case absorbed according to the meaningfulness of that which is factually experienced. It is as though ongoing experience constantly took shape and came to manifestation through the merging in the present situation of two horizons: the sphere of intelligibility (ipseity) and that of meaningfulness (the world). As we have seen, this correlation is what forms the structure of appearance.

In facticity everything is meaningful, and we are completely engrossed by the ever-changing contexts of meaningfulness which we constantly encounter. As already noted, the dividing line between the act of experiencing and what is experienced is blurred. Indeed, the sense of the content is an integral part of the very structure of the meaning of experience; so as I experience the world, the latter manifests itself in all of its concreteness in each of the contexts of meaningfulness I encounter. The situational actualization of concrete life according to a given tendency, and in relation to what it encounters, and the expression of life in terms of actions, gestures, manifestations, or performances in view of given goals are thus accompanied by the mobilization of particular lived experiences and hence specific configurations of one's own world, while other experiences remain in the background.

When viewed from this perspective, there is no need to invoke a "seething cauldron" of mechanically regulated impulses and desires and to call this the "unconscious," in order to realize that there is an activity of the mind outside our waking consciousness which might cause a parapraxis (Freud 1939). There is no need to resort to mechanisms that explain why certain lived experiences can pass from a conscious to an unconscious state, or vice-versa, as a consequence of processes and causal links that are ultimately explained by an unconscious impulse or desire (i.e., instincts and/or—more recently—motivational systems).

The relation between these two supposed dimensions is akin to that between the warmth of the sun and the heating of a stone: it is a cause-effect relation. But if I fail to perform an action—e.g., if I forget my keys when leaving the doctor's—I should account for the phenomenon of forgetfulness, rather than hypothesize that this is caused by unconscious desires! The fact of letting oneself be concerned with a given meaningfulness, of being absorbed by certain facts, implies that other tendencies and motivations will slide into the background, becoming irrelevant and hence indistinct, and that meaningfulness will vanish from the horizon, that they will be postponed and displaced by the ongoing engagement, without having to invoke any unconscious repression.

The fact of going to the doctor's and forgetting my keys on his desk when I leave, then, does not conceal any unconscious desire to return there—as the cause of my forgetfulness—but simply shows that after my visit, I was probably still occupied with the doctor's opinion or that I was perhaps anticipating how to reach the nearest pharmacy in order to get my prescription. The presence of the keys is part of a distant background that remains unthematically present without addressing me because I am absorbed by what the doctor told me or by some other matter. I am in the street, heading toward my car to drive back home, when I cannot find my keys: the set of keys that simply was of no concern for me, which did not exist for me, when I left it on the doctor's table on my way out. The fact that I must drive back home leads me to relate to the keys again—I now consider the keys thematically again, in terms of their presence. So there is no need for the inventions made by Freud and his followers in order to account for forgetfulness, if we understand this phenomenon in terms of the transformation of the presence of a thing—from thematic and immediate to non-thematic—within my concrete horizon.

On the other hand, as Boss (1977) explains, "it is another kind of forgetting when I lose track of the name of someone I know and cannot come up with it. This type of forgetting manifests a deficiency in my perceptive retentive capacity and, as such, constitutes a special form of my otherwise intact relationship to a thing. In this case it is a deficiency in my ability to relate myself to a person with such and such a name" (p. 117).

5.7 Trauma and Nachträglichkeit

The same kind of interpretation provided for parapraxes is to be found in the psychoanalytical view of the unconscious erasing of traumatic memories, entailing the repression of the impulses deriving from the trauma in the conscious sphere: a repression which, according to Freud, has damaging effects that give rise to all the various forms of neurosis (Freud 1920).[12] A certain event, labeled a traumatic event, is thus seen to engender an overexcitement of the nervous system—which the latter "has difficulty in abolishing through motor reaction" (Freud and Breuer 1892–1895, p. 156)—capable of breaking through that wall which protects the mind from particularly dangerous excitations. This process would be accompanied by painful inner representations, the repression of which is seen to lie at the basis of pathogenic effects. It is as though, after the trauma, a residue remained within the breached wall, encapsulated within the psychic apparatus, capable of exercising an enduring pathogenic effect (this residue being the repressed representations). This first conceptualization—which was developed with a closed organism in mind (a "living vesicle," as it is called in "Beyond the Pleasure Principle"), protected against excitations by a kind of shield—remains the foundation of all other conceptualizations.

Psychoanalytic treatment would therefore consist in reactivating these representations by tracing their causal links back to their source, so as to engender a kind of catharsis capable of integrating the foreign body, according to the well-known Freudian adage "remember, repeat, and work through." The reintegration of the past would be favored by the therapeutic setting, which would provide a new and safe container for processing the event.

This view of a protective wall,[13] which, in different forms and with various nuances, shapes all psychoanalytic literature on trauma (Garland 1998; Fonagy et al. 2002), came to be reinterpreted within the framework of the theory of attachment in terms of the capacity of emotional regulation. Being aware of situations that present potential risks, a child's caregiver will protect it from affect states which it would be incapable of regulating on its own. Hence, a child's early relations with its caregiver provide a kind of filter against harmful stimuli, a protective wall, while at the same time laying the foundations for the child's system of representation of its own inner states (mentalization).

The individual capacity to modulate affect states, which is to say, to build the wall *(der Reizschutz),* would be connected—throughout one's life—to the mentalization system, which in turn would depend on the person's affective reciprocity with his or her caregiver (mentalized affectivity). The trauma experienced would shatter this barrier, which evidently may already be more or less undermined by the quality of one's attachment (e.g., secure, insecure, disorganized)—so much so, that

[12] Freud's casual use of the term trauma to describe "any excitations from outside which are powerful enough to break through the protective shield *(den Reizschutz zu durchbrechen)* … with a breach in an otherwise efficacious barrier against stimuli" (Freud 1920, p. 29)—which has become en vogue once again through the ambiguous category of post-traumatic stress disorder—points to a quantitative view of affections which has its roots in Aristotle (Met. Delta 21, 1022 b 15–21). However, alongside a quantitative definition of affections, in Chapter 21 of Book Delta of the *Metaphysics*, Aristotle puts forward another three definitions, pertaining to quality: according to the movement of alteration, according to the act of changing, and according to one's disposition.

[13] The term *Reizschutz*, which is usually translated as protective wall or shield, literally means "protection against stimuli."

some people have spoken of "attachment trauma" (Schore 2002; Solomon and Siegel 2003).[14]

If, as a consequence of abnormal attachment, a child experiences an altered development of its ability to access painful mental states, it will learn to dissociate itself from them, as it cannot restrain accompanying them. Over the course of his adult life, a patient who has not acquired the ability to think about personal painful emotional experiences, in certain circumstances will associate the recent traumatic event with the inner representation he has repressed in his childhood, thereby giving rise to a symptomatology. As noted by Stolorow (2011), developmental trauma is viewed as the instinctual flooding of an ill-equipped Cartesian container.

As in the case of the forgotten set of keys, when it comes to the erasing of traumatic memories, the psychoanalyst explains the experience, not in relation to the person experiencing it and the circumstances in which it occurs, but rather by invoking causal links. On the other hand, the psychoanalyst finds himself at a loss when it comes to explaining the weaving and unweaving of memories and hence the remodeling of the personal history of a patient.

In relation to this historical aporia faced by psychoanalysis, a lively debate has sprung up in recent years on the term *Nachträglichkeit*, which Freud uses with different nuances at different stages of his copious production and which is fully articulated in his book *Moses and Monotheism*. Without wishing to enter into this psychoanalytic debate, it seems to me that the confusion surrounding this term points to the difficult underlying problem—reflected in all of Freud's work—of the polarization between scientific explanation and hermeneutical interpretation: two cultural traditions which Freud sought to bring together in the early twentieth century.

This confusion is particularly evident in the conceptualization of the meaningfulness of biographical phenomena, which are initially understood as motives (i.e., causes of a particular kind) and then slowly but steadily reinterpreted by Freud in terms of causal links. The cultural dialectic which underlies this structural

[14]The research on attachment which one of the authors of the present book conducted with Guidano in the early 1990s (Guidano 1991; Russiello et al. 1995; Arciero 1994, 2002; Arciero and Guidano 2000) was aimed, not so much at interpreting the difference between various forms of reciprocity on the basis of the level of security/insecurity of early relationships, but rather at grasping—within the context of a non-representational epistemology inspired by Merleau-Ponty—the various forms of attachment as a source of emotional organization and the narrative capacity of the individual as a means of emotional articulation and regulation. By making the relational system coincide with that of emotional organization, this approach opened up a new research horizon that placed affectivity at the center of every relationship and hence implied a radical contextualization of all the aspects of human psychological life. This perspective, which was developed together with Guidano, pointed to the need to study individual differences. On the other hand, by generating different forms of misattunement and misregulation in continuity with early experiences, the endurance of the same forms of emotional organization throughout one's life may be seen to lay the foundations for the understanding of various personality disorders (Arciero and Bondolfi 2009). This perspective also helps understand the findings made by some recent studies on emotional experience among street boys living in extreme adversity in post-conflict Sierra Leone (Ardizzi et al. 2013, 2015).

confusion of psychoanalysis, and whose origins are still largely unexplored, continues to guide the contemporary tendency to redefine—*après coup*—this Freudian term. This point is clearly summed up by Dahl in a paper on the topic, as its abstract suggests: "The two temporal vectors of *Nachträglichkeit* are discussed: the first consists in a causal process which operates in the direction of the time to come against the background of a factual experience; the second consists in a backward movement which enables an understanding of the unconscious scenes and fantasies which occur at the level of primary processes" (Dahl 2010).

On the one hand, we have the reconstruction of one's history, with a causally determined sequence of meanings; on the other, we have a hermeneutic investigation of a hypothetical psychic system, the unconscious, which is invented—and hence assumed—by the theory itself, whose operational rules guide all possible interpretations of experience. Thus what remains unquestioned is the very foundation of the problem, the incoherent articulation of which stands at the basis of the whole psychoanalytic edifice. Clearly, to grasp motives as causes is to refuse to acknowledge that each individual act rests on something—a motive—that is acknowledged by the person performing the act, on the basis of his being engaged with that which addresses him (Arciero 2002).

To further elucidate this point, let us focus once more on what a psychoanalyst would describe as the erasing of a traumatic memory.

Let us suppose that one of the two partners in a couple suddenly informs the other that their relationship is over: an unexpected and unsought-for separation. Let us also suppose that, in order to cope with the pain of separation, the forsaken partner immediately starts a new relationship—precisely as a means to deal with the effects of the painful event. The memories and, with them, the emotional manifestations related to the person's bereavement gradually vanish. Are we to regard this "strategy" of the patient as a form of dissociation? Furthermore, if the patient's ability to think about his personal experience is actively inhibited, is this the result of the associating of the recent traumatic event with a repressed representation pertaining to an altered form of early reciprocity?

But if one of the modes of forgetfulness implies that the patient does not appropriate the painful event and avoids it instead, this does not mean that the capacity to access the event is inhibited. Forgetting here may simply be a consequence of the fact that the more a given experience is avoided, the less the person realizes that he is avoiding it. The experience remains unthematically present in a distant background, and such distance amounts to more than just retention: it is an active position with respect to the past.

The dissolving of a memory, therefore, does not imply its repression into the unconscious; rather, the very opposite is the case, in the sense that—to return to our example—the reorienting of one of the two partners toward a new relationship takes place through active avoidance and hence through the presence of that "traumatic" experience which gradually vanishes, as the individual becomes involved in a new relationship: the experience in question becomes unprominent the moment in which it is no longer necessary for it to be present. However, it still remains available, since it can be remobilized in terms of sense by biting into a madeleine, having a morning dream, or simply thinking about it—in other words, by thematizing that content which is unthematically present.

On the other hand, while the avoidance of unpleasant emotions related to the separation continues to be unthematically present, every negative variation affecting the person's present relationship corresponds to the opening up of an access to those painful feelings which this new condition ought to have helped him cope with. For example, if the person in question entered into a new relationship as a means to face the sense of emptiness connected to the perception of loneliness caused by the separation, every resurfacing of this sense of existential loneliness during his present relationship will correspond to the reemergence of the sense of emptiness and of the symptomatology which may spring from it. Just like when I realize that I have forgotten my keys on the doctor's desk as I am heading toward my car to return home, in this case too a certain situation will reopen an access to a past experience; but unlike in the case of the keys, it might be difficult to integrate this past experience within the present meaningful context. Clearly, this does not imply that symptoms are the price paid for the repression of experiences, as psychoanalysis claims; rather, what it implies is that the possibilities of meaning pertaining to the person's present life condition do not allow him to grasp, and hence thematically integrate, a past experiential condition that therefore proves foreign to the individual, even though he perceives it as his own. While this state of affairs does not correspond to a symptom in itself, it may lead to the emergence of one. So it is possible to argue that the genesis of a disorder lies in the relation between the prereflective sphere and its narrative reconfiguration (Arciero 2002, 2006; Arciero and Bondolfi 2009).

5.8 Life's Relation to Itself: Symptoms

According to the perspective outlined so far, symptoms may appear when a new mode of access to oneself emerges in relation to ongoing everyday life events. This new existential position shapes the way in which we are situated with respect to ongoing circumstances and, at the same time, our possibilities of relating to our past experiences. However, it is incapable of assimilating significant past experiences in a way that is coherent with our current condition. Certain contexts of intelligibility, therefore—corresponding to configurations of the domain of the proper which is constantly reshaped through our encounter with new circumstances in life—are, so to speak, excluded from any possible integration and renewal according to the sphere of meaningfulness opened up by events. A gap thus emerges within ourselves, a form of self-estrangement, a sort of break between ongoing experience and the underlying motives and tendencies on which it rests; and it is precisely this discordance that engenders the sense of passiveness, of suffering, which patients speak of.

This intimate estrangement may act as the basis for the mechanism of symptom formation, which therefore represents an attempt to achieve self-integration and self-cohesion (narrative reconfiguration) that ultimately confirms—and often intensifies—the inner discrepancy that engendered it in the first place. In other words, the mechanism of symptom formation takes shape as a kind of repetitive system which, activated by given conditions, will regenerate the elements from which it originates.

As the expression of a mechanism, each symptom points to an experience of the self which constantly produces itself in the same way and imposes repetition upon the motility of life. From this perspective, psychological symptoms and, more generally, disorders may be seen as a limitation of one's personal freedom, of the possibilities of sense; as such, they are always accompanied by a certain degree of suffering. The previous example may help clarify this perspective.

The person who within a short time embarks on a new relationship in order to cope with the feeling of emptiness caused by his separation will gradually discover, as he gets to know his new partner better, that the latter is not quite as he imagined (new position of access to oneself). This will lead to the resurfacing of the feeling of existential loneliness, of emptiness, which the new relationship was meant to assuage. Consequently, the development of the new relationship engenders not only a change of the person's perspectives, and hence of the existential dimensions related to them, but also—more generally—a change in his current position with respect to the original projects of the couple: a different opening up to the future. At the same time, this new horizon redefines the person's relation to his past, causing situations and feelings connected to the previous separation to resurface—memories that seemed to have vanished at the beginning of the individual's engagement with his current partner and which now instead increase his feeling of emptiness.

Intertwined with the person's present life conditions, emotional states connected to past situations resurface which, once removed from their original contexts, are experienced as something that is suffered: as an affliction the understanding of which is excluded from our present being-there (gap between ongoing experience and the background made up of contexts of intelligibility). This past manifests itself as something unrelated to the individual's present life condition, as something foreign to himself—since it cannot be thematically integrated within his ongoing existential position—but at the same time it is perceived as something that is his own. These emotional states perceived as something that is suffered may never be articulated, thereby engendering an emotional condition that cannot be appropriated within the subject's current perspective, and may therefore trigger symptoms.

The sense of emptiness which becomes increasingly present and foreign to the person as loneliness takes hold within the couple may engender a state of fear, which he will attempt to face by focusing on the bodily aspects that accompany it, thereby anticipating the state of illness. In other words, the person who feels almost a sense of estrangement with respect to his own perception, as he cannot integrate its sense within his ongoing life situation and thus becomes alarmed, will transform the body through which this perception takes shape into an object to be monitored (mechanism of the symptom). Consequently, in every condition in which emptiness re-emerges, causing fear, bodily modification grows more acute for the patient and with it the certainty of suffering from an illness, which in turn—through a vicious circle—fosters the fear that engendered it in the first place. This mechanism may lie at the basis of a hypochondriac disorder with panic attacks.

It is clear here that each symptom always provides a double indication: on the one hand, the factual condition in which it emerges (the lack of correspondence between the person's expectations with regard to his relationship and the actual present situation); on the other, the life situation within which it originated (the previous separation and sense of emptiness), and which is thus preserved by being reiterated, yet not assimilated. What also becomes clear, then, is the meaning of Merleau-Ponty's intuition that "the movement toward the future, toward the living present or toward the past and the power to learn, to mature, and to enter into communication with others are all somehow blocked by a bodily symptom; existence has become entangled" (PhP, p. 192). Symptoms bring the motility of life, which reveals existence in its bodily actualization, to a halt; at the same time, through the repetition of the mechanism, the movement by which the world opens up reaches a standstill. This process takes a very different form in the case of a traumatic experience, as we shall see (p. 175 ff.).

Symptoms, then, require interpretation, a restitution of meaning, the care of self; and this care deconstructs the mechanism (reduction and destruction), identifies the gap on which it rests, and reintegrates the disunity of experience—interpretatively retrieving the unsaid—within a cohesion of sense (construction).[15]

Evidently, connecting in vital terms the experiences that lie at the origin of a disorder cannot mean simply recalling the trauma which represents their "cause" nor explaining them according to a theoretical approach. Rather, it entails newly articulating these experiences as a vital part of oneself in a sense, based on a possible view of the future.[16] The first step consists in looking back toward the modes of being that accompany the patient's behaviors (the unprominent correlates of factual life), so as to bring them to light, opening up to inquiry the domain of phenomena which the psychopathological condition of the person obscures to a substantial degree (reduction). This first stage includes the deconstruction of the symptom and of all those obstacles that prevent the understanding of the concrete situation in which meaning takes shape (destruction). **In the previous example, this first stage consists in defining the actual situations in which the malaise emerges (the**

[15] The treatment of a symptom, therefore, also has its starting point in the patient's experience of the mechanisms that underlie this gap.

[16] Historical reality coincides with virtuality: the disclosure of possibilities that call one's actual condition into question.

sense of emptiness), defusing the underlying mechanism (the emptiness-fear circle) and hence tracing the experience of emptiness back to the situations it is connected to.

By following these guidelines, we can bring the problematic experience to light and provide a concrete outline of it, so as to grasp its origin and the contexts related to it. This enables us to define a field of inquiry by clarifying the ongoing life situation that governs its understanding. **In our example, guided by an analysis of the exacerbation of the problematic experience through the development of the patient's present relation, we may grasp the person's sense of emptiness, connected to his condition of loneliness and freed from the vicious circle of fear, as an indication that points to his separation and the end of his previous relationship.** The articulation of this access thus outlines the perspective within which meaningful moments can be disclosed—and with them the possibility of turning toward the past and of being addressed by it (construction). **In our example, these moments emerge through an analysis of the end of the patient's previous relationship and of the ways in which he faced this separation.** The establishment of this interpretative condition leads to the development of new directions and hence to a renewal of ipseity in the present: a renewal which cannot be confined to the acquisition of knowledge detached from concrete being-there (according to the theoretical approach), but which rather co-constitutes ipseity itself. In other words, the past is appropriated through a new relationship, opening itself up—once again within ipseity—to becoming.

5.9 Life's Relation with Itself: Memory and the Living Appropriation of the Past

As a disorder manifests itself, it becomes increasingly evident how the fact of experiencing such and such a thing, each time according to certain motives and tendencies, always implies the relation which life has with itself, i.e., that self-familiarity—or self-estrangement, as in the case of symptoms—which becomes associated with a particular ongoing experience; that basic character of actual life, whereby experiencing the world means encountering oneself.

While this aspect largely slips into the background in everyday life, it clearly emerges in the way in which I remember myself. In this case—unlike in everyday life—the relation between the current experience of recollecting and the past event that is being evoked becomes visible, because memory creates a distance between myself and that which I experienced. This relation concerns the content of the experience I had at the time, the past experience which is still familiar to me.

When I turn to the past, the sense of the event I experienced—as opposed to the objects, situations, people, atmospheres, and circumstances it comprises—suddenly becomes accessible to me. By moving within the framework of the sense thus disclosed, I can retrace situations down to their most concrete aspects. In other words, within the memory of lived experience, I can immediately grasp not just what I experienced at the time—the particular set of circumstances in which *I found myself* and which makes the past accessible to the present—but also the sense of the completed experience, a sense which is immediately familiar to me and manifests itself within the present context of intelligibility.

Whereas memory has to do with a sort of definiteness of the past, I *newly find* myself in lived experience, not through the observing of past experiences, but through an interpretative context that grows out of the current present; and it is through this access that I come to experience my past in an immediate way. Once again, through the things I come across I encounter myself: my lived experience manifests itself to me, as a phenomenon. In this respect, memories reveal themselves according to specific aspects within the current sphere of intelligibility; hence, they give themselves to the extent that I experience their meaning. Therefore, through the enactment of ongoing experiences, I achieve or lose a certain degree of intimacy with my past, with the content of my lived experience. In visualizing a personal experience, I am not remembering it but, in this visualizing way of being there where it is, I am in some relationship to it. In this respect, having oneself means becoming familiar with one's proper world, and hence with that context of intelligibility within which I myself live and which I re-actualize on the basis of the circumstances at hand.

On the other hand, the fact of directly grasping myself as something familiar within the content of memory reveals the concealed and unprominent presence of a certain motive and/or tendency within that sphere of intelligibility that is not apparent in factual life and is usually absorbed by the present circumstances. What such an experience reveals is that every behavior is immediately oriented in the light of life itself, in all of its fullness, and hence by one's personal story: a living story, a story that lives in concrete experience in terms of the intimacy which life has with itself in each of its relations. This story is newly mobilized at every moment through the availability of lived experiences in actual occurrences, which lend concrete life its fullness. My personal story is thus understood on the basis of occurrences and measured in the light of my life developments, as an expression of my very existence.

The difference between memory and the living appropriation of the past becomes particularly striking in the case of the rereading of a text. Let us start from the considerations on rereading made by Paul Valery in a famous essay on Stendhal, published as the preface to the reprint of *Lucien Leuwen*, 30 years after Jean de Mitty's first edition: "I have just finished reading a *Lucien Leuwen* that is very different from the one I loved so much thirty years ago. I have changed and so has the novel. I should state right from the start that this second *Leuwen*, which corrects, expands and improves upon the first, revives and intensifies the charming memory of the old reading—but I will not disclaim the enjoyment I experienced back then" (Stendhal 2002, p. 1263). Two different life stages mark the distance between the old reading and the present one. Valery provides the context for that charming memory: the literary Tuesdays at Mallarmé's home and the walk he would take with Jean de Mitty after each soirée, when the two of them would stroll down Rue de Rome to the radiant center of Paris. The poet then sketches his interests at the time: his enjoyment of Stendhal's lesser-known novels and the fascination he felt for psychological processes, for the uniqueness of meaning. Each memory, therefore, is situated within a specific circumstance in the past, as the visualization of a thing experienced at a certain time. Such is the phenomenon of memory.

It was in this period that de Mitty offered Valery the slim volume of *Lucien Leuwen*, and within this context of occurrences, the poet's own inner world manifests itself: "As a young man I used to place love so far above and so far below … Perhaps I had good reasons to feel so intimately affected by those indefinable qualities; on the other hand, I was amazed that this was the case." Here the poet evokes that bygone world in which he, a somewhat prudish young man, had been charmed by a primitive and imperfect *Leuwen*—for such had been the circumstances!

A distance of 30 years stands between the charm which the poet experienced in the past and his familiarity with the novel he had first read in his youth, a familiarity ensured by his present access to the book. Valery, the literary egotist who has found his way, now regains the past enjoyment of reading—a sweet, living experience—only by renewing his grasping of himself: "because I could not stand—and still cannot stand—that a literary work may deceive me to the point that I can no longer clearly distinguish between my own personal feelings and those conveyed by the writer's artfulness. I see the pen and I see the person holding it. I pay no attention, I do not need his emotions. I only ask him to tell me about his means" (p. 1264). Valery here exists in some relationship to his past!

This second phenomenon is very different from that of memory, as it means having that lived experience within reach in the present. The experience of reading a text which I had in my youth gives itself according to a current sphere of intelligibility that determines the aspects according to which both the experience in question, and the circumstances in which it took shape, become accessible. What emerges is the way in which a lived experience from the past becomes familiar to me, how I *newly find* myself within it. The very context of occurrences, within which reading disclosed Valery's own world, is grasped through the motivations characterizing the ongoing conditions. It is evident, therefore, that while access to these occurrences represents the connection between past and present, it depends on what themes and problems present experience poses in the light of the foreseeable future. Any grasping of life relations—the fact of having oneself—cannot amount to the collecting of past occurrences, and less still to a hermeneutical *Nachträglichkeit*, whose direction is determined by the phantasmagorias of an analyst faithful to his a priori; rather, it must be oriented by life itself, in all of its fullness.

In this respect, a person's relationship with his or her past is always preserved by becoming a story. This relationship, renewed through actual occurrences, points to a being-there that concretely exists precisely by virtue of its relation to what it conceives and chooses as its story. This being-there gives itself a story within which to live through the way in which it takes charge of its own past on the basis of the events it encounters. This perspective allows us to understand the following claim by Heidegger: "The proper *organon of understanding* life is *history*, not as the science of history or as a collection of curiosities, but rather as lived life, how it goes along in living life" (GA 58 1992, p. 193).

As it flows by, experiencing the world which it is attuned to, and which enables it to find itself on every occasion, our life articulates itself as intelligibility or the lack thereof, as the acquisition or loss of familiarity with itself, with the environment it lives in, and with others. Within such an encounter with a self that constantly renews itself, generating meaning, what also takes concrete shape and acquires relevance is experiences, which may be of greater intensity and therefore define one's perspective while at the same time disclosing a new horizon of expectations. In this sense, in the situation at hand, I will grasp and single out a particular field of meaningfulness with respect to the overall context of my factual experience, on the basis of which I will experience myself in relation to how I deal with my own world. We here catch a glimpse of that peculiar interweaving of an experienced past, more or less sedimented according to certain changing inclinations, and the horizon of expectations defined by it—a tension which is resolved again and again through the enactment of ipseity. What we have here is the endless dialectic between sameness and ipseity, mediated and configured by a narrative that makes it possible to integrate the diversity, variability, discontinuity, and instability within a coherent story, engendering what has been described by Ricoeur as a person's narrative identity. Ricoeur thus argues that the essential difference between the narrative model and all other models of connection lies in the status of the event which, by entering into the flow of the narrative, in each case connects the person to his or her own story.

5.10 Trajectories and Turning Points

This distinctive intertwining of different existential dimensions, which become sedimented through emotionality, coincides with the background of immediate accessibility and intelligibility which each human being carries as self-cohesion: a structural connection which the encounter with occurrences enables the person to reshape at each moment by engendering a direction, orientation, and meaning—but also a personal story in the making—that may become stabilized, establishing new tendencies and motivations which may change over the course of the person's life.

The circumstantial quality of everyday life determines the way in which our relations with ourselves, the surrounding world, and the people we deal with intertwine, unfold, change, and/or become stabilized. Every situation, then, corresponds not just to a range of vantage points that lend life meaning but also to the possibility of following a trajectory, of keeping within a certain stable and meaningful context; and in everyday life, we always preserve a certain tendency, an existential rhythm according to which we constantly make things take a certain course. Within this framework, an emotionally situated individual action (within the context of the praxis of living), directed toward something and with a given aim in view, acquires a sense that is far more complex than the aim of the action itself, insofar as the determinations which it acquires depend on the referential context within which it is generated (Arciero and Bondolfi 2009).[17]

I go to work, devote a lot of energy to my profession, occupy myself with botany, read nineteenth-century French literature, play golf, cook dinner, and, more generally, live with a person I love, have a family and life project, and worry about my teenage daughter, while my wife is thinking of changing job, my mother requires assistance, and so on. In other words, I am always directed toward something, or rather toward a "not yet," according to a weave of perspectives that change over time and lend my existence meaning but also a certain familiarity and, with it, an intimacy with certain other people, a solidity, a sense of being settled somewhere, a dwelling place. Levinas notes: "The subject contemplating a world presupposes the event of dwelling, the withdrawal from the elements (that is, from immediate enjoyment, already uneasy about the morrow), a recollection in the intimacy of the home" (1991, p. 153).

This is the subtle stability that colors everyday life and of which our attitudes and habitus are but one manifestation. This corresponds to prominent ways of realizing ipseity, which generate tendencies to lend meaning to the things we encounter—tendencies that are never permanent. Together with these tendencies and the modes of enacting them, problematic spheres take shape that may lead to new orientations and new forms of fulfillment. Each life transformation therefore coincides with a moving away from certain nexuses of intelligibility and accessibility, from particular configurations of familiarity, and at the same time with the emerging of new availabilities that may be mobilized in view of new tendencies. The fact of drawing upon the experience of oneself during a transformative process implies a new self-intimacy, which structures itself through ongoing experience in view of possible new horizons of sense. On the other hand, if the way in which ipseity expresses itself in each case is through a situation, it is evident that the most significant transformations over the course of our life will be associated with the way in which we actually grasp the material content we encounter. Once again, life "addresses itself and answers itself in its own language" (GA 56/57 1999, p. 42).

Various circumstances can change the course of our life, altering our personal stability and lifestyle, thereby affecting the interpenetration of the various levels of existence (the world of the self, the surrounding world, the collective world). These impermanent saliences which life gives itself as it unfolds represent its modes of manifestation, which stabilize or change as forms or styles of familiarity.

One real-life example—a tragically common one in these years of economic crisis—will help illustrate this point: I lose my job. Consequently, my economic condition changes and, with it, my possibility of accessing certain social contexts and of spending time with certain people, my perception of my own admissibility

[17] In this sense, an action, to the extent that it implies the intention of being recognized for what it is, may express a certain illocutionary force—during a conflict in a couple, for instance—that has very little to do with the energetics of vitality affects, as argued by some researchers working on the motor system (Di Cesare et al. 2013).

and personal worth, the emotional dynamics between myself and my wife, the way in which I relate to my children as a father, my career projects and how much I intend to invest in them, my balance of efforts and rewards, the satisfaction I derive from what I do, the way I relate to my desires—and so on.

In other words, in the search for a new stability, motives and tendencies recombine, engendering new meaningful contexts and problematic areas, new specific areas of self-understanding, which constitute a renewed direction of sense. The life of men and saints alike is marked by such turning points and more or less sudden and momentous reorientations, which lend life a new direction.

Echoing a theme—that of extreme situations—dear to the young Jaspers, Poeggeler writes: "Clearly, we interpret the table we use in our everyday life on the basis of the 'what' that is known to us about it; but man is not the mere realization of a 'what', since this 'what' discloses itself in the great moments of life and, furthermore, must always be determined anew" (1995, p. 297).

Actual cohesion of the sense of self takes the form of a changeable situatedness, manifesting itself in ever different ways in contexts of occurrences—albeit within the framework of directions of sense pertaining to different periods of one's life. But, at the same time, it is precisely through this circumstantial quality that the cohesion of the sense of self becomes accessible: through it, the domain of the proper reveals itself, acquiring greater intensity, while lending definition to the surrounding world and the social one, which therefore always manifests itself in a fluctuating way. As has been repeatedly stressed, this intensification corresponds to the current enactment of certain tendencies (or the emergence of new ones), and hence to a certain life trajectory, which in each case is motivated by its very flowing. So each person reveals himself to himself within particular occurrences, and in a particular context of events, and this unique disclosure at the same time represents the way in which life addresses itself, by taking charge of it—or not. Facts therefore unveil that the encounter with that which becomes present reveals to the individual the meanings to which he can open himself up in that particular moment of his life. At the same time, the very same events will motivate—or fail to motivate—the person to act in accordance with the meanings he perceives. It is evident, therefore, that motivation does not coincide with—and cannot be traced back to—an inner, objective, and all-encompassing system governing the individual and his world; rather, it only occurs in response to what the person grasps as being meaningful within his world.

The experience of the encounter with oneself, which takes shape through the horizon of sense produced by coming face to face with a situation, is a historical phenomenon. To put it in different terms, history—being-there historically—is an experience which takes shape again and again in concrete situations, in view of a given horizon of expectation. The perception of being alive, which in each case manifests itself through its enactment in such or such a way, is experienced in a historical sense through this actualization: the phenomenon of history, therefore, emerges as an act that experiences itself through its specific realization.

The fact of being myself in this or that situation, then, coincides not with any precise "me," with the "now moment," (Stern 2004) but with self-possession and self-understanding: it is a form of self-possession which has nothing to do with self-perception and less still with self-observation or my own self-image. Rather, it takes shape as participation in lived experience, in the past, according to a present mode of access and hence according to a mode of grasping.[18]

As regards this familiarity which each person establishes with himself, and which changes from one period of life to the next, the lived experiences which I gain access to again and again by being alive cannot exist before me as objects amassed in a repository called the past. In each individual experience across the different seasons of my life, I am fully myself in one particular circumstance: when an occurrence leads me to encounter my own self. It is through this ever-situated restlessness of existence—*kinesis tou biou*—that life determines itself by choosing its possible directions and, without ever withdrawing from itself, prereflectively notices itself in each case: for in each case the experiences I have are my own. The world of ipseity is marked not by cinematographic inwardness, but by a unique and intimate self-involvement—the immanent historicity of life—which takes shape again and again in specific situations.

The sphere of historicity primarily coincides not with the reflective access to or inner perception of one's experiences, but with the kind of familiarity which every human being has with himself: with his own past, which is experienced according to an intelligibility that is constantly renewed in accordance with a web of tendencies and motives subject to the changing circumstances. Clearly, the way in which life notices itself also depends on how lived experiences manifest themselves, how the weight of the past makes itself felt: the more the past determines a person's present life—by circumscribing the situations he or she can experience—the more the weight of the person's being-there proves momentous and the experience of grasping oneself becomes an objective experience. Thus one's own past is constantly perceived as a burden, as a duty.

This aspect is what Heraclitus may have in mind when, in the remarkable fragment 119 D.-K. (Eraclito 1978), he uses a striking expression—which in ancient Greek carries an absolute meaning (Arciero 2006)—to argue that "a man's way of life determines his fate, while his fate determines his way of life.[19]"

The same theme, which Heidegger describes as a key element in early Christianity and as a value for all men, is developed by St Paul in his reply to the Thessalonians: life

---

[18] Heidegger grasps this experience as an intuition: an original retro- and pre-conceptual formation.

[19] Fragment 119, *Ethos antropoi daimon*, revolves around the terms *ethos* and *daimon*, character and destiny. Most commentators (Zeller and Mondolfo 1961; Kirk et al. 1983; Kahn 1979; Marchovich 1978) adopt Snell's reading, according to which the destiny of a man is determined by his character.

If we instead read the fragment as a noun phrase (Arciero 2006), i.e., as a "nominal predicate, with neither verb nor copula, regarded as the standard expression in Indo-European, whereas the verbal form would have been the third person of the present indicative of 'to be'," then it "does not express any factual data, but establishes a timeless and permanent relation which acts as an authoritative argument" (Benveniste 1966). In other words, *daimon* cannot be considered as predicative of *ethos,* but rather carries an essential value, insofar as it expresses an integral part of the human being. What makes this such a remarkable fragment, then, is the speculative tension springing from the simultaneity of *ethos* and *daimon*, which coexist in parallel yet distinctly: the simultaneity of the one and the many. From this perspective, it is even more interesting to note that if in the same sentence we had the verb *esti*, the third-person singular of the present indicative of "to be," the sentence would refer to present situations: it would find its content in the situation it refers to.

is actually unfolding, and this facticity "must be historically assumed and directed toward an instant that is not determinable" (Pöggeler 1995, p. 297). A person who has faith will take charge of his life by acting at every moment in the enduring light of hope.

"Noticing oneself," then, does not consist in any final determination of the content of one's own story; rather, it coincides with the "vital process of the loss or attainment of familiarity with concrete, lived life itself" (GA 58 1992, p. 165). In the very act of unfolding and taking shape through the range of references offered by the situation at hand, experience partakes of the underlying web of lived experiences, which is therefore part of the integral unity of experience. In this respect, the interpretation of the ongoing situation (*die verstehende, die hermeneutische Intuition*) carries with it a certain provenance while at the same time being directed toward a range of references in view of which we encounter each occurrence. Hence, the sense of familiarity with oneself, the sense of self, emerges and takes shape in conjunction—and in conformity—with a sense of enactment. In other words, one's own story becomes accessible within the framework of a present context of intelligibility, and life transformations disclose the possibility of mobilizing the world of the self in relation to new tendencies.

Bibliography

Arciero G (1994) I and Thou: notes on the phenomenology of reciprocity. Paper presented at the meeting "Ethology and Knowledge" -Nov. 1994 University of Siena-Published in "Conoscenza e Etologia" (1996) Melusina ed

Arciero G (2002) Studi e dialoghi sull'identità personale: riflessioni sull'esperienza umana. Bollati Boringhieri, Torino

Arciero G (2006) Sulle tracce di sé. Bollati Boringhieri, Torino

Arciero G, Bondolfi G (2009) Selfhood, identity and personality styles. John Wiley & Sons, Chichester

Arciero G, Guidano VF (2000) Experience, explanation, and the quest for coherence. In: Neimeyer RA, Raskin JD (eds) Constructions of disorder: meaning-making frameworks for psychotherapy. American Psychological Association, Washington

Ardizzi M, Martini F, Umiltà MA, Sestito M, Ravera R, Gallese V (2013) When early experiences build a wall to others' emotions: an electrophysiological and autonomic study. PLoS One 8(4):e61004

Ardizzi M, Martini F, Umiltà MA, Evangelista V, Ravera R, Gallese V (2015) Impact of childhood maltreatment on the recognition of facial expressions of emotions. PLoS One 10(10):e0141732

Baltes PB, Lindenberger U, Staudinger UM (2006) Life span theory in developmental psychology. Wiley, New York

Barbaras R (2007) Le mouvement de l'existence: études sur la phénoménologie de Jan Patočka. Éditions de la Transparence, Chatou

Benveniste E (1966) Problèmes de linguistique générale, 1. Gallimard, Paris

Boss M (1977) Existential foundations of medicine & psychology. Jason Aronson, Northvale

Caputo A (2015). Emozioni come infezioni? La critica heideggeriana al contagio emotivo di Scheler. In Dibattista (Eds.) Emozioni infettive. Mimesis, pp 127–155

Dahl G (2010) The two time vectors of Nachträglichkeit in the development of ego organization: significance of the concept for the symbolization of nameless traumas and anxieties 1. Int J Psychoanal 91(4):727–744

Di Cesare G, Di Dio C, Rochat MJ, Sinigaglia C, Bruschweiler-Stern N, Stern DN, Rizzolatti G (2013) The neural correlates of "vitality form" recognition: an fMRI study. Soc Cogn Affect Neurosci 9(7):951–960

Eraclito (1978) Frammenti, a cura di M. Marcovich, La Nuova Italia, Firenze

Erickson EH (1982) The life cycle completed: a review. W. W Norton, New York
Fonagy P, Gergely G, Jurist E, Target M (2002) Affect regulation, mentalization, and the development of the self. Other Press, New York
Freud S (1920) Beyond the pleasure principle. Hogarth, London
Freud S (1939) Moses and monotheism. Hogarth, London, pp 7–140
Freud S, Breuer J (1895) Studies on hysteria. SE, 2. Hogarth, London
Garland C (1998) Thinking about trauma. In: Garland C (ed) Understanding trauma: a psychoanalytical approach. Duckworth, London, pp 9–31
Greisch J (2000) L'Arbre de vie et l'Arbre du savoir.: Le chemin phénoménologique de l'herméneutique heideggérienne (1919-1923). Cerf, Paris
Guidano VF (1991) The self in process: toward a post-rationalist cognitive therapy. Guilford Press, New York
Hayek FA (1952) The sensory order: an inquiry into the foundations of theoretical psychology. University of Chicago Press, Chicago
Heidegger GA 56/57 1999
Heidegger GA 58 1992
Heidegger GA 61 1985
Jaspers K (1919) Psychologie der weltanschauungen. Springer, Berlin
Kahn CH (1979) The art and thought of Heraclitus. Cambridge University Press, Cambridge
Kahneman D (2011) Thinking, fast and slow, New York. Macmillan
Kierkegaard S (2013) Kierkegaard's writings, XI: stages on life's way, vol 11. Princeton University Press, Princeton
Kirk GS, Raven JE, Schofield M (1983) The presocratic philosophers: a critical history with a selcetion of texts. Cambridge University Press, Cambridge
Levinas E (1979) Totality and infinity: an essay on exteriority, vol 1. Springer Science & Business Media, Dordrecht
Levinas E (1991) Totality and infinity: an essay on exteriority. Kluwer Academic Publishers, Dordrecht
Marcovich M (1978) Eraclito: frammenti. La nuova Italia
Patočka J (1976) Le monde naturel comme problème philosophique. Martinus Nijhoff, La Haye
Patočka J (1995) Papiers phénoménologiques. Editions Jérôme Millon, Grenoble
Pöggeler, O. (1989–1995). Heideggers logische Untersuchungen. In Martin Heidegger: Innen- und Aussenansichten: Forum für Philosophie Bad Hamburg (pp 75–100) trad.it. In Poggi S, Tomasello P (eds) Martin Heidegger. Ontologia, fenomenologia, verità. LED Edizioni Universitarie. A cura di. Milano, 1995
Popper KR, Eccles JC (1986) Self and its brain: an arguments for interactionism. Springer, Berlin
Quesne P (2004) Les recherches philosophiques du jeune Heidegger, vol 171. Springer Science & Business Media, Dordrecht
Richardson WJ (1988) The place of the unconscious in Heidegger. In: Hoeller K (ed) Heidegger & psychology. Review of Existential Psychology & Psychiatry, Seattle
Russiello F, Arciero G, Decaminada F, Corona R, Ferrigno L, Fucci M et al (1995) Stress, attachment and skin diseases: a casecisco Varelaal. J Eur Acad Dermatol Venereol 5(3):234–239
Schore AN (2002) Advances in neuropsychoanalysis, attachment theory, and trauma research: implications for self psychology. Psychoanal Inq 22(3):433–484
Solomon RC (ed) (2004) Thinking about feeling. Oxford University Press, Oxford, p 76
Solomon M, Siegel DJ (2003) Healing trauma. WW Norton, New York
Stendhal (2002) Romanzi e racconti 2 Vol. I Meridiani. Mondadori
Stern DN (2004) The present moment in psychotherapy and everyday life (Norton series on interpersonal neurobiology). WW Norton & Company, New York
Stolorow RD (2011) Trauma and human existence: autobiographical, psychoanalytic, and philosophical reflections. Taylor & Francis, New York
Strawson, G. (1999). The self. In Gallagher, S., & Shear, J. (1999). Models of the Self. Charlottesville Imprint Academic.
Weimer WB (1977) A conceptual framework for cognitive psychology: motor theories of mind. In: Shaw R, Bransford J (eds) Perceiving, acting, and knowing. Erlbaum, Hillsdale
Zeller E, Mondolfo R (1961) La filosofia dei greci nel suo sviluppo storico. La nuova Italia, Milano

6 Personal Stories and Psychotherapy

6.1 The Theoretical Approach to Personal Stories

Scientific psychology and the forms of psychotherapy inspired by it have always tackled the problem of personal stories on the basis of a priori principles usually borrowed from the natural sciences. By molding psychotherapeutic practice on such principles, it has been possible to bring the actual unfolding of life under some general laws and to dilute the process of individual maturation into a hodgepodge of invariants. This has forced the dynamics of life to conform to a set of ideal norms in the light of which the actual experience of the living person may be converted to suit the observer's ideal: an ideal through which to reduce history to a logical chain of objective facts.

The most straightforward example of this mode of operating concerns those therapies that are designed to repair a malfunction and which draw upon the patient's personal story simply in order to contextualize the ongoing dysfunctional conducts and the behavioral schemas which are seen to lie at its basis. Treatment, in this case, takes the form of a continuous attempt to fix up the patient's behaviors and cognitions, according to a series of techniques that enable the therapist to correct the dysfunctional patterns, and to manage the therapeutic relationship in accordance with given models of mental functioning.

For the clinicians of all persuasions who investigate the theme of personal stories, the theoretical approach to psychotherapy translates into a search—within ongoing and past experiences, for those simple elements, mechanisms, structures, complexes, and systems—capable of accounting for the patient's current condition in terms of more or less remote causes. One telling case is that of the theory of motivational systems (attachment, intersubjectivity, sexuality, etc.). According to such theory, (a) individual actions are suggested by motivations; (b) motivations derive their strength from the fact that agents tend to follow innate systems of behavior; (c) these systems provide the laws and hence principles connecting the motivations for action in individual cases (Von Wright 1971). Once disclosed, these

G. Arciero et al., *The Foundations of Phenomenological Psychotherapy*,
https://doi.org/10.1007/978-3-319-78087-0_6

principles become a means to access one's own story and experience and, in certain cases, a way of shaping one's own destiny.

This approach to personal stories coincides with the one endorsed by Hempel (1942) to show that there are no unique facts in stories but that, as in the natural sciences, particular cases can only be understood through general laws (Arciero 2002). The most interesting aspect of this approach is the way in which the patient opens up to his own story according to the prescriptions of the theoretically oriented therapist. Given that a person's relationship with his own story is articulated through its enactment and hence according to a given sense, for the theoretically oriented therapist, this relationship takes shape instead, in ways that are determined by his theory of reference. It is in the light of such theory, for instance, that the patient will perform self-observation exercises by focusing on problematic experiences, evaluating and neutralizing negative patterns of thought, or recognizing how irrational ideas affect his mood, behavior, physical condition, etc.

The patient's relationship with his own ongoing experience or past thus becomes the correlate of another relationship: that with the therapist's theory—a theory which guides the patient's grasping of his own lived experiences. According to this perspective, the patient's past exists for the therapist as an objective element which may be grasped as the object of a series of theoretical determinations that specify the modes of defining the sense of one's lived experience. These determinations may be based on the organization of personal meaning, motivational systems, traumas, unresolved conflicts, family constellations, mental models, narrative patterns, or personality structures. In any case, what matters is not the specific approach adopted, but the way in which the patient's personal story is conceived, namely: as a causal chain of past facts revolving around central elements that may be grasped as objective data of consciousness. In keeping with Descartes' 6th rule (1998), these may be described as the simplest things (*quid sit maxime simplex*).

In this sense, the actual concreteness of one's personal story is subordinated to general principles that govern it, thereby dehistoricizing it: for the historical domain is analyzed by referring to given a priori.

Consequently, the effect which the grasping of one's own lived experiences—according to preestablished determining conditions—has on the development of one's current life is limited to the heuristic extension of the same theoretical principles to different spheres of life. The principles in question are thereby articulated in an increasingly broad manner, to the point of extending the therapist's theory—which shapes the patient's relationship with his own history—to the ongoing critical situation, which is explained in the light of them. Different forms of psychotherapy acknowledge this broadening as a "growth" of individual consciousness and hence as a therapeutic act.[1]

As the enactment of the patient's relationship with his own past is carried out, according to the theoretical motivations of the therapist who establishes this

[1] From this perspective it is easy to understand why constructivist therapies in particular—obsessed as they are with the idea of an all-encompassing theory of knowledge—have sought to develop a model for therapeutic change by drawing upon epistemology.

relation, the person concretely seeking to grasp his own history is never really involved. The relation between the past and the present ipseity is detached from concrete existence, insofar as the very condition for grasping one's personal story is the determination of lived experiences in the light of a series of theoretical principles that shape this reflection. Hence, the grasping of the past cannot influence concrete being-there in such a way as to renew it; rather, it only touches upon present ipseity within the framework of the theoretical knowledge which has determined that grasping, thereby consolidating the reflective distance between the I and the me—or, as Kant would put it, between the transcendental and the empirical.

The acquisition of theoretical knowledge and its application to the objective sphere of lived experiences, in other words, do not enable the renewal of the context of access to oneself and hence of ongoing ipseity but will only neutralize every possible novelty through the same theoretically oriented trajectory. Indeed, this mode of operating—already known to Hellenistic philosophical schools—derives its reassuring and soothing quality from the fact that the theory according to which the gnosiological self is structured makes it possible to control the disturbing character of the future by "knowing" its sense in advance, thanks to supratemporal principles.

From this it follows that the "gnosiological" enactment of the relationship with one's own story, conceptualized as a therapeutic act, does not change the concrete context of access to one's own experience, becoming a factor of stabilization instead. At times it may even turn into a destiny, which takes shape according to the coordinates of the theory adopted by the therapist, who is usually the one to establish the criteria governing the process of self-inquiry.

6.2 Therapeutic Care and the Theoretical Relationship

There is another, implicit aspect of the patient-therapist relationship which is of crucial importance. If "therapeutic" knowledge of oneself is attained by interpreting experiential data according to a given theory, the investigation and explication of the patient's experience is bound to be subjected to the therapist's opinion. As the repository of all theoretical principles, the therapist directs, evaluates, rectifies, and corrects the sense which the patient grasps in his own experience.[2]

What we have here is the very core of the therapeutic relationship, which—as we will see in Chap. 11—is characterized by an ancient theoretical structure rooted in the methods and forms of individual spiritual guidance practiced by almost all the philosophical schools of Antiquity (Rabbow 1954; Hadot 1995, 2014). The same structure shaped the experience of confession for over a 1000 years, down to Suarez's idea of confession at a distance, by letter, and—even more strikingly—Descartes' *Meditations*, a solitary, written expression of the author's work on himself. Yet, as Foucault rightly noted, there is one crucial difference here: whereas from the time of the earliest Christian monks to the sixteenth century, it was believed

[2] All the various forms of training and supervision go in this direction.

that there could be no access to the truth without sacrifice (i.e., without foregoing self-attachment and hence without working on oneself), from Descartes onward the idea took hold that anyone could access the truth simply by establishing a clear and distinct relationship with oneself capable of disclosing it (Foucault 2016). According to such a view, one opens up to the truth not by ascetic effort, but through knowledge or, better, through a science capable of grasping the truth of the soul. So when the soul is dimmed, when the fact of no longer being in possession of oneself makes it necessary to shed light on oneself, self-examination draws upon the ancient structure of confession. However, it does so in such a way that experiences are verbalized, not within the framework of a relationship of obedience to a spiritual father, but according to scientific principles capable of revealing the truth that lies behind all the confusion.

The same metaphysical core has come to distinguish all forms of scientifically oriented psychotherapy: to some extent, the act of entrusting oneself to another person's care always goes hand in hand with an explicative work that is oriented by the principles of a natural science of the soul and which is intended to shed light on one's true self—by separating it from that which dims it and prevents one from grasping it.[3]

Personal experience, therefore, is interpreted in accordance to the theoretical principles upheld by the therapist, who acquires the role of a guide for the patient seeking to discern a new configuration of the self in keeping with the therapist's theory.

Based on this mode of accessing the self of the other, the exercising of therapeutic care consists precisely in promoting the distinction of those movements of the soul that prove meaningful in the light of the therapist's scientific ideal and in bringing them to manifestation: a process no different from crafting a table from a log. The same *techne*, rooted in the Greek way of conceiving man's being, shapes the effort to interpret lived experiences. Each theory thus establishes a new mode of relating to the patient's experience of himself—a new technology of the self—based on the idea of knowledge that lies at its basis. On the other hand, as a consequence of the constant interpretative focus on those objective and simplest elements of life, what takes shape and solidifies in the patient through verbalization work is a self that stems from the application of the very same "therapeutic" technology to the person's lived experiences. This self is the outcome of the enactment of the relationship with one's personal story in terms of knowledge: a process directed by a theoretically oriented interpreter who, faithful to his guiding principles, is capable of discerning the sense of certain experiences more clearly than the patient who has lived through them—precisely because he possesses the principles required to grasp this sense. What follows from this is the constitution of a new mode of accessing

[3] What deserves a separate mention is mindfulness psychology, in which the therapeutic practice takes place in exactly the opposite way from conventional Western psychotherapies, insofar as treatment occurs through a process of deindividualization—implemented through a meditative practice—rather than through a regaining of the self. Nevertheless, the techniques for transformation are much the same (see Foucault 2013, p. 95).

one's own lived and ongoing experiences that is entirely determined by the idea of knowledge that lies at its basis: a gnosiological self whose degree of articulation varies according to the degree of distinction and verbalization attained over the course of the therapeutic relationship.

Each theory establishes the parameters for what may be regarded as change (and for its stabilization), which therefore consists in the consolidation of the theoretical relationship with one's own past experiences. Change is seen to amount to the realization of a "new self"—regulated by the therapist's theory—that establishes the criteria according to which experience is stably grasped and structurally freed from existence. This product—the self-subsistent table at the end of the manufacturing cycle—becomes a stable object and may thus be investigated and brought into relief according to the rules of an evidence-based practice: a practice whose only merit is to certify the theoretical grasping of experience and hence the distance from actual experience of the approaches based on it.

In therapeutic practice this is particularly evident in the case of patients who have chosen to change therapist and adopt a new approach. Even when they are dissatisfied with the outcome of their past treatment, these patients continue to access their mode of being according to the theoretical grids provided by it, to the point that a seasoned therapist can easily infer what a priori principles guide these individuals' relationship with their own experience and what current their previous therapist subscribed to.

6.3 The Access to Other People's Stories and the Possibility of Therapeutic Care

If a person's relationship with his own past (*Gehaltssinn*) takes shape through the enactment (*Vollzugssinn*) on the part of a someone who de facto exists and each time, starting from a concrete context, realizes this relationship according to a sense of reference (*Bezugssinn*), it is evident that one's personal story—and hence the way in which a person has his past—depends on how the individual accesses his own past experiences.

As we have seen, when a person's relationship with his own experience is defined according to some kind of theoretical determination, the understanding of past experiences never extends to concrete existence so as to contribute to renew it, since the past is only grasped within the framework of a reflective objectivation and is therefore seen as something removed from actual being-there. Lived experience, in other words, manifests itself as a passing-by, as a process that unfolds within the psychic domain in time in such a way that it only takes shape and becomes visible under the reflective gaze of a disconnected I that fixes the salient elements of this process: the invariants. Lived experience gives itself as a present object that is perceived by the eyes of the mind.

Once again, what appears between the folds of the practical science of therapeutic care is that conception of the motility of life typical of ancient ontology which, guided by the paradigm of production, reduces every "not yet" to the constant presence of a graspable object open to theoretical manipulation. However, existence is not available as an object: it is never given in a permanent, certain way; rather, it is constantly in the process of being actualized. In this respect, "the 'how' of life lies in its being at stake in the world, 'on a day to day basis'" (GA 61 1985, p. 133).

In this being absorbed in everyday life, the way in which life reaches itself in given situations at the same time coincides with the possibility of accessing the self: with the way in which life is understood on the basis of occurring circumstances. Hence, in each case the sphere of intelligibility manifests itself in the way in which life is grasped and possessed (or not) via those circumstances that shape its aims and projects. The experiences of the self which emerge again and again—and which have an instable prominence in relation to the course of our life—more or less explicitly provide those relevant aspects that distinguish the way in which we feel situated with respect to other people, within the world and within the framework of our own existence.

This leads us to a crucial point: to grasp the motility of life as a self-enclosed phenomenon is to miss a key element, namely, the impermanence of being on a journey, which constitutes its essence. In other words, that (theoretical) way of grasping life betrays the very nature of life, as it fails to take account of the motility of the experience and event of living: the incompleteness of our being-there, which in each concrete life context manifests itself as the constant pursuit of meaningful possibilities.

If, by contrast, we attempt to fully understand the motility of life "from within", i.e., in the light of the experiences made by individual human beings who conduct their lives each in his or her own way, we will perceive motility not as the alteration of that which endures—as a passage from what is no longer present to something else that comes into presence—but rather as the enacting over and over again of our relationship with what we encounter: a passage which begins anew at each instant.

Again and again, life finds itself in a given situation and from this situation projects itself into an imminent future which is yet to come. Hence, life in each case accesses itself as it unfolds, without the need for an I to turn back and reflectively grasp itself as an object. Since the way in which we relate to our own lived experience is through its enactment by a concrete being who factically chooses his own path starting from the situation at hand, and who on the basis of such context in each case accesses and appropriates (or fails to appropriate) his own self historically, it is evident that our relationship with the past is indicative of the degree of intimacy between our actual being-there and our personal story.

If lived experience reveals itself in relation to the way in which life notices itself, to the way in which it intuits itself in this or that circumstance, the concrete meaning of one's personal story has to do with the way in which it is actually experienced and hence with the way in which the relationship between ipseity and one's own past experience is factically articulated. On the other hand, if a person's relationship with those experiences which were significant for him or her brings the self into play, enactment comes with an "updating" of the meaning of past experience, which thus *becomes* alive, and, at the same time, with a renewal of the ipseity that takes charge of that past.[4] The meaning of the past is grasped on the basis of the living present (motivational context) and in each case its appropriation has an effect on ipseity. As lived experience is appropriated, it *comes to be* again and again in ongoing experience, thereby contributing to forming it and lending cohesion to one's own life.

In this respect, a person's relationship with his or her personal story is co-directed toward concrete ipseity: an ipseity which takes charge (or fails to take charge) of the past by preserving it through the enactment of its ongoing being-there—hence, never permanently so—and thus chooses to live within the story it is capable of taking on and bringing forth again and again. Lived experience does not flow before the eyes of the mind as film images do before the eyes of a spectator, yet the present I resonates with it in its meaningfulness and is inseparable from its occurrence: lived experience is appropriated according to the sense which is proper to it (*Ereignis*).[5] If we grasp being-there in its historicity and hence "from within," it is clear that the motility of life is not determined by the fact of thinking or reflecting about something; rather, it takes shape as a "someone" that generates sense through the encounter with things themselves in the light of a horizon of expectations.

The fact of grasping another person's experience from within points to a completely new access route which, if it is to disclose the sense of this experience without destroying it, must require us to somehow go along with this experience as it unfolds. In other words, an understanding of the other can neither be achieved via explanation nor via description—less still via empathic resonance. The experience which the other has of himself cannot be explained or described as though it were an object within reach or under my gaze, since the other is there, in relation to me in the various circumstances of everyday life. I meet the other at work, on a train, and in the street; the other is together with me on a project team, in an association, and in the struggle for a common good. The other is at home, travels, and comes back late; the other is in memories, lessons learned, time lost, hopes, desires, and horizons. The other is always with me, in the particular situation *I find myself in*; and in each of these worlds, the other is another face, as Levinas has shown: the face of the other relating to me.

Starting from inside the womb, being with the other always, in every moment, constitutes a determination of what one is like; and it is precisely this relationship in terms of one's position—expressed through countless modes of sociality—that makes it impossible to refer and reduce the experience of the other to one's own.

On the other hand, we cannot understand the other through empathic resonance. As we have seen, empathy is a methodological requirement within the framework of an approach based on the myth of interiority and leads from the isolated solipsism of a self-referring I to the one-to-one solipsism of the I-you relationship. This is referred to nowadays as the second-person perspective (Varela 1999). However, grasping the experience of the other does not mean building a bridge between two interiorities based on affective identification with the other's lived experience

[4] This is like being directly confronted with a past experience.

[5] What is meaningful is "primordially living and experiential" (GA 56/57, p. 107).

(*Einfuhlung*) or the transposition of oneself into the other's life (*Hineinversetzen*); rather, it means situating oneself with respect to the other, based on the fact that one already stands in relation to the other as another human being: an "other" in the same home, city, society, or history—in other words, in a shared world.

What we have here is an "other," then, and not a "you," understood as a second I that stands before me and onto which I project my modes of understanding myself or which I grasp by simulating its modes of being (Gallese 2005, 2007a, b). This is not an "other" that can be grasped by reducing it to the inner dynamics of a self-enclosed I; nor is it an "other" in which I can grasp typical behavioral traits or patterns.

The other is himself an ipseity and I can grasp him as such in the light of our coexistence in a shared world. To quote Patočka (2003), "originally we grasp neither ourselves nor the other as two Is, but rather as components of the situation … we – the other and ourselves – reciprocally discover one another within the unity of the meaningful situation" (p. 105–106). In this respect, being with the other—with attitudes ranging from solicitude and intimacy to indifference and unfamiliarity—is always a determination of how I am in a particular situation, i.e., of the way in which I make the experience my own, on the basis of the fact of *finding myself* with the other in a world in which the two of us coexist.

Existence as a shared experience is the foundation which ensures "that there is a *being-for* and *-against* and *-without-one-another* right to the indifferent walking-alongside-one-another" (GA 20 1988 Deu. p. 332 p. 241). It is precisely through this constant feeling of being supported by an emotionally attuned body that, from the very start, ipseity is never a subject but a being-together-with; and it is only through this form of belonging that one can be an I. Saying "I" always indicates a configuration of one's being-there—in the world with others, always in a specific situation—that can only be accessed by the person who says "I" (Husserl 2006, First Investigation §26, Sixth Investigation §5, Heidegger GA 2 1977).

Clearly, from this perspective, being empathic is but another possible position with respect to other people's lived experience, as are being intrigued, annoyed, or frightened—and all other modes of understanding of the other, which are made possible by the experience of existing. From the very start, no matter how it manifests itself, this experience is always also a way of perceiving oneself (*Bezugsinn*) and a particular mode of enactment (*Vollzugsinn*) vis-à-vis the mode of being of the other (*Gehaltsinn*) who relates to us. The experience of the I, you, us, and them can only take shape starting from this foundation. It is evident therefore that, as a unique mode of relating to the other, the therapeutic relationship is inextricably bound to the way in which the experience of the you is conceptualized.[6]

[6] On the other hand, to conceive the dialogic relation between the I and the you, not as something based on the fact of being with the other, but as the cornerstone of the entire social framework—and, as some have suggested, of the therapeutic relationship (Hermans and Hermans-Jansen 2004; Hermans and Gieser 2012)—means refusing to acknowledge the defining character of ipseity, which essentially coincides with being with the other, before and beyond any I and you. In other words, ipseity does not become a being with the other after one has understood that it exists in the presence of another entity that shares its disclosing nature; rather, ipseity always coincides with

6.4 The Therapeutic Method and Relationship

When it comes to therapeutic practice, the relationship just outlined raises two crucial and interconnected questions. First question: if being with the other is always a determination of one's own mode of being, what mode of relating to the other is therapeutic care associated with? Bound up to this first question is a second one: since being with the other always implies a specific way of grasping the other's experience, to what particular mode of understanding does the therapeutic relationship correspond? It is evident that the question of method, when framed in such terms, cannot be separated from the question of how the other is considered within the context of that particular relationship that is therapeutic care and hence of how psychic disorders are conceived.

If the other is understood according to the natural-scientific method—by recognizing in the other's experience what may be reduced to a priori rules—the you is systematically stripped of all personal experience. This may occur to different degrees, ranging from the reduction of the other to a biological, instinctual-biological, behavioral, cognitive-behavioral, or computational object, for example, to the acknowledgment of the other as a you on the basis of personality categories or motivational systems, achieving an understanding that still remains a form of reduction to the I—to an I: that of the therapist, who thanks to his knowledge understands the other better than the other could ever understand himself.

It is evident that in these forms of therapy—ranging from psychoanalysis to the various constructivist and cognitivist currents—the experience of the you is that of a subject to be interpreted in the light of the criteria and categories deriving from one's theory of reference. At times, in actual practice, when these criteria and categories fail to fully account for the meaning of the patient's lived experience, as a last resort many practitioners turn to banal empathic sensitivity, combined with a fair dose of common sense! Thus the relationship with the other emerges as an educational relation and, in certain cases, as the taking charge of someone, as an act of providing assistance that relieves the other of the weight of self-responsibility. A more refined version of the relationship of assistance is the therapeutic relationship conceived as a secure base: as the construction of a new vessel that replaces the defective one developed over the course of the patient's early relations with his caregiver, thereby enabling him to acquire the ability to think about painful emotional experiences.

In a therapeutic context, understanding the other primarily means answering his plea and being open to what it verbally expresses as an indication of the other's feelings and actions. Therapeutic care is the attempt to access those aspects of this plea that cannot be verbalized, but only expressed as a form of suffering. If this is so,

being together with the other. The fundamental fallacy of the dialogic approach to psychotherapy, which undermines its entire conceptual framework, lies in the fact that it reduces the occurrence of experience, which is always mine, to I-ness. In this case, the I-you dialogue, the we, takes shape as the go-between for two encapsulated subjectivities that are inherently opposed to each other: two self-contained entities of which we know nothing, including their possible mutual relation.

then the experience of the you is synonymous with the acceptance of an otherness that imposes itself in its privation and suffering and which resorts to therapeutic care as a means to change its condition: "to straighten his life out with words" (Plutarch 1927, 8.42).

Heeding the plea and providing therapeutic care with words do not merely imply the kind of solicitude, consideration, and benevolence deriving from the passion for healing but also the capacity to orient oneself through speech toward that which lies concealed behind the request for therapeutic care and remains ungraspable for the patient: that which, as we have seen, usually lies at the basis of the disorder.

The succession of clues provided by the narrative woven by the patient, and which in most cases culminates with the symptom, cannot point the listener toward well-defined circumstances, since the experiential origin of the condition in which the patient finds himself is bound to escape him. The symptom occurs despite oneself and proves stronger than oneself; often patients—but also the DSMV and all other manuals before it—do not even grasp its connection to the circumstances in which it took shape. Then, one of the basic problems underlying the suffering that the therapist faces is the fact that the lived experience gets separated from the ability to account for it, to give sense to it. So the patient has a particular experience that he cannot give words to, that he cannot recognize, or perhaps he gives a completely different meaning to this experience, a meaning that is not in consonance with the meaning of the experience itself.

While in everyday casual conversations, but also in the recovery of experiences, in saying "I" the speaker—who knows in advance what his statements refer to—can orient the listener toward his unique situation through an indicating meaning; and while the listener, in turn, can re-enact the situation evoked and the occasion of expression which the meaning points to (Husserl 2006, First Investigation §26), the indicating meaning provided by the patient somehow always remains "pointless."

The account which patients give of their present condition of suffering, then, points to a moment in the past (in which the symptom emerged) which typically has no meaningful connection with the symptomatology displayed. The symptom emerges through the establishment of a mechanism that keeps the meaning of the lived experience separate from its appropriation; and it is precisely this gulf that strengthens the disorder. This is why patients' accounts generally fall short when it comes to their symptoms.[7] For instance, in the case mentioned in the previous chapter, the woman who in relation to a modification of her involvement with her partner experiences the emergence of certain emotional states as an affliction, perceiving them as a bodily alteration (e.g., tachycardia), will perceive the condition she is in as being unrelated to the present situation and removed from past circumstances. In other words, those emotional states perceived as physical symptoms are experienced and suffered as something alien and—as we have seen—may trigger a mechanism, producing a disorder.

[7] Janet was the first to realize—albeit only in relation to hysterical phenomena—that symptoms are produced through the establishment of a mechanism between two levels that continue to engender it by remaining separate.

6.5 Therapeutic Care and Unappropriated Narratives

Typically, as patients' accounts are polarized by their affliction, they certainly do not point to the prereflective meaning of their experience; on the contrary, it might be argued that they conceal it. For instance, in the case of that patient of ours who has described a detailed sequence of her cardiac difficulties, the "pathology" she is experiencing and relating is unconnected to the actual conditions within which it emerged and even more so to any historical origin. Failing to recognize the conditions triggering his disorder, a patient seeking therapeutic care "to straighten his life out with words" will produce an account of his own experience as though this had just sprung out of nowhere.

Evidently, this is the kind of patient who yearns for sense and hence easily falls prey to the soothing and reassuring effect achieved by the theoretical perspective of many forms of psychotherapy through the standardizing and anticipating ordering of the future. But this is also a particularly vulnerable patient, who can more easily be exposed to the charisma of the healer and hence to the kind of abuses perpetrated in the name of therapeutic care.

A "pointless" account of the sort just outlined hinders one's understanding of the experience at its basis, since the verbal reconfiguration made of it by the patient loses any connection to the way in which meaning has prereflectively taken shape. Hence, an account of this sort de facto prevents one from retracing the patient's experience to its origin: to that context of circumstances in which the occurrence of ipseity became detached from its narrative reconfiguration, thereby leading to the formation of the symptom.

For instance, in the case under scrutiny, the description of cardiac difficulties that stands at the center of the accounts and everyday life of the patient certainly does not allow us to grasp the way in which the patient perceives herself when she feels without the other—nor does it allows us to understand within which existential horizon this perception fits; rather, the patient is simply reporting that in certain moments of the day or week she suffers from tachycardia.[8] In this respect, the indicating meaning that the patient provides through her narrative is "pointless": it does not lead the therapist to re-enact the situation and occasion of expression to which the meaning refers and hence does not orient him to a fulfillment of the experience indicated.

In everyday life an assertion conveys the way in which a certain concrete experience is pointed out. By following the direction revealed by the assertion, it is possible to distinguish that which is being indicated, what the assertion has in view. But while an assertion points to a given experience, and by grasping it freezes it within a certain linguistic meaning, so to speak, at the same time, it strips the meaning of its context and turns it into something present.

[8] To simply dismiss incongruities of this sort as self-deception is to assume that the patient is implementing a maneuver (negation or rationalization) to avoid facing the element of truth in her account, as opposed to examining this personal history from the perspective of how the patient relates to her own past.

So while an assertion indicates how to grasp a certain experience, at the same time, it covers the field or web within which the experience in question took shape, thereby concealing the access to its modes of direct actualization. It may be argued that every assertion is nourished by its simultaneous and necessary capacity to veil the prereflective condition that anchors it and out of which it emerges. For instance, the act of mentioning a desk—the famous example provided by Heidegger in one of his early lectures—conceals the different meanings which this object has for the professor, his students, the cleaner, or someone who has never been in a lecture hall.[9] In the case of symptoms, however, the patient's account completely obscures that prereflective meaningful structure which at the same time it presupposes. Nowhere does assertion prove more dangerous than in the production of symptoms (Caputo 1984, 1988). To quote John Caputo, "Interpretation is always threatened by the pull (*Zug*) of fallenness, the withdrawal *(Entzug)* of authentic self-understanding. The opposite pull (*Gegenzug*) of projection, which makes authentic interpretation possible, is always liable to degenerate. It is just this threat that is posed when interpretation passes over into the 'assertion'" (Caputo 1988 p. 73).

An assertion completely loses the disclosiveness of the original understanding to which it refers, amplifying the patient's unfamiliarity with his or her own experience—which is therefore suffered—through the establishment and endurance of a vicious circle between the actual happening of experience and its symbolic representation. This difficulty of accessing personal experience therefore manifests itself in the form of a symptom, which—as the Greek etymology of the word suggests—occurs together with the activation of the mechanism (i.e., the triggering of the vicious circle).[10] On the other hand, a symptom indicates that the motility of life has come to a halt: through the repetition of a mechanism, it reduces the possibilities of signifying that which the individual encounters, and hence to access a shared world, thereby hijacking the person's freedom and opening up the domain of suffering.[11]

Most of the problems that lead patients to undergo treatment revolve around the subtle and constantly ongoing interplay between prereflective meaning and its linguistic appropriation. In the footsteps of Aristotle's *Poetics*, Ricoeur (2010) emphasizes how the construction of a plot—which combines a range of different factors (agents, aims, means, situations, outcomes, expectations, etc.) into an intelligible whole—constitutes the mediating element between the need for concordance in one's narrative and the acknowledgment of discordances. Through its reconfiguring power, language makes it possible to compose, and hence also to reinforce or ignore, the congruence between different experiences, thereby creating the possibility to alter the dynamic between experience and its symbolic reconfiguration.

In the case of the aforementioned cardiac symptoms, then, the prereflective experience at their basis, instead of being revealed and appropriated through an adequate

[9] Consequently, the analysis of meaningfulness, which is to say of how the desk is understood by each person, coincides with the study of the access to it, which is achieved by re-enacting it.

[10] Etymologically, "symptom" describes a fact that coincides—and occurs together with—another fact that is its effect or marker.

[11] It is on the basis of this that psychopathology structures itself as a science.

narrative, is only reconfigured on the basis of a sense that excludes it. As already noted, the account loses its reference to living meaning and through a vicious circle engenders an experience that produces the same account again and again, amplifying the distance with the meaningful structure which it presupposes in a self-referential and recursive fashion.

On the other hand, the reconfiguring of events may force the unstable structure of discordant concordance which distinguishes the narrative plot, creating concordance where discordant elements are present. In this case the narrative invents a new meaning, de facto allowing the experience at its basis to remain unappropriated. In other words, the distance between experience and its narrative appropriation is produced through the weaving of a narrative that is so incongruous—or perhaps oblique, one might say—with respect to prereflective meanings that, although it does not explicitly manifest a symptomatology in itself, it may engender other conditions of malaise. Hence, it represents a different form of impediment compared to the simple production of a symptom.

This is, for instance, the case with patient X, who after a 7-year conjugal relationship and three children born 2 years apart finishes weaning her newborn and finally gets back to work, only to discover that she has never really loved her husband.

This discovery marks the beginning of a marital crisis, which the patient attributes to her husband's behavior, his everyday attitude toward her, and his having forced her to marry, undergo three pregnancies, quit university, etc. Some time after the beginning of the crisis, the woman falls for a married colleague of hers and begins an affair made up of sporadic encounters that completely take over the patient's sense of her own life. The dilemma which leads the patient to seek treatment is whether she should leave her husband or not—a nagging doubt, accompanied by constant anxiety, with fluctuating behavior toward both partners.

The patient, who has turned to therapy to resolve her dilemma, perceives this experience as being completely unrelated to the situation in which it took shape, namely, the glaring fact that she got back to work. Returning to work after 6 years and three pregnancies totally changed the patient's existential horizon and sense of her everyday life; her condition as a woman, wife, and mother; her perception of her own femininity and capacities; and her way of relating to her own family of origin, money, everyday things, and stage of life. It is within such a context that the patient falls in love with a colleague she has known for years.

Equally unrelated to this transformation of the patient's life is her reprocessing of her personal life. This revolves around a narrative of self whose protagonist is the patient herself, a young, fragile, and inexperienced victim who has fallen prey to an attractive man—her husband—and is forced to quit her studies and job in her father's company in order to conceive three children against her will. The patient no longer realizes that she chose to get married despite the opposition from her parents, who had advised her to wait, given her young age. The patient has forgotten that it was she who urged her husband to have some children, while he would have waited longer. The patient also misses the fact that it was her own choice to quit her studies and job with her first pregnancy. Memories, then, instead of being appropriated are resignified according to the perspective provided by the ongoing circumstances.

Through the new mode of access to the past that emerges together with the transformation of her existential conditions, the patient rearranges her own memories into a new narrative yet does not appropriate them or engage with their sense. Hence, she fails to grasp the fact that, now as much as in the past, she is always making her life choices in opposition to someone, or at any rate in view of someone, rather than in view of herself. In other words, the patient does not grasp the fact that when she is required to decide for herself, she searches for a support and center of gravity because she feels at a loss. Hence, the woman's narrative reconfiguration of the past in the light of the new possibilities opened up by her affair with a colleague does not lead her to investigate what significant factors are driving her in that direction (motivation) and thereby to renew her present ipseity by appropriating this perception of herself (her feeling disoriented when required to decide for herself); rather, this narrative reconfiguration redefines the web of events to the point of obscuring it and removing it from the domain of the proper. In such a way, this mode of being continues to determine the patient's orientation.

The motility of life is here suspended at a different level from that of the mechanism of the symptom. The distance from oneself is created and maintained through an unappropriated narrative: a narrative, that is, which does not thematize the mode of being that lies at its basis so as to make it its own, but which rather conceals it. In the particular case under consideration, the fact that the patient portrays herself as the victim of a charming and overpowering husband prevents her from dealing with her difficulty of taking charge of herself in the first person. This distance from oneself can be generated again and again over the course of one's life, through the creation of different narrative reconfigurations that repeat the same underlying motif while continuing to conceal the experience that determines them, without ever remobilizing and taking charge of it.

All this is clearly visible in the case of our patient, who articulates the same difficulty of taking charge of herself in different ways at three different stages of her life: she chooses to marry against her parents' advice, to have three children against her husband's will, and to leave her husband while blaming him for the separation. In such a way, the patient remains distant from herself, so to speak: she does not deal with herself; she does not appropriate herself by establishing her own existence according to the repetition of the same dominant mode of self-perception. In our works on neurotic psychopathology (Arciero 2002, 2006; Arciero and Bondolfi 2009), we have distinguished different personality styles marked by particular tendencies, identifying the rules governing their manifestation on the basis of recurrent traits that, aside from accounting for the repetitive stabilization of given modes of being and related personality disorders, link them to the emergence of different neurotic configurations.[12]

When one is confronted with such recurrent modes of reconfiguring life events, as is often the case in clinical practice, it is tempting to extend them to mankind as a whole, and hence to scientifically interpret man according to a psychopathological model, in terms of causal invariants: in terms, that is, of recursive patterns for the ordering of personal experience, patterns which remain unaltered over the course of one's life and thus lend a sense of stability and continuity to one's experience. This permanence may be seen to extend to symptoms.[13] The origin of these invariants will then be sought in processes of development, including early forms of reciprocity. This is the burdensome legacy which, in various forms, has been dominating the world of psychotherapy since Freud. Those therapists who embrace this legacy in their practice explain life histories in terms of recurrent patterns. In doing so, they make a destiny of recursiveness and the non-appropriation of the reconfiguration that accompanies it. In other words, they freeze the patient's life in an endless repetition, with each therapist accounting for it according to his own theory.

On the other hand, an approach that is the exact reverse of the one just illustrated, and which is adopted by narrativist therapies in particular, consists in envisaging the treatment as a "re-authoring" of the patient's personal story, accomplished together with the therapist. Treatment, in other words, consists in transforming the patient's narrative in such a way as to create new identities through the resignifying of the experiences it reconfigures, as though it were enough to alter the patient's narrative to change his or her personal story. The distance between the two—which is brought into relief by MacIntyre (1981) when he states that stories are lived before they are told—becomes apparent in the case of hysterical fugue states. These are characterized by the fact that instead of reconfiguring lived experience, the narrative is produced as though it were a film starring the patient in the lead role, a film disconnected from the person's actual life. It goes without saying that this way of crafting experiences coincides with a pathological condition.

As in the case of therapies of natural-scientific inspiration, the outcome of this rhetorical operation carried out by the narrativist therapist is always the same: to distance the patient from his own experience, preventing him from dealing with it.

But how can language have such power?

---

[12] Our aim was to grasp the connection between normalcy and psychopathology by showing how the various neurotic syndromes may be traced back to the fossilization of modes of experiencing.

[13] In Kraepelinian biological psychiatry, this perspective is pushed so far that symptomatology is reduced to chromosomic determinism.

6.6 Orienting Assertions

As we have seen, the distance that becomes structured through the narrative is already created through assertion, which constitutes the very foundation of language as discourse, as argued by Benveniste (1966, 1974). If ipseity is constantly realized in the encounter with this or that thing, and if it is on the basis of such an encounter that language articulates the meaning of ipseity through discourse, an assertion not only stably preserves the topic dealt with by lending it a consolidated meaning, but this immobilizing of sense within the assertion at the same time freezes the possibility to access the way in which the understanding of an experience originally developed.

In the case under consideration, for instance, the meaning conveyed by the patient's assertion, the dilemma of whether she should leave her husband or not, no longer concerns the situation within which it took shape, but rather its transformation into something that is on hand, something which may be passed on, expressed, and manipulated through language and which therefore turns into an unresolvable doubt—and hence into an assertion which is no longer connected to any other fact. Heidegger describes this peculiar element as the fading of meaningfulness. The philosopher explains: "The content of factical life experience falls away from the existence relation towards other contents; that which falls away remains available; the availability itself can, however, for its part fade as a sense-character of the relation and pass into that of mere usability" (GA 59 1993, p. 26–27). If we here detach the assertion from the prereflective structure it presupposes, not only does meaning become an entity that is on hand, but language as a whole can be viewed as an enclosed system—in the manner of De Saussure.

It is once again evident, therefore—if only in a more subtle way—why assertion is dangerous: the indicating meaning, reduced to a meaning fixed in the assertion (doubt), prevents the therapist from fulfilling what it says. The patient is asking the therapist to solve her dilemma as though she were seeking to resolve a mathematical equation.

If, however, we understand assertion as the means by which lived meaning is verbally expressed, then that account which constitutes an obstacle for getting back on the trail of oneself at the same time also represents an indication; in other words, it points to a direction to be followed in order to retrieve the prereflective experience that is what in any case supports it, i.e., the range of implications that lie at its origin. It is precisely for this reason that the self-narrative which therapeutic care deals with or attempts to restore—by steering clear of the Scylla of naturalistic determinism and the Charybdis of the narrativist play—rather ought to enable the patient to take control of his own experience, so as to decide what to do with himself. This would also appear to be suggested by Heidegger (GA 59 1993), when he claims that "The enactment is in each case and necessarily such that what has been self-worldly meaningful *becomes* once again in it; (…) I appropriate my own past so that it is had again and again for the first time and that I myself am always affected anew by myself and exist in renewed enactment" (p. 84).

In the case we are considering, for instance, the patient's dilemma of whether she should leave her husband or not is, at the same time, an obstacle to be deconstructed and a formal indication to re-enact those life changes which culminated in the marital crisis. The re-enactment and appropriation of the meaning that comes to light along the meaningful trajectories traced by formal indication suggest new possible outcomes for the patient's present situation. From this perspective, interpretation—understood as the re-enactment of experience—requires the analysis and explication of the dynamics of meaning in the situation at hand. It is evident that this kind of research, which aims to grasp a unique life in its movement, must keep to such movement.

The establishment of the interpretative condition, then, leads to the development of new directions and hence to a renewal of ongoing ipseity: a renewal that cannot be confined to the task of knowledge-acquisition detached from concrete being-there, according to the theoretical approach, which captures experience in order to standardize and silence it. Rather, this renewal co-constitutes ipseity itself by restoring its movement and suggesting possible directions. The past, in other words, is appropriated according to a new relation (i.e., to a new sense of reference) by opening up oneself once again to becoming.

The fact that the narrative may be completely discordant with the circumstances it reconfigures does not change the fact that it still indicates the experiential space from which it emerged, making it possible to open up, and hence to retrace, the experience it is directed toward by enabling one to regain it in the light of one's current life conditions. The logic governing the understanding of a given experience is given by the way in which this experience was originally meaningful. This regaining of experience through re-enactment does not consist in the mere recollection of something past; it cannot be grasped through introspection, but rather newly emerges through the remobilizing of the conditions that have generated it and which find completion in reactualization. With the re-enactment of experience, something actually new occurs that prospectively reflects itself in everyday life in terms of the generation of meaningful possibilities while also retrospectively projecting itself in terms of the stabilization of the "effectiveness" of the interpretation achieved. The movement forward, therefore, always also amounts to a movement of retrieval.

6.7 Formal Indication and the Therapist as Re-enactor

Regaining access to those contexts which lie behind the accounts presented by the patient implies two movements: (1) the grasping of content as an indication and (2) the retrieval of those experiences targeted through indication by means of their re-enactment.[14]

The first step in any word-based therapy is to start from what is being communicated—the content—and trace it back to those experiences which support it and to which the patient's discourse is pointing: meaning, as prereflectively generated in

[14] The third movement—the sense of reference—corresponds to the passion for healing.

its contextuality and within the framework of the implications through which it has taken shape. This implies a way of comporting oneself toward another person's experience in order to grasp it in its happening—i.e., according to the ways in which a human being *finds himself* with respect to this or that concrete situation—and hence to retrieve the meaningful trajectories through which the person in question, and no one else, has found a position with respect to given events. What clearly emerges here is not just the positum of psychology—personal experience (ipseity)—but also an indication as to how to access that prereflective and non-thematic dimension, which can only be understood as a specific way of comporting oneself: a "going with" the patient's living appropriation of his past that enables one to grasp the very genesis of those experiences of the other that are still unspoken and unexamined. To quote Dahlstrom (2001), "there is a prereflective manner of encountering things, access to which is, nonetheless, possible in a reflection, indicating a foregoing self-disclosure of the manner of encounter (not be confused with some sort of inner self-awareness)" (p. 235).

The task of the therapist as the re-enactor of the experience under investigation, therefore, consists in determining the meanings of the patient by retrieving and executing tendencies that are faithful to the situation. Hence, it coincides with the disclosure of a context through which to grasp exactly what it means for the patient to be situated with respect to this or that event belonging to a given stage of his life—the period toward which the indication orients the person. So it is through the capacity to submit oneself to the patient's plea by interpreting that which it points to that therapeutic care takes shape as something that the healer leads to, and provides, by relieving the suffering that the patient is unable to take care of himself. The practice of therapeutic care therefore corresponds to a way of comporting oneself with respect to the meaning of the experience under scrutiny. It emerges as a specific way of grasping the other's experience within the framework of a relationship that addresses the patient's plea for help, through a passion for investigation and an interest in questioning him in accordance with the actual unfolding of circumstances.

Because it is only by re-enacting original, non-thematic experiences that these can be grasped and appropriated, this kind of inquiry cannot be guided by a priori principles. From this perspective the sense of a narrative, of an assertion, but also of a symptom, corresponds to a formal indication, which is to say, to a signal which is empty—i.e., merely formal—before it is re-enacted and which points to an original behavior, the meaning of which can only be retrieved by retracing it. This attitude rules out any theoretical slip, so to speak: any attempt to grasp experience through categories of meaning stripped of the assumptions, prejudices, context, and setting of the interpretation. Finding oneself in a given situation—the most basic fact of existence—always corresponds to an expression and manifestation of one's mode of being oneself: it is in specific situations that, as it discloses itself pre-thematically, ipseity encounters itself with others in the world and together they come to manifestation. The one cannot be separated from the other.

Differently from the theoretical approach, formal indication becomes not an object fixed by the gaze of the mind, but a clue as to where to look, a disclosing of the way in which the meaning that subsequently trickled out, solidified, and finally faded in words, dimming in the symptom, actually originated. Precisely for this reason, formal indication cannot specify the object of its research but only point to it and provide a reconnaissance space while at the same time suggesting to the person seeking to understand it a certain direction and the need to follow it if he is to grasp its origin. Dahlstrom (1994), who calls this aspect of formal indication the reversing transformational function, emphasizes that the search for original meanings is not a form of contemplative reflection, but something that forces us to change our approach to experience, which by being reactualized and hence appropriated engenders a renovation of current sense.

This function acquires a key role in psychotherapy and constitutes the foundation of the study of personal stories. The analysis of the patient's lived experience does not investigate the past in search of the remote causes (or imaginary traumas) of a chain of facts that may explain the person's ongoing experience; less still does it search for the repetition of patterns that dissolve present experience in the regular recurrence of theoretical elements. Rather, what makes the analysis of lived experience so crucial is precisely the task which it imposes: taking care of what has been significant in the patient's personal story forces ipseity to take a position with respect to those lived experiences which are still a vital part of one's own tendencies. The transformative function of word-based therapy, therefore, consists in the possibility to find a new orientation in one's present life, by using the patient's story to reactualize the meaningful perspectives it conceals. This movement forward is therefore also a movement of recovery which becomes a new movement forward, according to a generative circle which is nothing but a "re-grasping inclusion of previous interpretations" (GA 61 1985, p. 163) within the interpretative context that has been attained.

In keeping with an ancient hermeneutic principle, formal indication discloses the movement of the person's interpretation according to a direction that goes from what is evident to what is obscure, unexamined, and unexpressed. Hermeneutics, which thus emerges as a counter-movement to the "natural" pull of the withdrawal of the experience of being alive, has to do with the recovery of those intimate tendencies and concealed motivations—tendencies that, once renewed via appropriation, contribute to lending shape to one's existential perspectives. Indeed, it is only by re-enacting those experiences which had vanished behind words that the therapist—faithful to the situation under investigation—strives, together with the patient, to ensure that what is being formally indicated may be appropriated in its own terms. This co-retrieval of what was meaningful, and is still unspoken, inevitably changes the person who has turned in that direction by renewing his or her perspective.

In the light of this view, we can agree with Patočka's conclusions with regard to the way in which temporality becomes time with respect to the future. Patočka (1988) argues that in order for this to happen—i.e., in order for temporality to become such in the light of the future—"what already exists (viz. the past) must no longer count as an authority which determines all possibilities; the condition (viz. for temporality to become time with respect to the future) is the possibility of non-being which, by emerging, sharpens one's gaze as regards that which it is possible and necessary to pursue" (p. 106).

As the experience to be understood has meaning in the original context of actualization for the person who lives it, and cannot be grasped "from the outside" or analyzed in the light of "what is simplest," or indeed assimilated through an act of empathy toward one's own experience, the experience must be articulated in such a way that its meaning will emerge according to the way in which it was originally had. The relation to the patient's story must therefore be developed according to an understanding way of looking at things, or as an observing understanding of the other person's experience, in agreement with its actual realization. This inevitably implies a retrieval of the original access to this experience, a retrieval which—as we have seen—corresponds to a way of approaching the other's experience by attempting to grasp it as it takes shape. The grasping of phenomena through formal indication, then, is "nothing else than the explicit, factically genuine execution of a tendency that is faithful to a situation, a tendency that, while developed and acquired (in scientific research, knowledge), is implicitly there in the object at issue itself (factical life)" (PIA 171).

Before further investigating these fundamental questions, let us briefly return to the problem of method. If grasping another person's experience means accessing the prereflective domain for the therapist, how can he retrieve the immediacy of sense, rediscover the way in which a situation originally presents itself to the person experiencing it, and regain this access?

The method of an a-theoretical science, and hence of a psychology based on it, should allow us to newly grasp and preserve the genuine self-showing of the prereflective living experience. It was no doubt this problem which Natorp had in mind when denouncing the limits of reflective phenomenology, a problem which he unsuccessfully attempted to solve through the method of reconstruction. It is necessary, therefore, to specify in what way we grasp how each person experiences this or that situation, prereflectively. To achieve this, a method cannot be confined to the remote analysis of objectified life experiences.

A person's relationship with his own story constantly takes shape through its actualization within the present contest. It is this relationship which must be examined in order to bring to light those lived experiences which prevent the patient from grasping his own history and, more generally, prevent his present ipseity from gaining full possession of itself and of renewing itself. To grasp experience as it takes shape, it is necessary to reactualize it by "going along with" the enactment of the other person's lived experience according to its proper meaning, which is to say, according to its sense of enactment. In such a way, we can explicate sense without destroying it, in "sympathy" with its flow. However, this is not enough, because in order to actually grasp the full experience of a self, it is first of all necessary to grasp how concrete ipseity encounters (or does not encounter) the domain of the proper. Therefore, the therapist's interpretative involvement in the enactment of another person's lived experience coincides with the bringing to light of this person's experience, according to the distinctive sense possessed by a given experience within the coherent framework of his life. The therapist's interpretative involvement, in other words, takes shape in keeping with the constitution of the patient's experience. An explicating understanding of this sort coincides with the experience of a living experience. In order to remain such, it constantly strives to keep to the original experience, a-theoretically. As any theoretical stance engenders a reflective domain, and thereby transforms the other person's experience into reflected-lived experience, it no longer grasps experience in its immediacy but only observes it.

Therefore, not just for the explicative method of the natural sciences but also for that of descriptive reflection (developed by Husserlian phenomenology), it is impossible to disclose the domain of the immediacy of lived experience. Von Herrmann is right to stress the fact that, unlike the theoretical approach, "hermeneutical seeing and understanding does not modify the lived-experiences in lifting them out of the living enactment of lived-experience … instead, in going-along with the immediate living-experience, it shifts it from pre-phenomenological unexpressedness into phenomenological explicitness. Only in this way does hermeneutic understanding not deprive lived-experiences of their own character; rather, it preserves that character in order to reveal it as such" (2013, p. 79).

It is evident that the kind of interpretative understanding which is faithfully directed toward living experience, as this actually manifests itself to a person, without any theoretical mediation, corresponds to the first methodological principle of phenomenology: the return to things themselves, as these appear in themselves. The question of the access to life itself, and of a method capable of ensuring the scientific rigor of the study of the prereflective sphere without entailing any form of reflection, is the direction which the young Heidegger derived from life and presented to phenomenology. Heidegger starts by elucidating the concrete situation of access and through a deconstructive retrieval of the original sources of meaning seeks to record the conditions for the emergence of those categories through which life is understood in its hidden motives, tendencies, and unexpressed explanatory paths (GA 58 1992).

As we have seen, reverting to the sphere in question implies not the direct grasping of lived experience as though this were an object of sensory experience, but rather the deconstruction of the overall context of references, which has produced the explicative direction fixed through the narrative (*Destruktion*) and its recomposition, as a direction that is determined by partaking of the vitality of experience according to its flow. In this respect, the past discloses itself, or becomes foreign to us, in the light of the capacity we have to access it in the present, because it is the concrete situation of living that brings out the motives leading us to engage with our own history, to regain familiarity with ourselves.

Phenomenology thus repeats the original familiarity of life, bringing it to manifestation. Just as hermeneutic phenomenology is an a-theoretical science capable of making the domain of lived experience visible in itself, phenomenological psychotherapy retraces the genesis of configurations of sense, by retrieving the story of their production by a specific individual over the course of his or her existence.

Unlike in phenomenology, the genesis of sense which is the focus of phenomenological psychology always concerns a well-defined sense, a particular history, a unique life, and a positum. The interpretation of concrete life—and hence its narrative explication—is therefore bound to revolve around categories that articulate the way in which the life in question already understands itself in actual contexts, in ongoing situations. The narrative investigation of factical experience brings to light and stabilizes that experiential context which the constant slipping from one situation to the next, from one context to the next, which characterizes our concrete life, prevents us from articulating. As noted by Ricoeur, through the act of plotting, the succession of events that make up the story are gathered into a constellation (the narrative plot) which more or less successfully expresses the overall contextuality of unarticulated references, which are an integral part of the experience made by the person, lending them a new unitary configuration and a renewed intelligibility.

Retracing one's life does not mean reflecting on it as though it were the object of a preestablished form of knowledge; rather, it means borrowing meanings from it—so to speak—in order to verbally express them. Thus, for instance, a present problem will open a window into the past, indicating a direction of sense which is to be re-enacted starting from the situation under scrutiny, so as to grasp that specific sphere of life in which it took shape. In this respect, the different themes characterizing the successive periods in a person's life represent contexts of sense that must be newly explored starting from the situation under investigation. In this movement, performed together with the therapist, what is already mine—as the experience of my own life—becomes genuinely my own through the narrative appropriation of my still unarticulated experience. The linguistic expression which is grafted upon a person's suffering and acting verbally conveys the person's experience of living and reverts to it, according to a constitutive circularity between intuition and expression and understanding and interpretation.

What also takes shape within this domain is a destructive or deconstructive process that consists in finding a way to access the origin of experience in its tendencies and motivations: phenomenology must be capable of grasping experience as it takes shape.

This is precisely the insight reached by the young Heidegger, and it is in the light of it that we are to understand his response to Rickert's criticism that the *Lebenphilosophie* was incapable of distinguishing between living life and thinking about it. Heidegger realized that phenomenological practice consists in "going along with" (*Mitgehen*) lived experiences, as a repetition of life: that it is not a reliving, but a *categorial research* in which life's re-collectability (*Wieder-holbarkeit*) "simultaneously brings its evidence to fruition" (GA 61 1985, p. 88, quoted in Crowell p. 125).

It is difficult to understand how this phenomenological mode of partaking of the vitality of experience—by going along with its flow—may be subsumed within the broader notion of reflection, as suggested by Crowell (2001) and Zahavi (2003). This is especially true if we understand the term in question as the mode of grasping phenomena typical of theoretical experience. "With reflection we adopt a theoretical attitude," the young Heidegger writes—meaning that what was merely a lived experience becomes a lived-observed experience with reflection (GA 56–57 1999, p. 178). The risk is therefore to "equate the intuitive phenomenological view with the intuition of objects" (p. 237). What emerges here is a significant difference compared to Husserlian phenomenology. According to Husserl, the meaningfulness of a given object is given together with its sensory perception (as a sort of excess). Hence, being must be given and must be given as the object of an intentional consciousness. This phenomenology thus takes the shape of a science of acts of consciousness, which evidently is not concerned with the question of what makes it possible to grasp the being of a thing as an intelligible meaning—i.e., of how that giveness (being) of the thing is itself given.

This problem of conflating phenomenological intuition with the intuition of objects is all the more acutely felt in Husserl's phenomenology because the fact of turning toward the essence (genus, species, individual) makes phenomenological intuition coincide "with the grasping of relations of order" (GA 58 1992, p. 240), which amounts to the intuition of an object.

The distinction between reflective phenomenology and hermeneutical phenomenology, therefore, does not concern—and hence does not rest upon—the methodological use of reflection, as Zahavi argues; rather, it implies a particular way of grasping living experience, which cannot be understood if it is grasped according to the theoretical mode of accessing things.[15]

Clearly, if ipseity is envisaged as something other than a thing, then the sense of being this ipseity cannot be an object of knowledge. The defining feature of the relation with oneself which distinguishes ipseity is the fact that ipseity can only relate to itself by enacting existential possibilities. Hence, it would be more appropriate to speak not of reflection but of refraction—a refraction which is constantly taking shape. As stressed by Patočka (1988) "the primary form of this relation is not an orientation toward the object, but the practical grasping of a given possibility … it is a *reflection* clearly guided by an idea of being which amounts entirely to its praxis" (p. 93—see too Patočka 1998 lecture 12).

6.8 The Therapeutic Approach to the Other's Relation with the Ongoing Situation: The Care of Self

Based on what has been argued so far, the sense of a person's relation with his own past may be grasped according to a theoretical or hermeneutical approach, depending on whether the person's story is grasped as something which has objectively occurred, and is therefore removed from the present, or as the concrete enactment of the person's relation with his own lived experience, such that his story is constantly renewed by the enactment itself. These two different ways of conceiving the past imply two different methods for the study of a personal story, which correspond to two different ways for the therapist to enact his approach to the patient's lived experience and, conversely, the patient's own relation with his own experience.

[15] The problem, then, concerns *theorein* and its primacy, and even the deconstruction of Descartes and Husserl (GA 17)—polemically emphasized by Crowell—cannot be grasped if not in the light of formal indication.

On the one hand stands the determination of lived experience on the basis of preestablished principles and laws, on the other the practice of appropriating the past on the basis of factical life (and hence in the light of the possibilities of a unique life), by going along with the directions of its actualization and explicating its modes of being. These two different experiences of the relation with the past reveal two different movements with respect to factical experience. The first approach, the theoretical one, fails to access this experience, because it fixes it in general terms under the observer's gaze, so that what it grasps is no longer the experience at all, but rather a reflected objectification of it.

The second approach operates within the actualization of the lived experience according to the intentions of living experience and without betraying its essence. It brings about the explication of the motility of this experience, retracing the motives and tendencies that it conceals and from which it draws its vitality, so to speak. One might say that in the former case, experience is construed as an object in general (and then increasingly specified until it becomes present in flesh and blood), while in the latter, it is construed as a phenomenon which can only be grasped if one keeps to the way in which it comes into shape, manifests itself, and hence occurs by co-re-enacting it.

What emerges here, in a more complex form, is the problem we encountered in relation to the famous example of the table mentioned by the young Heidegger in his Freiburg lectures. A table may be grasped through a theoretical access, as a perceptible thing that is more or less rectangular, brown, made of wood, etc.—which is to say, as an object of the observer's direct perception which reveals itself to be a table under the categorial gaze. Alternatively, we can enter into the flow of the living experience of a postman, electrician, cleaner, or student, so as to produce an account of it that is in keeping with each person's sense of enactment. In this case, although everyone sees the table—which has a particular meaning in the context of my studio—it means something different for each person.

If knowledge, then, is an intentional behavior directed toward something—e.g., a table, seen from the point of view of the postman, electrician, cleaner, or student but also of the epistemologist, who sees it as a perceptual object, or of the geometrician or carpenter, who considers the quadrature or the type of wood—what motivation orients us toward grasping the table as something which gives itself in different ways, according to these different perspectives? First of all, however, we grasp the table, i.e., however we experience something, the thing in question always gives itself as a phenomenon and not an object. The aim of formal indication is precisely to grasp this diversity of meaning in terms of intentional referencing.

This diversity (of the meaning of the table), the content, has to do with the enactment of the way in which each person relates to the table—of how each person grasps the table and how the table is encountered. Therefore, we can only grasp the range of different personal perspectives if we go with the flow of those different ways of experiencing the table and of engaging with the table in that particular context and situation (the sense of enactment). Going along with the enactment of living experience, then, means understanding and explicating what a unique experience preserves.

It is evident that if we grasp the act of seeing the table (for each of the actors mentioned) according to the theoretical perspective—i.e., in terms of the perception of the table—this way of grasping it prevents us from understanding lived experience as it unfolds, as it actually manifests itself to the person having it. Making prereflective experience something theoretical means tearing it away from life, from one's personal story.

If, in keeping with its etymology, we understand the word "phenomenon" to mean that which shows itself (i.e., to each person), a phenomenon may be regarded precisely as the sense of enactment of a way of looking which cannot be reduced to perception, but which is rather embedded in the living understanding of the overall meaningfulness of what is encountered. Hence, it may be understood as the meaning which the table has in a given circumstance for the person gazing at it. One example would be looking at the table according to the meaningful trajectories traced by the electrician, and therefore as a way to reach the ceiling lamp, assessing the need to move it in order to work on the electrical system. In other words, the hermeneutic understanding of the way in which the electrician sees the table corresponds to an engagement with the enactment of that lived experience according to the sense proper to it. A lived experience can be encountered—so to speak—by the living experience of the interpreter, without any need for reflection, through an explicating understanding which does not shatter the original intentionality of life in order to present it objectively: for it leaves it intact. Therefore, the hermeneutic understanding of another person's lived experience coincides with an engagement with the enactment of the other person's lived experience according to the sense proper to that experience: a co-retrieval which implies a "sympathy," a conformity to the flow of the other person's lived experience, which evidently cannot be performed automatically. Rather, it corresponds to a (formally indicated) way of comporting oneself toward something which only attains completion through the enactment of the original experience together with the patient. Therefore, what formal indication gives access to is not an object, but that context of motivations for knowledge (i.e., the reasons why I know the table in one way rather than another) which the meaningfulness of the object points to.

The therapist thus methodically retrieves the actual experience of the patient's life in agreement with the three articulations that make up its full meaning—content, reference, and enactment—and according to which experience takes shape, in order to repeat it while explicating its sense without destroying it. This behavior of the therapist with respect to the patient's experience amounts to a way of approaching the relationship which the patient has with the situation under investigation—a relationship which is methodically retrieved as a sense of content, a sense of reference, and a sense of enactment and which is co-enacted according to the direction of formal indication. This method enables the explication of the patient's experience insofar as it articulates, and hence unravels, that lack of distinction between the act of experiencing and what is experienced which characterizes concrete experience: its *verbum internum*. This methodical retrieval therefore consists in an experience of the experience that is being studied; in other words, the therapist's enactment of the patient's experience brings into focus the patient's own articulation of the sense of the experience, so as to re-enact the latter and thus grasp its meaning.

The most relevant methodological aspect is that the patient's experience is not accessed through reflection—whose procedures cannot be methodically recorded (Quesne 2004)—but is instead guided by a kind of research that poses questions (questioning), creates doubts, and is enacted by relating to a content: the concrete experience of the patient, envisioned as a formal indication. The notion of phenomenon is therefore broadened as a principle and established as the foundation of the method. By paraphrasing what Theodore Kiesel writes with regard to the practice of philosophy, we might argue that the unthematic conditions structuring the decision to heal have to do with the situation of understanding and the fascination with questioning produced by the passion for healing.[16]

The content of the experience disclosed through questioning is not an object which appears in the light of reflection, starting from an empty generality that is filled according to progressive degrees of knowledge until a final specification is attained, thereby making the unique experience under investigation dissolve within a category; rather, the experience is regarded as a formal indication, as an intentional reference to a context of motivations for knowledge. The patient's experience, in other words, is not suddenly given to me, e.g., as a particular element of attachment (genus), whose type (species) I increasingly specify via analysis in order to bring the underlying motivation to light.

Differently from generalization, formal indication does not provide a horizon through which to determine the experience under investigation in advance, but indicates a context of tendencies and motivations for knowledge on the basis of which it is possible to gain access to the experience in all of its uniqueness. This means that the experience under scrutiny discloses an unprominent multiplicity of horizons of sense which attain determination through the experience itself; and it is precisely the disclosure of this context of tendencies and motivations that provides access to the experience as such, in its uniqueness. Heidegger can thus argue that motivation derives from the sense of reference of the interpretive attitude (understood as this unprominent and indeterminate multiplicity), that an object gives itself and is grasped in relation to such an attitude, and that the origin of the formal lies in the sense of reference (GA 60 1995, p. 94). On the other hand, the reference—the sense of reference—may be determined according to the content.

The content points to heterogeneous directions of sense which, while indicated, remain only implicit in the experience in question but which are then explicated and interpreted through the methodical retrieval of the sense of content of the situation, on the basis of an engagement in the re-enactment of the other's experience according to the various directions disclosed by formal indication (and therefore according to different senses of reference). For example, as the table points to the condition of

[16] "Having the situation of understanding and the passion for questioning" is for Kisiel (1993) the condition that structures the decision to philosophize (p. 234).

the electrician who sees it (a condition which remains implicit), in order to grasp the meaning which the electrician assigns to the object, it is necessary to disclose the situation according to his lived experience and retrace this situation together with him. The development and articulation of these trajectories, which point to the retrieval of the original meaningful contexts, have the present renewal of ipseity as their criterion—a criterion on which the enactment and hence the whole question-based research may be tailored. In other words, the guiding criterion of the interpretation is the retrieval of sense on the part of ongoing ipseity, in view of a more intimate fullness. The search for sense and the related questioning of the meaning of past experience are conducted precisely on the basis of this retrieval accomplished through the renewal of ongoing ipseity.

It is precisely by going along with the enactment of the experiences pertaining to the sphere under investigation, and by ensuring the deconstructive/constructive development of this process according to the orientations provided by formal indication, that the therapist can retrace the patient's experiences until he reaches the sense of the underlying experience and hence its appropriation: the lived experience which is already mine becomes proper to me. As Quesne notes, while "destruction means approaching the origin, the sense of enactment means attaining the origin" (p. 144).

Appropriation thus lends all the various situations an articulated unity of sense which corresponds to an increase in the fullness of sense. The formally indicated interpretation of what was meaningful reflects itself in actual life as the renewal of ipseity in the present: a renewal which enables one to open up existence to a new movement of understanding, restarting the interpretive circle. In this sense, phenomenological psychotherapy follows the ancient trail of the care of the self by focusing on the transformation of oneself through oneself.

6.9 The Therapeutic Enactment of Formal Indication

The therapist's understanding of another person's experience within the framework of therapeutic practice takes the form of a specific way of behaving with respect to the meaning of the experience under investigation—a looking "into" the experience that goes along with the production of the experience which is being progressively gazed at (and which cannot therefore be grasped in the light of anything general). If this is so, then the therapeutic method is structurally open to the determinations which concretely emerge from lived experience. Unlike in the case of the natural sciences, the method is not prescriptive but indicative, in the sense that it indicates the direction provided by those determinations: an orientation according to which it is possible to concretely co-enact the original search for sense. The method is therefore part of the execution of the interpretation insofar as it is developed through it.

In order to make this difficult passage easier to understand, let us return to the clinical case we have repeatedly referred to (Vd. Table 6.1).

This is the case of a patient—Giovanna, a 35-year-old divorced dentist with no children—who enters into therapy because of cardiac issues that emerge in conjunction with a wide range of conditions and which cannot be attributed to any ongoing medical pathologies. The relation between the appearance of the symptom, and the conditions which make it emerge, is utterly unknown to the patient when she turns to our practice. The patient's cardiac difficulties when she enters into therapy, and her detailed account of them, do not allow the therapist to re-enact the situations they point to nor to delineate the existential horizon within which they fit.

Studying the symptom and the mechanism which accompanies it therefore becomes the first step in the treatment.

The disorder, which suddenly appeared about a year ago, has a stable structure. It manifests itself in any condition in which the patient perceives her corporeal state as being physiologically "altered," in terms of fatigue, sleeplessness, tiredness, and hunger. At the same time, this corporeal alteration is also associated with particularly intense emotional conditions such as excessive anger, joy, fear, or anxiety or with certain characteristics, like a sense of emptiness and boredom. In all these situations, the fact of feeling "altered" engenders a state of alarm that triggers catastrophic thoughts of possible future illnesses. In turn, this increases the patient's alarm even further, creating a vicious circle which amplifies the condition of alteration. The course of the disorder has undergone a significant variation because, although it emerged a year ago, it is only over the past 6 months, after a high fever, that its corporeal localization has become more or less constant, multiplying the states of malaise.

This first step in the treatment is intended not just to define the composition of the symptom in such a way as to unravel its mechanism, but to enable—through its deconstruction—an access to those prereflective meaningful structures that the symptom obscures and presupposes. As the mechanism generating the symptom rests on the gap between lived experience and its appropriation, a gap which the symptom preserves, its disarticulation discloses the prereflective dimension which manifests itself through lived experience. The destructuring of the symptom therefore restores the possibility of reconnecting the meaning of a certain experience to the situations and occasions of the expressions which this meaning points to (Fig. 6.1).

We thus trace the symptom back to its temporal origin and to the indication of a context with respect to which issues can be raised and questions posed. This

Fig. 6.1 From the symptom to its origin

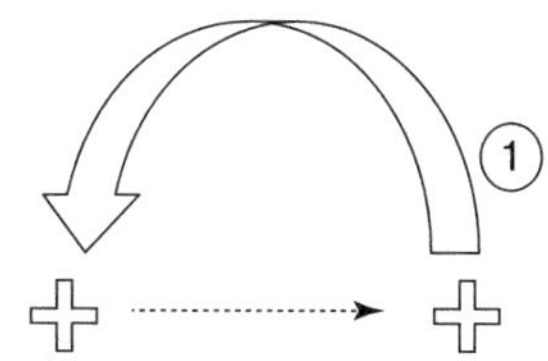

dimension that becomes the focus of the search for sense is the dimension of formal indication. It corresponds to a turning back toward the sphere of the proper (underlying motives and tendencies, or significations) starting from the problematic situation within which the symptom took shape, so as to grasp and retrieve the totality of implications supporting it: the prereflective meaningful structures.

This second methodological step, which follows the work conducted with the patient on the de-automatization of the symptoms—with the aid of self-observation exercises—releases some meanings trapped within the symptom, so to speak.

In the case under scrutiny, for instance, the feeling of emptiness experienced as an alarming alteration of the body which triggers a spiral leading to hypochondria and panic is discovered—through deconstruction—to be an indicator of how the patient feels in certain conditions which have to do with her affective loneliness.

This understanding of what the symptom concealed emerges through an analysis of the concretely delineated situation which triggered the first cardiac ailments a year ago. What lies at their origin is an apparently banal event: the patient's partner—Carlo, a 40-year-old divorced lawyer and father of two with whom Giovanna had begun a relationship the year before—informs her that he will spend Christmas Eve with his two children at his parents' house; Giovanna is not invited. It is within this situation that the sense of emptiness first emerges: it represents its meaning, the way in which ipseity takes shape. The sense of emptiness will become the lived experience which the patient is incapable of appropriating and will trigger the symptom and its subsequent development.

On the care path, addressing a problematic experience by investigating its sense discloses the direction to be followed in the attempt to appropriate the prereflective experience. In our case, the access ensured by the retrieval of the starting situation—**which is to say, by the retracing of the patient's alarming sense of disorientation at not having been invited to the Christmas dinner, something which she perceives as a form of betrayal**—enables us to delineate a field of inquiry, so as to thematize the underlying significations and the contexts related to them which constitute the prereflective domain within which the patient's distressing feeling of emptiness developed (Fig. 6.2).

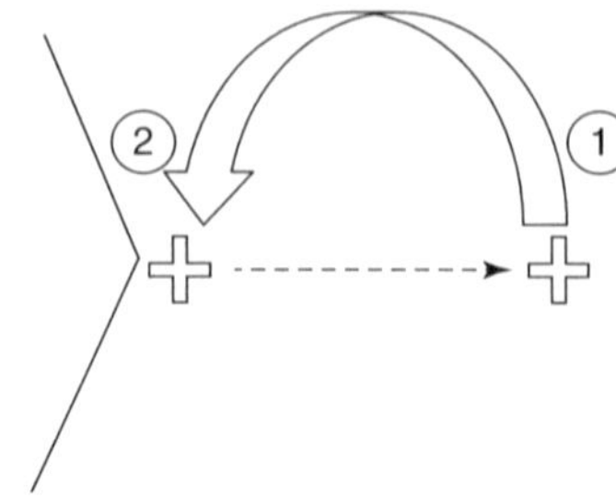

Fig. 6.2 From the origin of the symptom to a field of meaning to be investigated

> The patient's feeling of being at a loss and disoriented, which emerged in relation to the news about the Christmas dinner, reflects the underlying significations which this sense of emptiness points to.

It also becomes an opening—inevitably limiting the indicating perspective—through which to methodically trace the patient's experience back to the domain of the proper: the network of tendencies and motives which orient experience in advance and are transformed by it. It thus becomes clear that the past gives itself without any reflective mediation in its proximity to ongoing experience, which is therefore always *informed* by its temporal reference.[17]

A person's unique life journey is marked by the network of these prereflective structures of meaning and their corresponding domains, their relative stability, and the intertwining or vanishing of given tendencies and motives. The method therefore consists in returning from the experience which has taken shape in a given situation to the meaningful condition which precedes it and made it possible. The attempt to access the lost past from the present is undertaken through a double movement which simultaneously serves as a return, repetition, and retrieval of the living meaning of that which prereflectively precedes the experience (reconstruction, *Rückbau*) and as the critical disarticulation of the faded meaning to which we have access in the situation and which is broken down into a heterogeneous multiplicity of meanings (deconstruction, *Abbau*).

This passage is brought about through the practice of questioning, which focuses on the formal content that structures the underlying experience, whose composition comes to light in relation to the questions posed. This also reveals the most ancient etymological root of the term "therapy"—*therapeuein*—which fluctuates between being at the service of another person and supporting him or her, as the search for sense within this movement of destruction/construction and appropriation. It is precisely through the encounter between the therapist's capacity to pose questions—supported by his passion for healing—and the strength with which the patient is

[17] As previously emphasized (Chap. 4), ipseity is distinguished, not just by a sense of content, of enactment, and of reference but also by a sense of temporal cohesion (*Zeitigungssinn*).

capable of facing up to the task of "becoming a question to himself"[18] that the act of questioning discloses and newly mobilizes a theme and meaningful direction through which it is possible to appropriate the past.

In the case under investigation, for instance, the patient's problematic relationship is the starting point—revealed to us by the symptom—and as such lends direction to the method of analysis by disclosing a meaningful context. The news brought by the patient's partner that there is no place for her yet at an upcoming family gathering is perceived as a shock that brings out a new aspect of the relationship: a sense of disorientation which fills the patient with fear. A new meaningful element comes into play that will stand at the origin of the patient's symptomatology. This disorientation which takes shape in the situation at hand, becoming entrenched in meaning, provides a direction to be followed to reach the contexts at its origin: those tendencies and motivations—the prereflective sphere—that, while becoming congealed in meaning (as content which acquires sense through an occurrence), disclose new spheres of the domain of the proper.

Understanding through questioning plays a key role here: it follows the orientations suggested by formal indication.

In our case, what horizon of sense lies behind the information which the patient has received from her partner and which she perceives as a frightening disorientation?

The fact of perceiving the situation in question—i.e., her not having been invited to the Christmas dinner—as a betrayal, and hence of feeling profoundly alone for the first time since the beginning of her relationship, makes the patient feel emptied of all emotions, without a sense of self, and utterly lost. The patient starts fearing this new feeling, the feeling of emptiness, almost immediately and then starts fearing that she might be afraid of it.

What is the life context of this event? The patient began the relationship with her current partner a year before the episode. The two met by chance at a friend's house, just as the patient was leaving behind her the most important relationship of her life—a love story which had come to a dramatic end. The man with whom she had been living for the past 8 years, the man she had married and with whom she had done all the things that adults do together, had told her one day that he no longer loved her. Out of the blue, one day in May, he had collected all his things and simply walked out of the house—and her

[18] *Questio mihi factus sum*: Augustine, Confess. 4, 4, 9.

life—without providing any explanation. Suddenly the world had lost its colors. The stunned woman had faced some concrete changes in her life: wake up every morning, drag herself to work, return in the evening to an empty house filled with traces of him, eat in front of the television to dull her mind, and force herself to go out now and then, without knowing with whom or why. She had engaged in some casual sex two or three times just to shake off that sense of emptiness, to make sure that she still had a body, but in the end had felt nothing but disgust. It is in this life context that the patient met Carlo, at a dinner to which she was dragged by her best friend, who could no longer stand to watch her suffer.

Formal indication directs us toward this beginning in order to re-enact that domain of the proper which served as the backdrop to the sense of the patient's relationship down to the situation under investigation. It therefore discloses a sphere of content which is constituted by a range of directions of sense pertaining to the various situations it comprises.

In our patient's case, these are not just the abrupt end of the relationship, but also the interruption of what she regarded as an acquired position in her life—for the relationship with her husband over the last year, just before their separation, had given her the impression that she was about to enter a new stage of her life, that she had reached maturity! This transformation had filled her for the first time with the desire to have children and start a family, to change her career plans, and to redefine her position vis-à-vis her family of origin, engendering new aspirations and expectations. Then, with a sudden blast, all this had come to an end, and she had started obsessing over what had gone wrong, over the reasons for their breakup, and over the promises never kept and steps never taken, sifting through her memories over and over again, at a loss as to how she could make a fresh start.

The grasping of the sense of content of these underlying significations cannot be determined in the light of a priori principles that impose a univocal meaning, thereby once again obscuring the domain of the proper; rather, it discloses a situational dimension which is to be explored in all of its aspects. This is the prohibitive function exercised by formal indication, which in forbidding any univocal reference with respect to the experience under investigation indicates the formal sense of the content that must be disclosed by retracing the situations in which it has come to light.

It is interesting to note one crucial difference compared to the perspective developed in particular by Hayek and Polanyi, and according to which our actions and thoughts are governed by non-articulated rules—described as abstract rules by Hayek (1967, 1969) and tacit ones by Polanyi (1958)—that guide us without us realizing it and constitute a standard for correctness. Hayek writes: "What I mean to show … is that what I have called an abstraction is primarily such a disposition towards certain ranges of actions, that the various 'qualities' which we attribute to our sensations and perceptions are these dispositions which they evoke, and that both the specification of a particular experienced event, and the specification of a particular response to it, are the result of a superimposition of many such dispositions to kinds of actions, which result in the connection of particular stimuli with particular actions" (1978, p. 40). Like Pierce (1955), Hayek (1952) founded this view of his on evolutionary biology: he regarded dispositions as adaptations to environmental features and their identification as the activation of the kind of disposition adapted to them. According to this perspective, adopted by evolutionary epistemology, these rules are already a theoretical representation in themselves—albeit a tacit one—that is always passively available and is activated by a specific situation. Consequently, a univocal reference founded on biological objectiveness is used to grasp the meaning of an actual situation. In other words, one's understanding of the meaning of an ongoing experience is guided and justified by explicative principles based on evolutionary biology.

By contrast, from the perspective we are endorsing, the sense of content to which formal indication gives access—by disclosing the underlying experience with all its range of meaningful trajectories—corresponds to a dimension to be discovered, brought to light, and investigated through dialogue rather than conceived in the light of rules that are evoked by environmental stimuli and used to specify the person's unique experience. Thus the method of retrieval, which leads from a particular situation to the underlying signification, discloses the sense of content, yet only from a formal perspective, in the sense that the disclosure of that dimension coincides with given possibilities to determine the experience under investigation. Hence formal indication paradoxically discloses a sphere which is indeterminate in terms of content, a multiplicity of perspectives of meaning, which nonetheless may be determined within the concrete situation at hand and is therefore determinable according to a sense of enactment (see GA 61 1985, p. 20).

If in actual life we encounter a given occurrence in relation to indefinite tendencies and motives that orient our engagement in advance and in an indeterminate way, and if this indistinct directionality becomes actualized in one way or another in the concrete situation at hand (whose meaning becomes sedimented in words), moving from this situation back to the underlying prereflective conditions means retracing its enactment and hence newly disclosing the dimension of indeterminacy of the sense of content. Methodically exploring this dimension through question-based research means retracing—behind words—the multiple meaningful trajectories (sense of reference) it comprises by turning toward the original contexts, guided by formal indication (Fig. 6.3).

What all this means is that the sense of content of the experience and, correlatively, those underlying attitudes which guide the patient's understanding are methodically articulated and hence traced back to a sense of reference. The content,

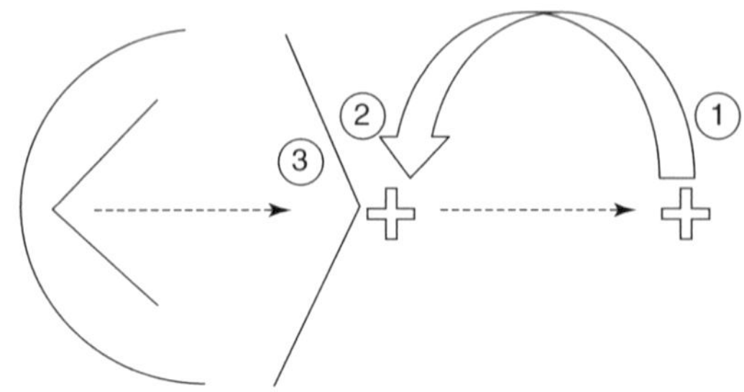

Fig. 6.3 From a field of meaning to the actual remobilizing of meaningful trajectories

in other words, is only grasped within the context of the attitude within which it gives itself and not in general—i.e., not according to any theoretical sense of reference. Thus heterogeneous, multiple meanings appear as we move back through the formal indication of the content from the situation at hand to the sense of reference. The heterogeneous meanings which take shape in relation to certain underlying attitudes are therefore traced back to the sense of reference, and these heterogeneous meanings find determination in the situation at hand and acquire a foundation in the flesh.

Like the flesh (see Chap. 8), the sense of reference gives itself as a multiplicity of meaning in the various situations in which the content takes shape, a meaning which evidently varies from situation to situation. This is what Quesne emphasizes in his valuable study on the subject: "Formal indication is the method to make something visible starting from the flesh or the original confusion of experience" (p. 138). In our view, this is where the key role of affectivity lies: it appears in its irreducible multiplicity and at the same time unity (Arciero and Bondolfi 2009).

In the case under investigation, therefore, by retrieving concrete experiences (sense of content), phenomenological construction/destruction leads us to disclose a perspective on the underlying condition and hence on the patient's position with respect to what she encounters (sense of reference).Within the framework of this methodically acquired preliminary orientation of the patient, the possibility emerges of grasping the original meaning for her of the encounter with her current partner and the way in which the relationship with him developed—and hence of the way in which content acquired meaning in the most meaningful situations over the year they spent together—by re-enacting this meaning according to all the various directions.

As if by magic, Carlo seemed to meet those expectations which the abrupt end of the previous relationship had broken and brought to light. Over the course of the year which the patient spent with him, all those dreams which the separation had shattered seemed to come back to life. From the very first evening, when he had showered her with politely seductive attentions, when she had found interest in conversation again and discovered a gaze that lit up when it met her own, she had felt something—something akin to a new hope. She had found herself indulging in this thought on her return home. This first date was followed by other, increasingly frequent and intimate ones filled with desires and promises. As the two of them opened up to one another, each taking a position with respect to the other, the memories and feelings related to the end of the previous love story progressively vanished. As the new relationship took shape, the role of this man in the patient's life seemed to be gradually displacing and replacing what had been her husband's position. Carlo had not just taken his place: he was more present, less aggressive, more accommodating, and much kinder. He was just as she wanted him to be. He had gradually yet steadily crept into her life, as though to fill a position which had remained vacant for a year. And as Carlo acquired this position, the prospect of resuming her interrupted plans emerged, her hopes gained new momentum, and she was filled with new motivation.

This sphere of meaning, in its various articulations, progressively appears as, together with the patient, we retrace her prereflective trajectories starting from the situations at hand. While the orientation provided by formal indication is motivated by a requirement—to search for meaning according to the content which it points to in its various manifestations—the enactment, or rather co-enactment, of the concrete experience restores the unity of meaning of the sphere which is the focus of the indication. As this research takes shape, along with the process of destruction, a renewal of meaning occurs, leading to the attainment of the original meaning, to the living appropriation of meaning (Fig. 6.4).

In our patient's case, this was a gradual process. She slowly realized that the place she had assigned to Carlo in her life was the one which her husband had left vacant; inevitably, her own position too had not changed. The life perspectives which the patient had created and imagined together with that man, and which had then been shattered by their separation, had remained as though concealed in a background that the loss of her husband had shrouded in mist, along with the emotions and situations that accompanied them. The patient could no longer find any meaning: at a loss, she wandered through life waiting

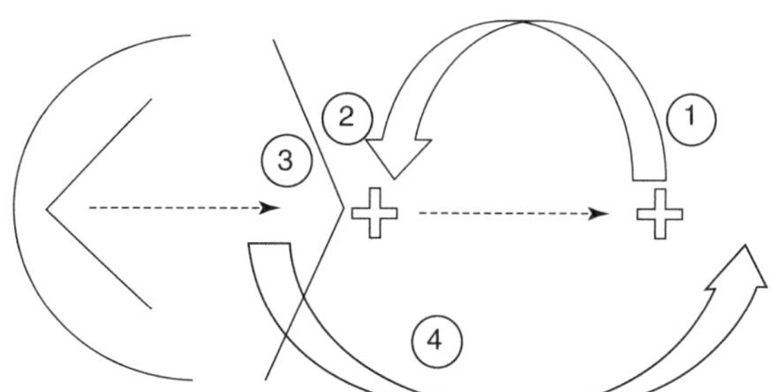

Fig. 6.4 From the remobilizing of meaningful trajectories to the progressive retrieval of the "meaning lost"

for the days to pass. When she started seeing Carlo, those horizons of sense gradually reappeared together with another protagonist. So as the past faded, what gained new momentum was a way of assigning meaning to her life that had never changed and had originally taken shape in her previous relationship with another man.

This is the sphere of meaning which the patient has gained possession of through therapy and which enables her to shed light on the emotional intensity related to the apparently banal episode of not having been invited to a Christmas dinner. Starting from this episode, the patient acquires a new understanding of the proliferation of memories related to her ex-husband, the emergence of doubts and questions concerning this relationship, and even the surfacing of a new feeling which she had never experienced before: a disorientation and loss of the sense of self, a feeling of emptiness. It was not just a matter of betrayal, or disappointment, or loss of meaning for the absence of her partner: all these meanings which seemed to have sprung out of nowhere pointed to a far more complex prereflective sphere.

The last stage of the method begins with the living appropriation of that dimension of signification which formal indication has revealed, by amplifying its features with respect to the domain of the proper in general, and which it had traced back to its origin, re-enacting it. This is the stage of the renewal of ongoing ipseity. The fact of taking charge of what therapy has brought to light implies that what was meaningful for one's own world becomes meaningful again through the opening up of existence to a renewed movement of understanding and, at the same time, to new horizons of expectation. Taking charge of oneself, then, means acquiring a new position, taking the initiative and setting out toward oneself. In this respect, phenomenological psychotherapy is based on practical enactment: it fosters the care of self, not in terms of knowledge acquisition—i.e., not according to a knowledge-based relationship—but as the transformation of the self through oneself (Conti 2018) (Fig. 6.5).

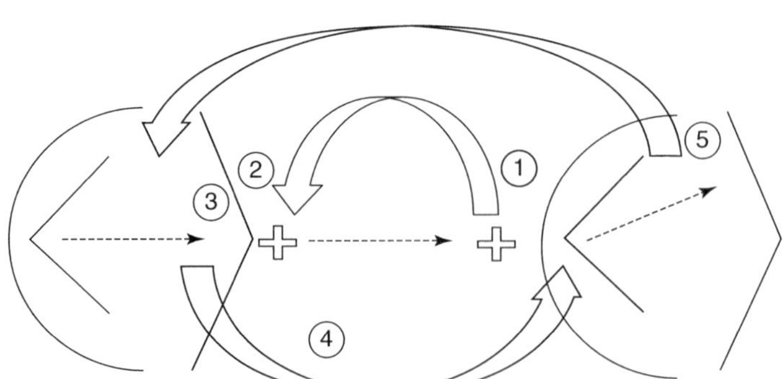

Fig. 6.5 From the retrieval of meaning to the taking up of a position with respect to oneself

At this stage, the feeling of emptiness and disorientation suffered by Giovanna as an affliction becomes an indication for her to ask herself whether her partner truly has a place in her life or whether he has simply taken up a function which belonged to someone else. At the same time, the fact of raising this issue calls into question the patient's own position, which has remained the same from her previous relationship, through her separation, down to the present. What emerges, then, is a new field of investigation, which concerns Giovanna's existential space without another person, and hence her way of perceiving herself and of managing such space, the motives and motivations behind her choice of her current partner, and the circumstances which led to the end of her previous relationship. As it unfolds, the analysis inevitably overexposes that feeling of the absence of all feeling—emptiness—as an emotional connection which holds together this overlap of different dimensions.

By following the double perspective disclosed by the analysis, i.e., the need for the other and at the same time his relative irrelevance qua other, the research is articulated through a process of destruction/construction—through the co-enactment of those meaningful trajectories which determine the development of the patient's relationship with her husband, with the aim of then retracing and reinterpreting the valences of her ongoing relationship, down to the emergence and stabilization of the psychopathology which Giovanna is seeking to treat.

Once again, the thematic appropriation of prereflective experience favored by the question-based research raises new questions, which influence the patient's capacity to take charge of such experience and hence the further enactment of concrete ipseity. The focus of this therapeutic method is not the knowledge of this or that aspect of the person; rather, it is precisely in the light of the detection of certain meaningful situations or personal characteristics—which are to be newly articulated and re-enacted over the course of the therapy—that the search for a new concrete position within actual experience is thematized.

The possibility of accessing her own world, and of appropriating dimensions of meaning through which the patient can understand herself with respect to that which she encounters, enables her to reintegrate what was meaningful for her through a renewed enactment, a new practical and actualizing position. The care path, therefore, is a transformative journey which—through the therapeutic relationship—leads from a given condition of limitation of one's own possibilities, and hence of suffering, to a state in which the person is capable of fully taking charge of the care of self, of his or her own freedom.

Table 6.1 The application of the therapeutic method

	Methodical movements
1	**Deconstructing the symptom and ensuring an access**
	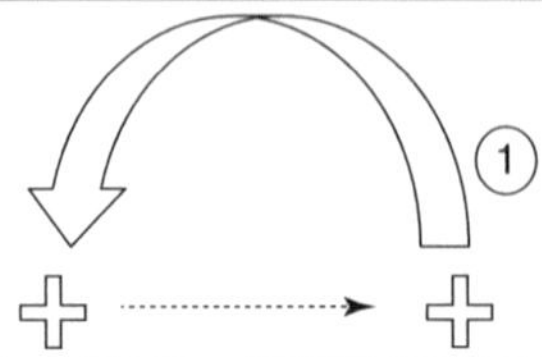
	"This first step in the treatment is intended not just to define the composition of the symptom in such a way as to unravel its mechanism but to enable—through its deconstruction—an access to those prereflective meaningful structures that the symptom obscures and presupposes." In other words, it is a "destructuring of the symptom, which brings out the experience trapped within it"
1a	**Giovanna's case**
	"Cardiac issues that emerge in conjunction with a wide range of conditions and which cannot be attributed to any ongoing medical pathologies" "It manifests itself in any condition in which the patient perceives her corporeal state as being physiologically 'altered'"
2	**Tracing the symptom back to its temporal origin while establishing a context of inquiry**
	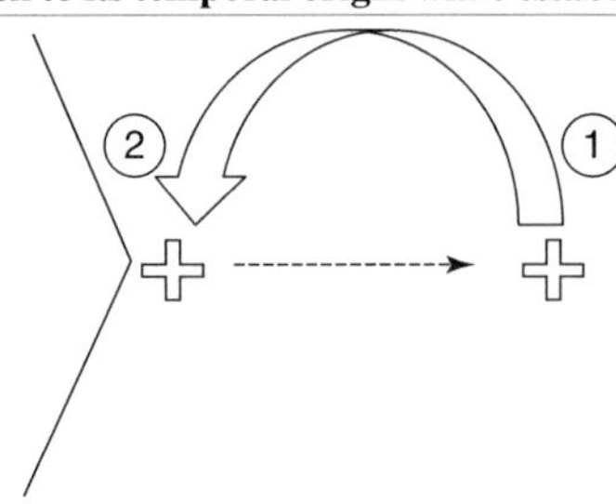
	"We thus trace the symptom back to its temporal origin and to the indication of a context with respect to which issues can be raised and questions posed. […] It corresponds to a turning back toward the sphere of the proper (underlying motives and tendencies, or significations) starting from the problematic situation within which the symptom took shape" "This change of perspective does not amount to the cognitive grasping of factically relevant modes of being as objects; rather, it consists in a disclosure of the field of investigation within which these modes of being can be grasped by going along with their sense of enactment" "On the care path, addressing a problematic experience by investigating its sense discloses the direction to be followed in the attempt to appropriate the prereflective experience"
2a	For Giovanna, "the feeling of emptiness experienced as an alarming alteration of the body which triggers a spiral leading to hypochondria and panic is discovered—through deconstruction—to be an indicator of how the patient feels in certain conditions which have to do with her affective loneliness." "This understanding of what the symptom concealed emerges through an analysis of the concretely delineated situation which triggered the first cardiac ailments a year ago. […] The patient's partner […] informs her that he will spend Christmas Eve with his two children at his parents' house; Giovanna is not invited." The feeling of emptiness engendered by this news "will become the lived experience which the patient is incapable of appropriating and will trigger the symptom and its subsequent development" (p. 22). This raises the following question: "what horizon of sense lies behind the information which the patient has received from her partner and which she perceives as a frightening disorientation?"

Table 6.1 (continued)

3	**An explicating understanding: retracing the disclosed sphere of meaning down to its prereflective sources**
	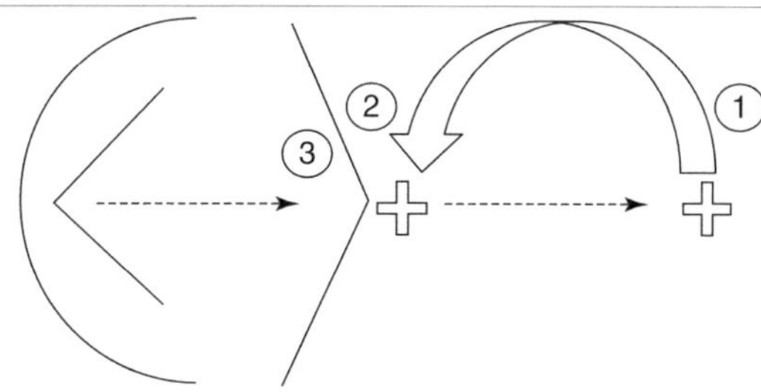
	Formal indication "therefore discloses a sphere of content which is constituted by a range of directions of sense pertaining to the various situations it comprises." However, "the grasping of the sense of content of these underlying significations cannot be determined in the light of a priori principles that impose a univocal meaning, thereby once again obscuring the domain of the proper." This means that "the sense of content of the experience" is "methodically articulated and hence traced back to a sense of reference. The content, in other words, is only grasped within the context of the attitude within which it gives itself and not in general"). It is a matter of "exploring together with the patient […] the sphere of meaning which has been disclosed, in such a way as to 'test' the experiential accounts provided […] down to their prereflective sources [the co-originary giving itself of the sense of enactment/content/reference]." "This passage […] brought about through the practice of questioning" makes it possible—through the alternation between deconstruction and construction—to "methodically trace the patient's experience back to the domain of the proper: the network of tendencies and motives which orient experience in advance and are transformed by it." An investigation of this sort "consists in an act of discernment with respect to the situational origin of meanings and the directions of sense they enclose." It "is not a technical or random matter—less still is it guided by an ideal of certainty established a priori; rather, it is the outcome of the fact of allowing oneself to be questioned by becoming involved in a search for meaning." It is therefore an "explicating understanding," typical of knowledge by acquaintance; and it "is precisely through the encounter between the therapist's capacity to pose questions—supported by his passion for healing—and the strength with which the patient is capable of facing up to the task of 'becoming a question to himself' that the act of questioning discloses and newly mobilizes a theme and meaningful direction through which it is possible to appropriate the past"
3a	"What is the life context of this event?" "Formal indication directs us toward this beginning [i.e., the beginning of Giovanna's relationship with Carlo] in order to re-enact that domain of the proper which served as the backdrop to the sense of the patient's relationship down to the situation under investigation." "It therefore discloses a sphere of content" which for Giovanna coincides not just with the abrupt end of her previous relationship, but with the existential horizon within which the very episode of her breakup is framed: the desire to have children and start a family, the changing of her career plans—"a situational dimension which is to be explored in all of its aspects." This "enables us to delineate a field of inquiry, so as to thematize the underlying significations and the contexts related to them which constitute the prereflective domain within which the patient's distressing feeling of emptiness developed"

(continued)

Table 6.1 (continued)

4	**The progressive retrieval of meaning**
	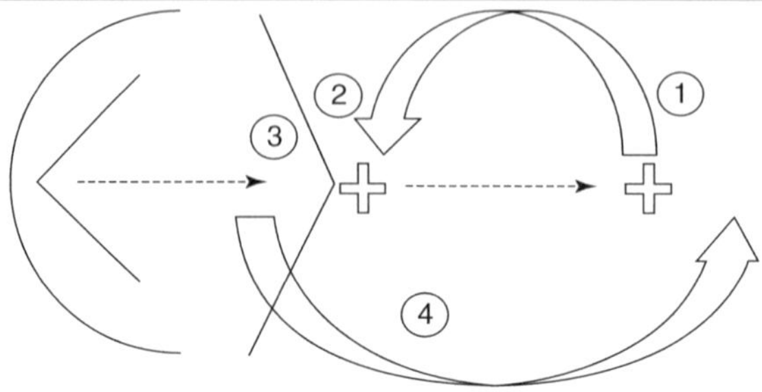
	"After having disclosed the unmanifest sphere of meaning, the investigation is conducted, according to the indications it provides, by following those shreds of meaning which point toward a range of different prereflective trajectories. Different possible paths thus emerge and, with them, different appropriate or inappropriate solutions for the articulation of the original sense." "By retrieving concrete experiences (sense of content), phenomenological construction/destruction leads us to disclose a perspective on the underlying condition and hence on the patient's position with respect to what she encounters (sense of reference)." Evidently, this "sphere of meaning, in its various articulations, progressively appears as, together with the patient, we retrace her prereflective trajectories starting from the situations at hand" Inevitably, "the presentation and thematic articulation of the sense of content and its appropriation" disclose "a new possible interpretation of the various situations under investigation." This "leads to the development of new trajectories and to a renewal of ipseity in the present," which marks the last movement in the practice of therapeutic care
4a	In relation to Giovanna, "the possibility emerges of grasping the original meaning for her of the encounter with her current partner and the way in which the relationship with him developed—and hence of the way in which content acquired meaning in the most meaningful situations over the year they spent together—by re-enacting this meaning according to all the various directions. As if by magic, Carlo seemed to meet those expectations which the abrupt end of the previous relationship had broken and brought to light." Through this re-enactment, Giovanna "slowly realized that the place she had assigned to Carlo in her life was the one which her husband had left vacant; inevitably, her own position too had not changed. The life perspectives which the patient had created and imagined together with that man […] had remained as though concealed in a background that the loss of her husband had shrouded in mist […]. When she started seeing Carlo, those horizons of sense gradually reappeared together with another protagonist. So as the past faded, what gained new momentum was a way of assigning meaning to her life that had never changed and had originally taken shape in her previous relationship with another man"

Table 6.1 (continued)

5	**The renewal of ipseity in the present**
	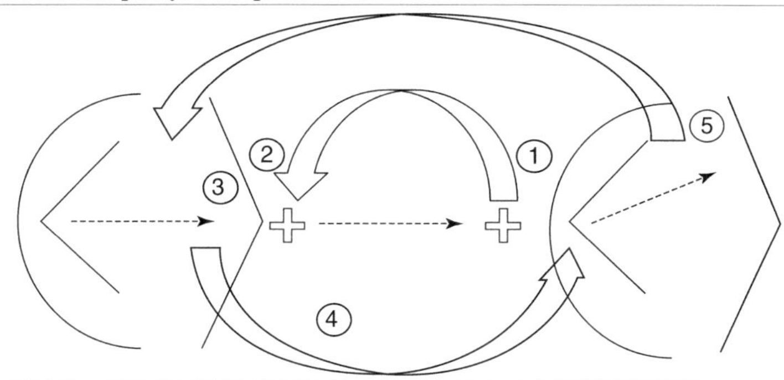
	"The last stage of the method begins with the appropriation of that dimension of signification which formal indication has revealed." "The fact of taking charge of what therapy has brought to light implies that what was meaningful for one's own world becomes meaningful again through the opening up of existence to a renewed movement of understanding and, at the same time, to new horizons of expectation. Taking charge of oneself, then, means acquiring a new position, taking the initiative, and setting out toward oneself" "The thematic appropriation of prereflective experience favored by the question-based research raises new questions which influence the patient's capacity to take charge of such experience and hence the further enactment of concrete ipseity"
5a	"At this stage, the feeling of emptiness and disorientation suffered by Giovanna as an affliction becomes an indication for her to ask herself whether her partner truly has a place in her life or whether he has simply taken up a function which belonged to someone else. At the same time, the fact of raising this issue calls into question the patient's own position, which has remained the same from her previous relationship, through her separation, down to the present. What emerges, then, is a new field of investigation, which concerns Giovanna's existential space without another person and hence her way of perceiving herself and of managing such space, the motives and motivations behind her choice of her current partner and the circumstances which led to the end of her previous relationship." This will require Giovanna to take a position with respect to herself (i.e., her own story) and what she wishes to do with herself (horizons of possibilities)
6	**The cornerstone of therapeutic care**
	"The focus of this therapeutic method is not the knowledge of this or that aspect of the person; rather, it is precisely in the light of the detection of certain meaningful situations or personal characteristics—which are to be newly articulated and re-enacted over the course of the therapy—that the search for a new concrete position within actual experience is thematized" "In this respect, phenomenological psychotherapy is based on practical enactment: it fosters the care of self not in terms of knowledge-acquisition—i.e., not according to a knowledge-based relationship—but as the transformation of the self through oneself" "In the analysis of the problematic experience, this movement implies 'siding with life,' that life which in encountering things remains intimate to the self—which in becoming involved with the world experiences itself"

6.10 Methodological Turning Points

At this stage, we can freshly address each of the questions posed in the first paragraph of the present chapter by framing them within the overall view we have developed so far. First of all, to what mode of relating with the other does therapeutic practice point to? Therapeutic practice is based on a relation which, as the etymology of the word (*therapeuein*) reveals, coincides with a connection between the idea of serving and that of searching or rather of serving as searching (Berti 2009). The therapist investigates the concrete situations in other people's lives, searching for an access and hence a perspective that will provide the right direction for him to find the sense which the patients have lost. But what does it mean for the therapist to search, as he goes along with the patient in the process of re-enacting lost experience?

With regard to the content of memory, we have seen how the sense of self-familiarity within the context of present circumstances discloses an unmanifest sphere of meaning. Memory, broadly understood as an unprominent sphere of meaning, provides immediate access to this dimension, albeit in a fragmentary way. For example, with regard to my trip to Egypt last year, I only retain the sense of the lived experience, but the situations, people, and objects it comprised fluctuate, emerge in isolation, and become blurred with one another, even though they all indicate the same sense. Memory undoes the existential web in which multiple moments of sense were articulated as a feeling of self-familiarity (the "myness" of experience). In fact, ipseity no longer actualizes the bond that held together in the original experience a range of different horizons, so that the content of the experience, while being directly available, requires a new sense and the renewal of one's relation with ipseity. This sort of paradox of memory—whereby elements which present themselves as being directly accessible are at the same time disconnected and distant—takes a radical form in the case of a pathology, where the content of an experience becomes foreign to oneself and endures as such through the establishment of the mechanism at the basis of a symptom.

How can one access that sphere of unprominent content which the symptom has concealed? How can one newly articulate the various fragments and set out after what has slipped away or has been forgotten? After having disclosed the unmanifest sphere of meaning, the investigation is conducted by following those shreds of meaning, according to the indications they provide, which point toward a range of different prereflective trajectories. Different possible paths thus emerge and, with them, different appropriate or inappropriate solutions for the articulation of the original sense and the living appropriation of the past.

Questioning is therefore a methodical movement, as it is only through the posing of questions that analysis discovers what it was searching for and discards inappropriate paths, tracing that unfolding of horizons of sense constituting the domain of the proper back to the original contexts. So as the perspective itself opens the investigation up in an indeterminate form, it gradually finds the right possibility of determination through the articulation and co-enactment of factical experiences, down to the origin of the sense of these experiences, enabling its appropriation and hence the renewal of the patient's actual existence.

While the concrete and contextual meaning connected to enactment vanishes with the sedimentation of experience, the sedimented experience endures as a trace—pointing to other traces and disclosing more profound trails, down to the original contexts—that must be followed in order to regain the spheres and situations in which the sense of the experience is concretely enacted. The search for and grasping of clues—which correspond to expressions of self—enable the posing of a series of question that allow the therapist to disclose concealed pathways and examine the traces and marks left in the clusters of experiences that progressively come to light, so as to reorient the investigation.

One is reminded here of that remarkable passage of the *Phoedrus* (274b–275c) in which Socrates recounts the ancient tale of Theuth, the Egyptian god who presents his inventions (*technas*) to King Thamus, in order for him to assess them before passing them down to the Egyptians. One of these inventions is writing: when introducing the letters of the alphabet, Theuth explains to the king that writing was invented as a sort of drug for wisdom and memory, to make the Egyptians wiser and better at remembering things. But the king, according to Socrates' account, replied that it is one thing to design a technique and quite another to assess its drawbacks or advantages. Although the sovereign realized that as the inventor of that technique Theuth would wish to defend it, he also realized that by limiting the exercising of memory, writing would only lead to oblivion: for those who would use it to recall things would no longer be drawing their knowledge from within themselves but only from exterior marks. This difference related to memory—understood as remembrance (*anamimneschomenous*)—is clarified a few lines later in the dialogue. After emphasizing the fact that this drug is not useful for memory but only for bringing things to mind, Thamus concludes that it makes pupils only apparently wise (*doxosophoi*), not truly wise (*anti sophon*). What, then, makes writing so disturbing? What makes it a point of disagreement between a god and the king of kings?

The answer is not slow in coming, as Socrates warns his listeners against the real danger posed by written discourse: that of severing the connection between the living and animated discourse of he who knows (*ton tou eidotos … logon zonta kai empsychon*) and its author—the person who in shaping the discourse must tap into himself. What this attack on writing brings into relief, therefore, is the sensible reference of each discourse to life, a relationship which writing breaks down.[19]

In this respect, the expression of experience in all its forms is a trace, a clue pointing to the contexts and situations in which sense is actually enacted. Freud argues that medical psychoanalysis "is accustomed to divine secret and concealed things from despised or unnoticed figures, from the rubbish-heap, as it were, of our

[19] This is no way conflicts with the fact that behind the attack on writing—and those who by using it seem to be omniscient but are in fact ignorant for the most part—there lies the major concern of the dialogue, namely, the vindication of the philosophical life against a life that merely seeks its (of writing) effects.

observation" (Freud 1955, p. 222)—an a priori which Freud masterfully confirms with a rather unlikely and exegetically flimsy analysis of the posture and possible movement of Michelangelo's Moses. By contrast to this view, in our case clues are not deciphered by means of generalizing conjectures that aim to explain the involuntary nature of the symptoms. Nor do symptoms point to a loss of control that allows incongruous acts to be performed as an indirect expression of one's unconscious—as in the case of slips of the tongue and, more generally, all parapraxes that psychoanalysis has sought to trace back to underlying aspirations or forces. The clues originating from a movement of life which we retrace are interpreted according to the directions provided by the existence of the patient, who in newly understanding them restores their flow together with the therapist. Hence, unlike in psychoanalysis, in our case "the capacity to trace apparently negligible experimental data back to a complex reality that cannot be experienced directly" (1986, p. 166)—one of the ways in which Ginzburg defines the evidential or divinatory paradigm—is co-enacted with the patient who, so to speak, has lost the traces of where he passed, the traces of self (Arciero 2006).

In discussing the remarkable study on the topic produced by Ginzburg—"who bravely opposed the paradigm of Galilean science" (Ricoeur 2004, p. 534 modified translation)—Ricoeur identifies those elements which enable a range of different disciplines to embrace the evidential paradigm: the uniqueness of the deciphered object, the indirect character of the decipherment, and its conjectural nature. The extraordinary insight provided by Ginzburg traces the evidential or divinatory paradigm back "to what is perhaps the most ancient gesture in the intellectual history of the human race: that of the hunter squatting in the mud to examine traces of his prey" (p. 169)—a paradigm also implicit in Mesopotamian and Egyptian divinatory texts. These traditions were transmitted to ancient Greece, and in particular to the Pythagorean school and its great physicians—Alcmaeon, Philolaus and Empedocles—thereby opening up and developing the vast field of conjectural knowledge founded on the examination of circumstances and governed by the goddess Metis (Vernant and Detienne 1974). This form of knowledge—centered on the opportune moment, on the capacity to seize the opportunity—is reminiscent, according to Philodemus, of the practice of medicine or sailing. Philodemus further associates it with *parrhesia*: a form of frank speaking which corresponds to a sort of therapeutic capacity characterized by the fact of telling the truth at the opportune moment to a given individual who is ready to accept it in relation to the condition which he finds himself in.[20]

This method of investigation, then, keeps to directly graspable phenomena with the aim of articulating the meaningfulness and contexts they disclose. The therapist sets out after lost traces together with the person who left them (Arciero 2006), in order to find their possible meaning. This relation of cooperation based on mutual trust coincides with the practice of serving another person, understood as the act of

[20] In the Epicurean milieu to which Philodemus belonged, *parrhesia* was transmitted from master to disciple and practiced in a group setting, in the presence of an overseer (Foucault 2001).

going along with him as he re-enacts his own experience, submitting to what it indicates.

This leads us to the second question: (2) to what particular mode of understanding does the therapeutic relationship correspond? This question is bound up with the first one because, if this therapeutic position with respect to the other consists in going along with him by re-enacting those prereflective domains that have been formally indicated, the method must ensure that the other's experience is grasped without betraying its sense, i.e., by preserving the situationality of the other's existence without turning it into a thing through generalization. The method must therefore be developed in such a way as to be co-originary with the content of the experience under investigation, and must therefore always unfold in a certain way, according to the life experience it seeks to clarify, by keeping to such experience.

In psychotherapy, the method (Vd. Table 6.2) we have been gradually articulating usually starts with:

(a) The destructuring of the symptom, which brings out the experience trapped within it.
(b) The second movement consists in disclosing the thematic field on the basis of the experience just released, by shifting the focus of the investigation from the facts of the world to the unprominent connections of the person's factical life. This process consists in turning back to the original contexts and situations through formal indication. In the analysis of the problematic experience, this movement implies "siding with life," that life which in encountering things remains intimate to the self—which in becoming involved with the world experiences itself. This change of perspective does not amount to the cognitive grasping of factically relevant modes of being as objects; rather, it consists in a disclosure of the field of investigation within which these modes of being can be grasped by going along with their sense of enactment and in which they can be explicated, in such a way that these modes of being will manifest themselves as the concrete involvement with this or that meaningful situation. If meaning is understood in terms of the sense of enactment starting from concrete life situations, each meaning may be seen to provide the original meaningful direction and the possibility to explicate it as the factical involvement with this or that circumstance.
(c) The third methodological movement, therefore, consists in exploring together with the patient—and according to formal indication—the sphere of meaning which has been disclosed, in such a way as to "test" the experiential accounts provided by identifying their roots, grasping their ambiguities, and discarding or selecting the paths they trace, down to their prereflective sources. The destruction or deconstruction which distinguishes this third movement, together with reconstruction, consists in an act of discernment with respect to the situational origin of meanings and the directions of sense they enclose. Starting from this disclosure of a formally indicated perspective, phenomenological destruction/construction makes it possible to bring to light concealed meaningful moments that can provide an orientation. If the way in which one behaves

Table 6.2 The movements of the therapeutic method

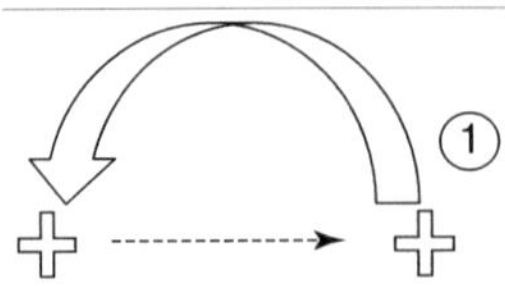	1. **Deconstructing the symptom and ensuring an access** This first step in the treatment is intended not just to define the composition of the symptom in such a way as to **unravel its mechanism** but to **enable**—through its deconstruction—**an access** to those prereflective meaningful structures that the symptom obscures and presupposes *From the symptom to its origin*
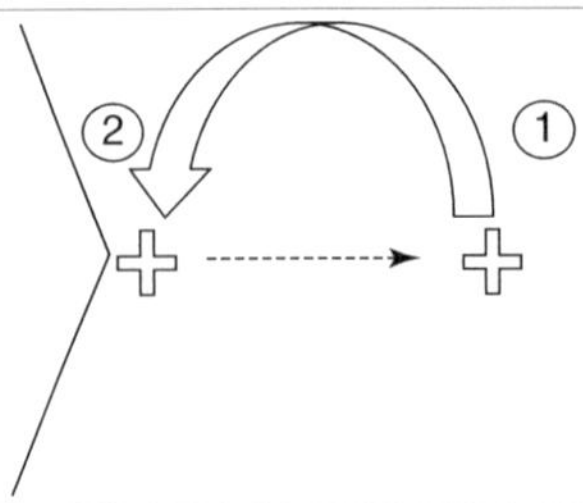	2. **Tracing the symptom back to its temporal origin while establishing a context of inquiry** Formal indication "therefore discloses a **sphere of content** which is constituted by a range of directions of sense pertaining to the various situations it comprises" *From the origin of the symptom to a field of meaning to be investigated*
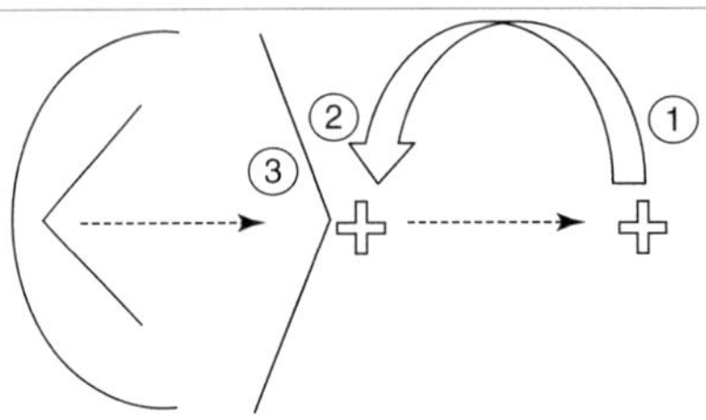	3. **An explicating understanding** It is a matter of "**exploring together with the patient** […] **the sphere of meaning which has been disclosed**, in such a way as to 'test' the experiential accounts provided […] down to their prereflective sources [the co-originary giving itself of the sense of enactment/content/reference]." "This passage […] brought about through the practice of questioning" makes it possible to "methodically trace the patient's experience back to the domain of the proper: the network of tendencies and motives which orient experience in advance and are transformed by it" *From a field of meaning to the actual remobilizing of meaningful trajectories*
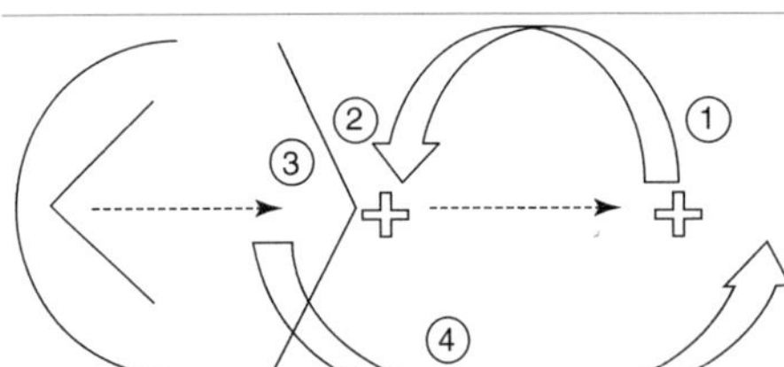	4. **The progressive retrieval of meaning** The "sphere of meaning, in its various articulations, progressively appears as, **together with the patient, we retrace her prereflective trajectories** starting from the situations at hand." This discloses "a new possible interpretation of the various situations under investigation" "which leads to **the development of new trajectories** and to a renewal of ipseity in the present" *From the remobilizing of meaningful trajectories to the progressive retrieval of the "meaning lost"*

Table 6.2 (continued)

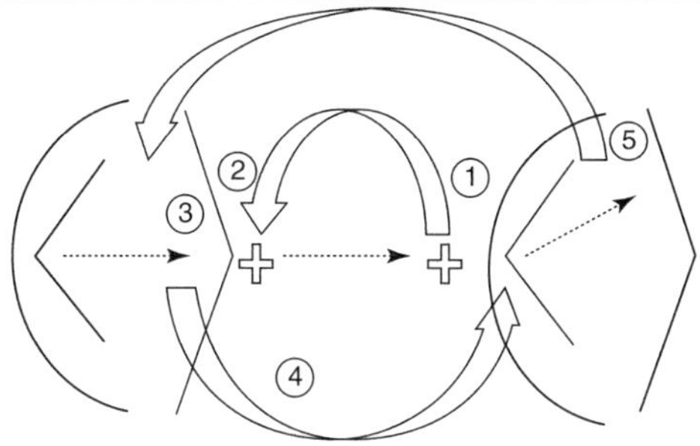

5. **The renewal of ipseity in the present**
"The fact of taking charge of what therapy has brought to light implies that what was meaningful for one's own world becomes meaningful again through **the opening up** of existence to **a renewed movement of understanding** and, at the same time, to **new horizons of expectation**. Taking charge of oneself, then, means acquiring a new position, taking the initiative, and setting out toward oneself"
From the retrieval of meaning to the taking up of a position with respect to oneself

toward others and in the world in a concrete situation and given moment is also essentially a way of behaving toward oneself, starting from this position, it is possible to explore sense relations, both according to the meaningfulness encountered within the world (the sense of content) and according to those motives and tendencies whose orientation the act of experiencing carries with it (the sense of reference). The patient's capacity to grasp these indications preserved by experience is not a technical or random matter—less still is it guided by an ideal of certainty established a priori; rather, it is the outcome of the fact of allowing oneself to be questioned by becoming involved in a search for meaning.

An intense participation in the investigation, then, is what enables the patient to gradually access the most invisible areas of life. The fact of following those traces makes it possible to clarify the patient's experience by tracing its sense back to the original motives. This process of moving back through an enacting understanding, following the direction imposed by traces, discloses, explicates, and articulates those horizons, until the overall thematic sense is attained.

(d) The presentation and thematic articulation of the sense of content and its appropriation represent the fourth movement of the method, which discloses a new possible interpretation of the various situations under investigation. The establishment of this interpretative condition leads to the development of new trajectories and to a renewal of ipseity in the present—a renewal which cannot be confined to the task of acquiring knowledge detached from concrete being-there (according to the theoretical approach) but which rather co-constitutes ipseity itself.

(e) We have thus reached the last movement of the method: the renewal of ipseity, meaning the acquisition of a new mode of access to the sphere of one's lived experience, and therefore—through the repetition of the movement—the possibility of newly articulating one's understanding. Thus every new interpretative context becomes a platform and a renewed indication for repeating that circular movement which leads to an even more in-depth understanding of a life story. We are once again following Heidegger's teaching: "The interpretation is only fully appropriated when by its 'enactment' it is understood that, turning back as it unfolds, it interpretatively informs what it has 'left behind' by in each case inserting it at the interpretative level reached; … this, then, is a re-forming and informing … which is nothing but a *re-grasping inclusion of previous interpretations* within the increasing strictness and simplicity of the interpretive coherence (*in die zunehmende Straffheit und Vereinfachung des Interpretationszusammenhangs*)" (GA 61 1985, p. 163).

Bibliography

Arciero G (2002) Studi e dialoghi sull'identità personale: riflessioni sull'esperienza umana. Bollati Boringhieri, Torino

Arciero G (2006) Sulle tracce di sé. Bollati Boringhieri, Torino

Arciero G, Bondolfi G (2009) Selfhood, identity and personality styles. John Wiley & Sons, Chichester

Benveniste E (1966) Problèmes de linguistique générale, 1. Gallimard, Paris

Benveniste E (1974) Problèmes de linguistique générale, 2. Gallimard, Paris

Berti E (2009) Nuovi studi aristotelici, vol 4. Morcelliana, Brescia, p 1

Caputo JD (1984) Husserl, Heidegger and the question of a "hermeneutic" phenomenology. Husserl Stud 1(1):157–178

Caputo JD (1988) Radical hermeneutics: repetition, deconstruction, and the hermeneutic project. Indiana University Press, Bloomington

Conti V (2018) Il chi della storia. Ph.D. Dissertation, Pontifical University Gregoriana

Crowell SG (2001) Husserl, Heidegger, and the space of meaning: paths toward trancendental phenomenology. Northwestern University Press, Evanston

Dahlstrom DO (1994) Heidegger's method: philosophical concepts as formal indications. Rev Metaphys 47:775–795

Dahlstrom DO (2001) Heidegger's concept of truth. Cambridge University Press, Cambridge

Descartes R (1998) Reguale ad directionem ingenii. Rules for the direction of the natural intelligence. A bilingual edition of the Cartesian treatise on method. Editions Rodopi, Amsterdam

Foucault M (2001) L'herméneutique du sujet. Cours au Collège de France 1981-1982. Gallimard-Seuil, Paris

Foucault M (2013) L'origine de l'herméneutique de soi. Conférences prononcées à Darmouth College 1980, Vrin

Foucault M (2016) Discours et vérité. Vrin, Paris

Freud S (1955) The Moses of Michelangelo. In: The standard edition of the complete psychological works of Sigmund Freud, Vol XIII

Gallese V (2005) Embodied simulation: from neurons to phenomenal experience. Phenomenol Cogn Sci 4(1):23–48

Gallese V (2007a) Before and below 'theory of mind': embodied simulation and the neural correlates of social cognition. Phil Trans R Soc Lond B Biol Sci 362(1480):659–669

Gallese V (2007b) Embodied simulation: from mirror neuron systems to interpersonal relations. Novartis Found Symp 278:3–12

Ginzburg C (1986) Miti, emblemi, spie: morfologia e storia. G. Einaudi, Torino

Hadot P (1995). Qu'est-ce que la philosophie antique?. (eng. Trans.) P. Hadot, What is ancient philosophy?, transl. by Michael Chase, Belknap Press, Cambridge, 2002

Hadot I (2014) Sénèque: direction spirituelle et pratique de la philosophie. Vrin, Paris

Hayek FA (1952) The sensory order: an inquiry into the foundations of theoretical psychology. University of Chicago Press, Chicago

Hayek FA (1967) Rules, perception and intelligibility. In: Hayek FA (ed) Studies in philosophy, politics and economics. Routledge & Kegan Paul, London
Hayek FA (1969) The primacy of the abstract. In: Koestler A, Smythies JR (eds) Beyond reductionism—the Alpbach symposium. Hutchinson, London
Hayek, F. A. (1978). New studies. In Philosophy, politics, economics and the history of ideas. Chicago University of Chicago Press.
Heidegger GA 2 1977
Heidegger GA 20 1988
Heidegger GA 56/57 1999
Heidegger GA 58 1992
Heidegger GA 59 1993
Heidegger GA 60 1995
Heidegger GA 61 1985
Hempel CG (1942) The function of general laws in history. J Philos 39(2):35–48
Hermans HJM, Gieser T (eds) (2012) Handbook of dialogical self theory. Cambridge University Press, Cambridge
Hermans HJ, Hermans-Jansen E (2004) The dialogical construction of coalitions in a personal position repertoire. In: Hermans HJ, Dimaggio G (eds) The dialogical self in psychotherapy: an introduction. Routledge, New York
Husserl E (2006) Logische Untersuchungen. Ergänzungsband. Zweiter Teil.: Texte für die Neufassung der VI. Untersuchung. Zur Phänomenologie des Ausdrucks und der Erkenntnis (1893/94-1921), vol 20. Springer, Berlin
Kisiel T (1993) The genesis of Heidegger's being and time. University of California Press, Berkeley
MacIntyre AC (1981) After virtue. Notre Dame, Notre Dame Press
Patočka J (1988) L'homme et le monde. In: Qu'est-ce que la phénoménologie?, a cura di E. Abrams. Millon, Grenoble
Patočka J (1998) Platone e l'Europa. Vita e Pensiero, Milan
Patočka J (2003) Il mondo naturale e la fenomenologia, trad. di A. Pantano Mimesis, Milano
Peirce CS (1955) Abduction and induction. In: Buchler J (ed) Philosophical writings of Peirce. Dover, New York, pp 150–156
Plutarch (1927) On listening to lectures. In: Babbitt FC (ed) Plutarch's Moralia, vol 1. Harvard University Press, Cambridge
Polanyi M (1958) Personal knowledge, towards a post critical epistemology. University of Chicago, Chicago
Quesne P (2004) Les recherches philosophiques du jeune Heidegger, vol 171. Springer Science & Business Media, Dordrecht
Rabbow P (1954) Seelenfuhrung: Methodik der Exerzitien in der Antike. Kôsel-Verlag, München
Ricoeur P (2004) In: Blamey K, Pellauer D (eds) Memory, History, Forgetting, trans. University of Chicago Press, Chicago
Ricoeur P (2010a) Time and narrative, vol 3. University of Chicago Press, Chicago
Ricoeur P (2010b) Power and Violence. Theor Cult Soc 27(5):18–36
Varela FJ (1999) The specious present: a neurophenomenology of time consciousness. In: Petitot J et al (eds) Naturalizing phenomenology: issues in contemporary phenomenology and cognitive science, vol 64. Stanford University Press, Stanford, pp 266–329
Vernant JP, Detienne M (1974) Les Ruses de l'intelligence: la mètis des Grecs. Flammarion, Paris
von Herrmann FW (2013) Hermeneutics and reflection: Heidegger and Husserl on the concept of phenomenology. University of Toronto Press, Toronto
Von Wright GH (1971) Explanation and understanding. Cornell University Press, Ithaca
Zahavi D (2003) How to investigate subjectivity: Natorp and Heidegger on reflection. Continent Philos Rev 36(2):155–176

Traces of Oneself and Healing 7

7.1 Experiencing and Narrating

We have repeatedly stressed how psychotherapy is a word-based therapy: it takes place through language. The therapeutic act takes shape through words: it aims to restore the patient's capacity to take charge of his experience by renewing the sense and movement of life which have been restricted by the disorder. We have seen how the therapeutic journey of transformation, which takes a symptom as its starting point, is guided by a method which first disentangles from the symptom the underlying experience that supports it and then subjects this to interpretation. For a patient, engaging with a problematic experience—be it a symptom or an unappropriated narrative—means opening up to the search for sense and hence allowing oneself to be challenged by the experience in question, so as to bring back to life a multiplicity of horizons concealed by the disorder. These are the horizons which formal indication enables the patient to bring to manifestation, so as to retrace their trajectories down to their original contexts through their co-enactment with the therapist and to fully appropriate the experiential dimension in question, renewing the sense of what has been significant for his own world.

The guiding thread of this process is clearly language, through which living experience is articulated and phonemically expressed, in the form of narrative accounts. The verbal expression of experience through a narrative makes the meaningful contexts concretely experienced by the narrator more or less present in the therapeutic conversation, according to the actual unfolding of the events and the style of the patient's factual experience. Following first Husserl's reflections (Husserl 2006, First Investigation §26, Sixth Investigation §5) and then Heidegger's ones on occasional expressions, we have emphasized this indicative aspect of assertion as a crucial point in the therapeutic conversation, insofar as it leads the listener toward the speaker's meaning. Hence, the objectivity of the narrative explication, co-conducted over the course of the therapeutic conversation, is not imposed by an a priori order but rather emerges within the framework of those actual contexts to which it refers.

G. Arciero et al., *The Foundations of Phenomenological Psychotherapy*,
https://doi.org/10.1007/978-3-319-78087-0_7

When I describe my relationship with my wife and the events that shaped it, this account has nothing to do with the explication of the same theme which may be provided, for instance, by an object relations psychologist, a psychoanalyst, or a cognitivist therapist. Each narrative explication unfolds according to prevalent tendencies and motivations, which for each person will vary depending on the circumstances—as well as, more generally, over the course of the individual's life—and which provide the guidelines and the direction of the narrative, so to speak. As previously emphasized, for the phenomenological therapist, the horizons of tendencies and motivations that punctuate and distinguish the different periods of a person's life—in a way that varies from individual to individual—also represent points of reference on the basis of which to newly articulate and retrace the series of meaningful contexts in the patient's personal story. In other words, in his narrative, the patient articulates and *relates* the meaningful contents he experienced in actual life situations on the basis of those guiding ideas which determine the development of his story. Narrating an experience is different from living an experience. It is evident that the context within which the narrative takes shape is not the same as that in which the experience itself was actualized. To quote MacIntyre, "Stories are lived before they are told" (1984, p. 212)—even though taking notice of one's own experiencing certainly does not amount to getting out of factical life.

The narrating of experience therefore represents a change with respect to the actual unfolding of experience—insofar as it is attuned in a different way to the current flow of events—and lends it an overall configuration which lived experience lacks. In an important passage from one of the densest texts from his early years, Heidegger writes with regard to the act of taking notice of experience and of its relation to factical experiencing: "What is decisive (i.e. in relation to taking notice) is that life, instead of advancing into tendencies of expectation and building life, gives form to the lived life from out of itself, stabilizing the meaning of the context of expectation and taking it as an explicit tendency" (GA 58 1992, p. 119). In other words, while human life spontaneously unfolds in particular situations according to given motivations and tendencies, in narratives, it becomes stabilized, in turn stabilizing the contexts of expectation. Hence, taking notice is an explicating modification of factical experiencing.

In the light of this perspective, we can understand Ricoeur's emphasis on the narrative status of events (1992), whereby events are distinguished from mere occurrences, in contrast with the perspective adopted by causal explicatory models. An event, then, will either carry a story forward, if it fits within this stable context without altering it, or mark a break, if it undermines the expectations created by the previous course of events.

In a narrative what is experienced is characterized as a well-defined whole with environmental boundaries, usually a beginning and sometimes an end. We might say that each narrative has a title, e.g., my holidays in Egypt last year, my relationship with my wife, my experience of working for an auditing firm, etc. The narrative explication, articulated according to the meaningful directions provided by living experience, lends unity to the meaningful contexts that have been concretely experienced, shaping one's lived experience and establishing one's expectations. In such a way, each narrative modifies the mode of access to lived experience, because the particular meaningful frames according to which ipseity manifests itself are oriented and hence stabilized by a guiding context of expectation.

This unifying power of narrative integration and the difference in terms of attunement between living experience and narratives emerge in a particularly clear way in that experience which culminates with the phenomenon of catharsis in Athenian tragedy. In his *Poetics*, Aristotle uses the term catharsis to describe the experience of the spectator, who feels pity and fear in a different way from the protagonists of a tragedy. As Michele Alessandrelli (2015) writes in a valuable study on the topic: "The moment the spectator sees protagonists on stage exclusively adhering to one meaning and losing themselves or tearing one another apart, he grasps that there are actually two or more possible meanings. The spectator is not hostage to the unilateral meaning which blinds the hero because he receives it from the μῦθος (i.e. from the plot of the story) in an enriched and integrated form. The μῦθος places this meaning within a wider context, in which it is possible to assimilate and bear it" (p. 17).[1]

In concrete experience, life is instead completely absorbed by a stream of events that follow upon and slip into one another, within a context of constant anticipation which is ceaselessly produced from one situation to the next and which determines the directions for its own development. Ipseity becomes such and attains manifestation together with the meaningfulness of the world, blurring the distinction between experiencing and what is experienced. The expression of factical life in a particular situation includes meaningfulness, for it encompasses the latter within itself and in doing so lends it form—and this form is an implicit sense-giving. The expression of life—coinciding with the way in which tendencies and motivations are actualized in a given situation with respect to the meaningful things one encounters—is therefore a trace.

What we have, then, are traces of oneself that are impressed upon life and which the narrative attempts to preserve by interpreting their meaning and orienting their vitality (Arciero 2006). Each trace thus points to a meaningful context and a significance that is enacted in the narrative. From this perspective, it becomes clear that a narrative can bring experience to manifestation by disclosing it yet also conceal it, becoming a source of illusion.[2]

A tension thus emerges between the way in which life generates form (the significance of an experience) in relation to the content it encounters and its linguistic expression: living meaning is verbally expressed—and hence articulated according to dimensions unknown to actual experience—as it becomes removed from life. While, on the one hand, the stabilization of meaning in a narrative lends a more or less enduring form to the heterogeneous multiplicity of tendencies and motives through which life expresses itself in relation to the meaningfulness nourishing it, on the other hand, it distances life from itself. And it is this distance which nourishes tragic catharsis.

[1] What we mean by *mythos* is "the composition of facts," which Ricoeur quite rightly interprets as the connection between facts (p. 59 vol 1) and which together with the *mimesis praxeos*—the imitation of action—constitutes the cornerstone of tragedy.

[2] We have previously mentioned how a certain way of telling one's story to oneself may itself be the source of a pathology.

7.2 The Therapeutic Effect of Words

In its self-sufficiency, human life—which factically unfolds giving rise to expression—contains a movement of dispersion, a movement away from oneself that leads to self-estrangement. We have already mentioned how this may occur when, through assertion, the meanings it refers to are fixed in words and made simply present, depriving them of their connection to actual life.

To return to an example mentioned earlier, the table—viewed from the point of view of the postman, the electrician, the cleaner, the student, the epistemologist, the surveyor, or the carpenter—is understood by each person in a different way, because each person has to do with the table for a different reason. Assertion conceals this particular way of having to do with a thing. The assertion that the table is sturdy, for example, a determination certainly deriving from the fact of having to do with the table, is a statement that concerns the table while at the same time concealing the genuine sense of having to do with it, which clearly varies from person to person.

Although the enunciatory character of the above assertion rests on the fact that each person has to do with the table in a particular way, this fact is concealed in the enunciation itself. Thus "the table is sturdy," not for the electrician who must use it as a support, or for the carpenter who sees it as a work surface, but rather as a sturdy thing which does not differ from any other sturdy thing. The table becomes a thing that is present and constitutes the object of the enunciation. In such a way, factical life sets out from expressive moments, from situations marked by the indistinction between experiencing and the things experienced, only to reach their reconfiguration through linguistic assertion. These moments crystallize in one's narrative, concealing the dimension of enactment from which they derive. Hence, if the means by which we express ourselves are a unique extension of factical life, what is required is a method enabling us to identify in those expressions meaningful and significant moments without crystallizing them: a method that preserves that uniqueness of life which comes to manifestation through its enactment in specific ways in life situations, instead of assigning meaning to experience on the basis of some theory.

As we have seen, the account of the symptom is the privileged means for concealing that movement which goes from expression to assertion. It is in a symptom that life most clearly reveals self-estrangement, as it traps prereflective experience within a vicious circle, stripping it of all meaningful connections through—and together with—its linguistic reconfiguration. A symptom, therefore, emerges through the interruption of the tension between significance and meaning, the stopping point of that motility of life (*kinesis tou biou*), which in the context of each person's life amounts to a dialogue between the expression of life and its surfacing in words.

So if the world gives itself to us as a relationship (since meaning is intrinsically connected to things, to entities) which takes shape through the expression of living being-there, and if it is through narratives that the articulation of sense takes place, by means of assertion, we become bound to the world—in one way or another. A symptom, which springs from the reduction of this tension to a mechanism connecting crystallized experience and the narrative which preserves it as such, exposes life to a sort of immobility in its lending shape to what it encounters. Psychotic delusion is the most obvious radicalization of this paralysis of meaning. The loss of freedom and ensuing suffering that characterize all disorders thus have to do with the difficulty to assign meaning to things, which inevitably impoverishes the patient's world.

Word-based therapy, therefore, has words—or, to be more precise, the obscurity of words—as its starting point. The opacity of words is what forces us to deconstruct the patient's assertion in view of the original relations it conceals. The path of healing sets out from an assertion as a trace, which in the crystallization of meaning conceals a reference to a multiplicity of contexts. Together with the patient, it then follows the trace back to the dimension of enactment which is being formally indicated. This procedure discloses the connection with specific situations and the way in which the factical encounter with meaningful contexts takes place, gradually elucidating and reestablishing the unity of meaning. Language, then, serves as an instrument for healing insofar as it is capable of leading us back to that relationship with the world, which it preserves in propositions and which may in each case be grasped as an enactment, as a phenomenon. As we have seen, this methodical movement of following one's traces coincides with the movement leading to the transformation of discourse and the renewal of sense. This is precisely why words have the power to straighten life out.

7.3 Words and Healing

Man, a living being endowed with speech, is always determined by the fact of being together with others, and speech is a fundamental mode of being with others. The therapeutic relationship therefore emerges as a unique way of relating to the other and as a particular way of conversing. The various psychotherapeutic currents conceptualize this relationship by establishing a series of a priori criteria for ensuring a good therapeutic outcome, first of all by defining how we are to relate to the other. These currents provide models for a disorder, change, relationship, treatment, and so on. However, none of them focus on the key aspect of the therapeutic relationship: what is the therapeutic potential of words associated with? How can we orient the articulation of understandingly being together with the patient, so that words may change the sense which the patient has of himself? This is the fundamental question that lies at the basis of all models, but which all models have ultimately sought to avoid. The only notable exception to this state of affairs is the work conducted by Jerome Frank and his daughter, who clearly identified the nature of the problem when, in the footsteps of the Aristotelian tradition, they associated psychotherapy with rhetoric and hermeneutics: "The assertion that psychotherapy operates in the realm of meanings raises questions as to which intellectual disciplines are best suited to examining its processes. In the United States the study of psychotherapy has traditionally been within the purview of science—specifically, behavioral science. We submit that psychotherapy also properly falls within the domains of rhetoric, 'the use of words to form attitudes or induce action' (Burke 1969, p. 41), and hermeneutics, the study of 'understanding and of the correct interpretation of what has been understood' (Gadamer 1975, p. xi)" (Frank and Frank 1991, pp. 52–53).

This way of accessing the problem also discloses a field of enquiry and area of study which to this day continues to hold the center stage and shape all research: the issue of common factors, i.e., those elements that are common to all forms of psychotherapy and are believed to account for its benefits (Garfield 1995; Grencavage and Norcross 1990; Hubble et al. 1999; Imel and Wampold 2008; Lambert 1992; Lambert and Ogles 2004).

We have seen how the practice of healing implies a mode of understanding, a specific form of listening, but also of questioning according to the directions suggested by the patient's story—in other words, a hermeneutics. Words, guided by the desire for healing, imply a specific way of expressing oneself aimed at moving the listener and hence a rhetoric. Within this context, rhetoric is understood not as a discipline designed to persuade—as sophistics is—but rather as one which has the function "of in any case identifying the appropriate means of persuasion" (Reth A 1, 1355 b 10) and is therefore connected to the capacity to encourage listening.

But what must the therapist's words—the words of a rhetor and hermeneut—target in order to mobilize the possibility of healing? This possibility does not translate into any one technique or model to be applied to the patient but rather differs from person to person since it is exclusively connected to the individual. One underlying motif of the present book so far has been the idea that the unprominent sphere of intelligibility—which constantly takes shape together with the meaningfulness of the situation and which changes in accordance with the transformations of life—at the same time coincides with a way of feeling emotionally situated and hence with an expression of oneself with respect to something. The emotional position of the living being with respect to existing circumstances is an integral way in which he finds himself in the world with others, as well as a way of being disposed toward himself in given situations. In addition to providing an orientation in the world, the emotional state elicited by an event manifests itself as a condition of self-affection.

In this respect, each person is given to himself in each situation through an affective knowing of himself which may be pleasant or unpleasant, light or burdensome: a kind of self-mastery which comes before all reflection and defines the emotional position of the living being in relation to what meets or might meet him. So we understand what we encounter through a certain way of feeling alive, an expression of the fact of being in the flesh at a given instant. In being affected by events—whether pleasurably or not—we orient ourselves according to what is desirable and more generally viable with respect to the existing circumstances, engendering fields of action and perception.[3]

Therefore, if emotioning corresponds to motioning from a certain context, in each case producing those possible new forms of effective involvement with the

[3] "In an emotional situation, being-there is always already conducted before oneself, is always already 'found', in the form not of self-perception but of an attuned manner of finding oneself disposed (*gestimmtes sich befinden*)" (GA 2 1977, §29).

environment that the online manifestation of different situations suggest (Arciero and Bondolfi 2009), this movement is always generated in view of something; and this something is that which is or is not feasible for each person.[4]

Human being-there is thus affected in a pleasant or unpleasant manner by that which one encounters; by orienting itself with respect to what is desirable, it determines how it can move, situate itself, and hence view things. Each person's unique life journey coincides with the unfolding of this movement from the cradle to the grave.

What we encounter, then, is not an objective reality, but rather something which is understood (*verstehen*) according to a mode of feeling alive (*Befindlichkeit*)—a mode of self-expression—based on a certain way of being disposed. This non-articulated meaning (*Rede*) can be brought to words and made explicit according to a given sense. Human being-there thus opens up in language, by paying attention (or not) to the meanings engendered within its own experience, so as to mobilize them—the search for sense takes shape through the possibility of opening up to words possessed by experience. Being endowed with speech does not only mean speaking to and being addressed by the other but also speaking and listening to oneself.

The relation between the expression of life—how living beings generate form in relation to the content they encounter—and linguistic expression, the articulation and interpretation of this content via words, constitutes the inner tension of life. In this respect, to live is to be in dialogue: a dialogue between the expression of being-there—which seeks to be vocally expressed and addresses itself in order to be heard—and oneself, as the reception of that which silently invites, summons and moves to the interpretation of sense. What emerges at the very heart of this silent dialogue—of human being-there—is therefore the decision to act in view of a given sense. In other words, the possibility of taking the initiative emerges through the joint functioning of the faculty of desire and the power to pursue certain existential directions according to a project rooted in the dialogue in question.[5] Symptoms interrupt this motility of life by halting the positioning of oneself with respect to the world through the repetition of a mechanism.

Hence, given that affective dispositions (*Befindlichkeit*) determine one's way of viewing and being open to the world, therapeutic discourse must focus on the affectivity of the patient if it is to change his position in the world and hence his way of looking at things. The directions orienting therapeutic practice cannot therefore be treated as a priori principles determining what shape the relationship must take; rather, they must be sought in those affective dispositions which orient the patient's sphere of intelligibility, since it is only by turning to those affective states that the therapist can elicit a change of perspective. The possibility of movement, then, always amounts to a reciprocal negotiation of one's position.

---

[4] Desire, which rests on the original desire to live, is therefore a principle of movement.

[5] It would be interesting to explore the birth and development of moral acting from this perspective.

7.4 Listening to Reason: Therapeutic Dialogue and Our Conscience

In a seminal 1993 paper exploring the theme from the perspective of therapeutic practice, Arnold Lazarus asks himself: "How is it that one of the first things that we all learn in 101 Psychology is that individual differences are paramount, that everyone is unique, but when it comes to treating patients some seem to assume that all come from identical molds?" (1993, p. 404). Before posing this question, Lazarus gives a witty, ironic, and accurate overview of the suggestions provided by various luminaries of psychotherapy. Thus Rogers and his followers, who claim to be promoting a person-centered therapy—in the belief that empathy and unconditional regard will result in personality change—actually offer the same identical therapeutic relation to all of their clients. According to Lazarus, the same lack of intersession or interpersonal diversity marks the work of other distinguished therapists of the 1990s, such as Ellis, Minuchin, Beck, and Sifneos. How can we take this irreducible uniqueness of patients seriously?

Regrettably, the answer which Lazarus provides is of the same sort as those of his learned colleagues, insofar as it boils down to a kind of commandment to be respected and imposed on the therapeutic relationship in order to safeguard interpersonal diversity. Lazarus thus prescribes the "authentic chameleon" conduct, referring to the therapist's attempt to modify his participation in the therapeutic process in order to offer the most appropriate form of treatment for the client.

However, as we have seen, the therapist who speaks oriented by and according to the directions suggested by the patient's affective states is far from being a chameleon: the (no doubt essential) fact that a therapist tailors his interpersonal style to different clients is not a "strategically" adopted attitude but rather has to do with the possibility of awakening in the patient an understanding of his own affective states, so as to renew their sense. The relationship generated on the basis of a process of listening and questioning is a form of involvement, which accompanies the flow of elicited experience and at the same time is guided by it.

Retrieving an experience and subjecting it to a new appropriation is an operation which is methodically oriented—not one which is left to chance—to the recitative creativity or mimetic capacity of the therapist or indeed to his merely persuasive attitudes or emphatic power. As the existential direction of living being-there is always given by the fact of being emotionally situated, every possibility of transformation goes hand in hand with the therapist's capacity to direct the therapeutic discourse toward those dispositions. In order to lead the patient to embrace a new perspective, the therapist's words must be capable of reaching this experiential foundation, so as to bring about a change of position on the patient's part. From this perspective, it is evident why the line between a therapist and a charlatan is so thin. After all, a patient's dispositions can be brought to light in many other ways, through methods ranging from seduction to adoption and from manipulation and provocation to suggestion. From our perspective, in order for words to awaken and newly mobilize affective dispositions, they must have such an impact on the person they are addressed to as to undermine and reorient the person's current interpretation of his own experience, by exposing him to the possibility of accessing the domain of the proper—the prereflective dimension. So the person who listens to the voice of the healer and allows himself to be questioned is first of all directed toward himself; and this is indeed bound to be the case, precisely because therapeutic discourse springs from the affective dispositions of the patient and is always addressed to them. The ultimate addressee of the therapeutic voice is the self of the patient who is capable of *understanding*. In this respect, that affective centrality in understanding and promoting change, which some therapists had emphasized in the early 1980s (Greenberg and Safran 1989; Greenberg and Johnson 1986; Guidano 1987; Mahoney 1984; Rachman 1981), acquires a new breadth and depth.[6]

Summoning springs from the *comprehension* which the self is capable of. The one summoned is the patient as a living being-there: through a dialogue with himself, the patient is called to himself by the need to listen to himself—by the plea that comes from himself and triggers a process of self-listening. In this respect, answering the summoning is always also a form of self-interpretation. The plea comes from that prereflective sphere according to which living being-there is in each case emotionally situated and understands the meaningfulness it encounters and which it is required to take charge of. According to Heidegger, this intrinsic possibility which being-there has of being called by itself is synonymous with conscience: "This inner voice of conscience is what properly discloses Dasein to me, to the extent that I am intimately with myself. This unique way of uncovering and disclosing that lies in conscience and prescribes for me the temporally particular possibility of my Being, this mode of resolute openness has the character of keeping silent. The voice of conscience speaks in silence" (Heidegger 1926, p. 286).

The therapeutic discourse therefore targets the patient's conscience, in the sense that its aim is to elicit a kind of listening which coincides with the patient's return to himself. The discourse is guided by rhetoric, not because it aims to persuade but because it mobilizes those means of transformation proper to each person. This return to oneself, an intentional referring which implies conscience taking shape as an actual enactment—and as a co-enactment over the course of the therapy—does not have any content but is an opening to oneself: a "lending an ear" to the silent voice which springs forth from one's emotional condition, revealing the situatedness of one's being in the world at a particular moment. As Ricoeur (1992) observes, this enactment, "prior to marking the capacity to distinguish between good and bad on the moral level and to respond to this capacity with the distinction between good and bad conscience, means attestation" (p. 422). Conscience, therefore, is not understood here as the judge presiding over the moral court in each person's soul or as a composite system produced by evolution which may be broken down into

[6]According to Aristotle, the irrational soul is made up of two parts: a vegetative part and a desiderative one, which unlike the former is capable of listening to reason. Listening and obeying to reason—as Aristotle explains—are here understood in the sense in which "we speak of 'paying heed' to one's father and friends, not in the sense of the term 'rational' in mathematics" (*Et. Nic.* I,13 1102 *b*—1103 *a* 5).

separate elements that function independently[7]; less still is it envisaged as the content of specific acts of judgment. Rather, conscience is understood as the intrinsic possibility for living being-there to be called by itself to *comprehend* its own expression, an unvoiced expression, which does not convey any content but which silently stirs and disturbs our being-there by bringing it face to face with a decision which it alone can make: to take charge of itself in view of something.[8]

The therapist's questioning of the patient—by investigating and obeying his story—therefore becomes a summoning if it awakens the patient's capacity to *comprehend* himself: to heed the voice of his conscience. In this respect, heeding the plea for a patient means deciding to be oneself, to bear witness to one's existence by accepting the weight and duty of it. Through the decision to take charge of itself, this reawakened self (which calls itself in its conscience) finds itself to be removed from and situated beyond the explanations which concealed and imprisoned it. In other words, the call—the summoning—is an expression of oneself which comes before one's voice and serves as its foundation: otherwise, words would be mere sounds and not the articulation of the opening up to the world of living being-there! This newly mobilized self-understanding is taken up and safeguarded through appropriation, which discloses a renewed self-mastery, new meaningful paths, and different life projects. The fact of heeding the silent call of the reawakened voice reflects the relation between one's corporeal being-there and that which one is listening to.

The patient who comprehends the therapist's words—the words of the other with a passion for healing—is thus called by his own being-there to take charge of himself by reorienting his own existential horizons. This is the ultimate aim of therapeutic discourse: the treatment helps "free" the patient, allowing him to grasp himself once again by bearing witness in a renewed way to his possibility of being himself. Therapy therefore takes shape through an interpretation of the ongoing situation, in an attempt to trace this back to the original phenomena by means of phenomenological destruction. Formal indication enables us to reach the sphere of underlying tendencies and motivations through discourse and to renew meaning through words. This retrieval of sense amounts to an attestation of the possibilities of living being-there which are most proper to it.

---

[7]A prototype for this may be found in the satirical "recipe" for conscience that Schopenhauer proposes when criticizing the popular understanding of the phenomenon: "Many a man would be astonished if he saw how his conscience, which seems to him such an imposing affair, is really made up. It probably consists of one-fifth fear of men, one-fifth fear of the gods, one-fifth prejudice, one-fifth vanity, and one-fifth habit; so that he is essentially no better than the Englishman who said quite frankly, 'I cannot afford to keep a conscience'" (Schopenhauer. *On the Basis of Morality*. §13, p. 127). This very prototype still constitutes the foundation of the contemporary view of conscience that considers the multiple states of the self and their coherence as the basis of the sense of personal integration. Putnam (1988), for instance, writes: "States appear to be the fundamental unit of organization of consciousness and are detectable from the first moments following birth" (p. 25). See also Bromberg (1998).

[8]What is also connected to this summoning is the acknowledgment of being exposed and of the lack of distinction between experiencing and that which is experienced—of that "*es gibt*" which prevents any solipsistic deviation. This is what inspired Heidegger's musings on nothingness and indebtedness in Being and Time—awakening as the assumption of the fact of being guilty.

7.5 The Therapist and the Chameleons

Answering the call from the prereflective sphere—a soundless expression—means opening up to oneself and listening to oneself, by taking charge of the temporally particular possibility of one's own being-there which this uncovering prescribes. The voice of conscience stirs us to a change of attitude. As the concrete enactment of the call, conscience summons living being-there to bear the weight and duty of being itself; and this appeal does not resolve itself in the uniqueness of a moment but rather takes shape through a perpetual renewal of interpretation.

The appropriation based on obedient listening therefore corresponds to a possessing of oneself which is re-enacted over and over again. It is in such terms that we can understand the young Heidegger's words in his review of Jasper's *Psychology of Worldviews*: "Conscience (*Gewissen*), here understood as the enactment of conscience, not as the mere having of conscience—*conscientia*—is a how of self-experience in accordance with its historical basic meaning" (GA 9 1976 p. 32).

From our perspective, then, the therapist's role has nothing to do with clever chameleons or with the application of a priori formulas to the patient—less still with the attempt to manipulate the relationship according to the guidelines imposed by a theoretical model. One of the most bewildering examples of this mode of operating is the approach which envisages the therapeutic relationship as one of attachment and the therapist as a universal caregiver. Rather, it is evident that if a therapist applies this theory, he becomes incapable of distinguishing between a therapeutic relationship and a relationship of attachment.

As the purpose of therapeutic discourse is to stir the patient's affective dispositions, it is evident that the therapist's words must be addressed to the patient according to—and starting from—the patient's experience: it is through its co-enactment that the therapeutic relationship takes shape. But in order for the voice of the therapist (driven by a passion for healing) to spark in the patient a stirring *comprehension*, it must inspire confidence. In other words, it is necessary for it to bear witness—within and through the therapeutic discourse—to the position which the therapist upholds and maintains for the unfolding and fulfillment of the treatment.

By once again taking our lead from Aristotle (*Reth.* II,1,1378 *a* 5–20), we can distinguish three conducts which promote the patient's confidence toward his therapist: equilibrium in dealing with life matters, rigor in managing the investigation and interpretation of experience, and goodwill, the intensity of which is given by the passion for healing. "Let it [scil. goodwill] then be taken to be the feeling in accordance with which one who has it is said to render a service to one who needs it, not in return for something nor in the interest of him who renders it, but in that of the recipient" (*Reth.*II,7,1385 *a* 18).[9] Within this particular relation which sparks self-listening in the patient, healing means helping the other to bear the responsibility

[9] It is important to note the close connection between this passage and the *Gorgias*. Socrates tells Gorgias that he who wishes to test a soul as to rectitude of life or the reverse must possess three qualities: "knowledge (*epistemen*), goodwill (*eunoian*), and frankness (*parrhesian*)" (487a). In relation to these references, it would be interesting to explore the relationship between *parrhesia* and reasonableness (*phronesis*), one of the key concepts in the Nicomachean Ethics.

and task of being himself—as opposed to subjecting him to predefined explanations with the aim of anticipating or prescribing ways of living.

If healing means letting the other be who he is, it is not by establishing and imposing bonds under the guardianship of a theory that the other is given the freedom to be but only by accompanying the unfolding of his experience through that form of serving which is investigating (*therapeuein*), guided by the other's existential possibilities and by his story. On the other hand, only if he is capable of taking charge of himself and of his own passion can the therapist ensure the kind of reciprocity which—to borrow Aristotle's words again—"very closely resembles friendship … but differs from friendship in not possessing the passion and attachment (*aneu pathous kai tou stergein*) for those it relates to" (*Eth.Nic.* IV, 5–6, 1126 *b* 20–25). The relationship is therefore one marked by reciprocity (Greisch 1994), albeit an unsettling reciprocity. It is real for a few hours per week and then virtual; it is necessary and sometimes unique but also without a future; it is close yet distant, polemical, difficult, delicate, and profound yet alien… and it is even paid for.

What the patient develops is a degree of intimacy with a stranger: a sort of passionate friend exempt from those feelings which extend beyond the treatment; a well-meaning yet at the same time balanced, frank, and rigorous friend who, driven by the desire to heal, leads him to grasp and accept himself, to the point of taking charge of his own freedom. This strange intimacy is the essential feature of the therapeutic relationship.

At this point, it is necessary to address a question which constitutes the very cornerstone of our analysis: where does this strange intimacy spring from?

7.6 Therapeutic Intimacy

Something common to all patients who come to therapy is the fact that they struggle to make sense of the enduring condition of suffering they find themselves in. Every explicative effort proves fruitless. The patients are incapable of finding release from their suffering. Frank (1961, 1995) has used the term "demoralization" to describe this state of affairs, pointing to an element shared by all patients: the perception of failure to tackle their symptoms.

The most immediate consequence of this experience is a decrease in the power to act and feel, accompanied by a reduction of the breadth of their relationship with the world. By stifling one's relationship with oneself, suffering limits the possibility of generating sense. This alteration of the motility of life perceived as a mode of passiveness by the patient—as the suffering of oneself—is the condition on the basis of which the anticipation of the future, increasingly burdened by the weight of suffering, is concomitantly dominated by the desire to be freed from it. All forms of psychotherapy, therefore, respond to this condition of demoralization by offering a new sense—usually informed by the theory on which the therapy rests—in order to replace the patient's ineffective explanation with a more adaptive one. So whereas the disorder appeared to be unavoidable and especially uncontrollable in the light of the explanation of it offered by the patient, the suggestion of a new possible explanation engenders the expectation that, if the treatment is followed through, it will be possible to resolve or at any rate control the disorder (Frank and Frank 1991; Wampold 2007, 2011). Different labels have been used to describe this process within the framework of the various different approaches: *response expectancies* (Kirsch 1985), *remoralization* (Frank and Frank 1991), *outcome expectations* (Greenberg et al. 2006), and *self-efficacy* (Bandura 1997).

If it is in view of this hope that the patient looks to the immediate future in search of relief, the healer—who, in accepting the responsibility to meet this expectation and take charge of the fragility which underlies it, finds a place in the patient's future—enters into his meaningful horizons, committing himself to a promise which meets that hope: "The patient—Ricoeur writes—'brings to language' his suffering in the form of a lament, which entails a descriptive component (this symptom...) and a narrative component (an individual involved in these or those stories); in turn, the lament becomes further defined as a request for... (for healing or—who knows?—for health or, possibly, immortality) and a request to... addressed as a plea to the healer. Grafted upon this request is the *promise* to be kept, once the suggested treatment protocol has been accepted" (pp. 229–230, Ricoeur 2001).

This is the basis of the therapeutic alliance established against suffering, a tacit agreement struck between the healer and the sufferer before any actual relationship takes shape between them. The two are brought together by a tension, by a way of being set in time, by a shared horizon which enables them to seal a pact—a healing alliance—which implicitly already defines a position of mutual acknowledgment: the sufferer relies on he who knows and who therefore promises healing, while the healer preserves his availability, meeting the requirements of he who is searching for a cure.

The immediate effect of the promise based on this mutual engagement is to disclose a horizon of hope and, with it, of new expectations which concern the actors who are involved and who share the common goal of healing. It is only starting from this condition that it is possible to understand why a good alliance formation measured in the early stages of treatment is predictive of outcomes: the stronger alliance, the better the outcome.[10] Different studies have emphasized how a positive alliance is one of the best predictors of outcomes (Horvath and Symonds 1991; Wampold 2001a, b, 2010; Norcross 2010; Bachelor 2013; Horvath et al. 2011; Johansson and Jansson 2010; Martin et al. 2000; Priebe et al. 2010).

It is hardly surprising, therefore, that the alliance is related to outcome regardless of the type of psychotherapy on offer (Carroll et al. 1997; Horvath and Bedi 2002; Krupnick et al. 1994, 1996; Wampold 2001b). The process of transformation does not rest on the distinctive ingredients that make up a specific therapeutic approach nor is one particular form of therapy preferable to another when it comes to tackling and resolving a certain disorder. Rather, the evidence suggests that the efficacy of

[10] The alliance refers to the quality and strength of the collaborative relationship between client and therapist. It is typically measured as an agreement on the therapeutic goals, a consensus on treatment tasks, and a relationship bond (Bordin 1976; Horvath and Greenberg 1994; Horvath and Luborsky 1993; Horvath and Bedi 2002; Horvath and Symonds 1991; Martin et al. 2000).

treatment is due to factors shared by all forms of psychotherapy. Thus the alliance and the pact on which it rests would appear to constitute the central core which integrates all other common factors at play (Imel and Wampold 2008), whose weight and influence depend precisely on the constant negotiating of positions between those party to the agreement.[11] Thus the alliance and the pact on which it rests would appear to constitute the central core—i.e., the hierarchically superior factor—which integrates all other factors at play (Imel and Wampold 2008), whose weight and influence depend precisely on the constant negotiating of positions between those party to the agreement. This is summed up by Rosenzweig's famous Dodo bird hypothesis, which asserts that when treatments are compared they yield roughly equal outcomes, thereby suggesting that the benefits of the various psychotherapies are due to some unrecognized factors common to all approaches: the common factors.[12] From the very first session, the actualization of the pact offers a response to the patient's feeling of demoralization because it sparks a new hope, mobilizing the client's strength to tackle his problems through the action of a possible therapy. For theoretically oriented therapists, what this means is providing a series of explanations for the patient's experience, according to a priori principles that make it possible to assign new meaning to it in line with the new explicative perspective adopted and to suggest or prescribe a coherent treatment designed to achieve more adaptive responses. Wampold (2012, p. 26) sums it up as follows: "Of course, each theory tells the story differently—irrational thoughts, unconscious motivations, unexpressed emotions, poor attachment histories—but each tells a hopeful story to the client: 'If you believe in this new explanation and follow the steps in this treatment, your problems will be manageable and life will be better.'" In common factor models, acquisition of adaptive explanations is central to change in psychotherapy (Anderson et al. 2009; Frank and Frank 1991; Imel and Wampold 2008; Wampold et al. 2006).

Evidently, while this attitude—which is shared by all theoretically oriented forms of psychotherapy—lends the patient's experience a scientifically certified meaning, restoring his mastery of control, it uproots experience from the actual meaningful context within which it is embedded. Thus, the possibility for the patient to access the original meaning of the experience is concealed. On the other hand, in

[11] These common factors were investigated by Lambert (1986, 1992), who ranked their importance by examining the relation between client progress and a range of variables. These variables typically included extra-therapeutic factors (e.g., spontaneous remission, fortuitous events, social support) (40%); expectancy, hope, and placebo effect (15%); specific therapy techniques (e.g., biofeedback, hypnosis, systematic desensitization) (15%); and common factors (empathy, warmth, acceptance, etc.) (30%). Wampold (2001a, b) further revealed that the differences among models accounted for only 1% of the variance outcome.

[12] The term derives from a seminal paper (1936) by Rosenzweig that used a quotation from *Alice's Adventures in Wonderland* to refer to the equivalence in outcomes. The quotation is from Chap. 3. A number of characters become wet, and, in order to dry themselves, the Dodo bird decides to issue a competition: everyone was to run around the lake until they were dry. Nobody cared to measure how far each person had run, nor how long. When they asked the Dodo who had won, he thought long and hard and then said, "Everybody has won and all must have prizes!"

line with the teaching of ancient physiologists, and with the tradition of the care of self (Foucault 2001), an explanation based on well-established principles—particularly if comprehensible and persuasive—has a soothing effect, as though this form of knowledge restored the power to govern the future by turning the latter into something available, controllable, and on hand.

7.7 The Pact and the Method

As already noted, the alliance takes shape through a constitutive asymmetry which manifests itself in the desire for and expectation of efficacy on the part of the patient, who puts the therapist in charge of his incapacity to accomplish something, with the assumption that through his work the healer will be capable of restoring his powers and possibilities. The common core of the alliance is thus a mutual promise, which is more or less tacitly made and which is founded upon the experience of suffering on the one side and on the exercising of the therapy on the other. According to this pact, the two parties commit to fulfilling their duties in order to accomplish the treatment. The whole course of treatment is designed to correct the initial disproportion generated in the patient's relationship with the therapist by the experience of suffering, to the point of restoring the patient's freedom—a step which marks the happy disbanding of the alliance.

Horvarth et al. (2011) is right to note that the therapist does not "build alliance" but rather "does the work of treatment in such a way that the process forges an alliance with the client" (p. 25). We might add that this is a working alliance, the directionality of which is given by the very aim of the treatment: the restoring of the patient's distinctive possibilities, of his freedom. This trust-based pact takes shape as the keeping of one's word and, reciprocally, as faith in the healer's words, as though they were the words of a friend.

This tacit pact, which is the foundation of all therapeutic relationships, accounts for an element that has been highlighted in more than a hundred studies: when patients are asked by clinicians what was helpful in their psychotherapy, they primarily attribute the effectiveness of their treatment to the relationship with their therapist (Sloane et al. 1975; Elliott and James 1989; Norcross and Lambert 2005a, b). On the other hand, various studies have noted just how fragile this pact is, how it contracts and expands, how it may grow and become stronger or be broken, and how over the course of the therapy it will be subject to interruptions, slumps, crises, erosion, and repair (Safran and Christopher Muran 1996; Safran et al. 2011). Finally, it is by virtue of this pact that the therapeutic relationship is the most common of common factors, the most common mode of convergence among professional divergences (Norcross 2010). This pact starts to take shape at the intersection between a desire—that of the patient—and a form of knowledge (*metis*), that of the healer, which becomes cemented as a reciprocal promise. It is then fulfilled and renewed as a pact of mutual trust, which serves as the foundation for those famous common factors that much evidence suggests are the key elements accounting for the effectiveness of psychotherapy (Lambert 1986, 1992; Wampold 2001a, b, 2007, 2010; Wampold and Bhati 2004; Hubble et al. 2010).

What emerges at the very heart of the pact, however, is a problem which we have already repeatedly touched upon and which now leaves no way out. If the point of intersection between the patient's suffering and the healer's knowledge is also where the working alliance is first established, what form of knowledge guides the healer? In order for the healer to be capable of grasping the patient's unique experience, what form of knowledge must he possess?

Bordin's conceptualization of the alliance (1979) has been particularly influential among psychotherapy researchers over the last decades. For Bordin, the quality of the alliance is a function of (1) the patient and therapist's agreement on the therapeutic goals, (2) consensus with respect to the therapeutic tasks, and (3) the development of a bond (mutual trust, liking, respect, and caring).

This widely shared perspective on the therapeutic alliance gives primacy and assigns special importance to the working alliance between patient and therapist. While always guided by the reciprocal promise on which the alliance is founded, the sense of partnership whereby each participant is actively involved in implementing the treatment gradually starts to take shape as an unfolding—and, for the patient, a disclosing—of personal experience, generally of a problematic sort. The patient's experience becomes the object of a shared task, which is oriented by the kind of perspective adopted by the therapist.

This task, which implies and at the same time lends shape to the working alliance, is carried out in the light of those theoretical a priori principles that constitute the therapist's frame of reference. Whatever the extent of the task set by the kind of therapy pursued, from the observation and modification of the patient's behavior to the search for childhood traumas, every form of therapeutic work is (or ought to be) theory-centered: as the patient reveals himself, the therapist resignifies his experience or corrects it according to the reference model(s) by which he is guided.

The explanation—whatever it may be—which is provided in the context of the treatment newly discloses the future in terms of the theoretical governing of the uncontrollable. Wampold (2001a, b) therefore goes so far as to argue that the truthfulness of the explanation is of no importance for the outcome of psychotherapy. This also helps understand why the power of the treatment rests on the patient accepting the explanation rather than on whether the explanation is "scientifically" correct. The equivalence between the effect attained by the explanation and the person resorting to it is related to the possibility which the patient accepts of resignifying his own experience according to the sense offered to him by the therapist in more or less confident terms. This outcome is shared by all the various theoretical approaches, which differ from one another precisely by virtue of the explanations they provide for certain psychological phenomena and for the treatments deriving from such explanations. This is the basis of the so-called Dodo bird conjecture.

The issue in question must have been quite clear to Rosenzweig, who wrote: "Whether the therapist talks in terms of psychoanalysis or Christian Science is from this point of view relatively unimportant as compared with the *formal consistency* with which the doctrine employed is adhered to, for by virtue of this consistency the patient received a schema for achieving some sort and degree of personality organization (…) Having in common this possibility of providing a systematic basis for reintegration, diverse forms of psychotherapy should tend to be equally successful" (2002/1936, p. 7). Treatment therefore consists in reintegrating the patient's experience by resignifying it in a coherent way, according to a systematic theory of reference, regardless of what that theory may be.

In the light of the explanations of the patient's experience, which the theory provides in order to bring about the "cure," a series of methods and means are deployed which evidently require the involvement and consensus of both the therapist and the patient. Hence, the particular approach used is unimportant as long as the treatment is accepted by the patient and fosters his active engagement. So while therapists of contrasting persuasions will differ in their emphasis on the central and enduring qualities of the goals for which they are willing to cooperate with the patient in view of his treatment, "different patients will be predisposed to find different tasks and goals meaningful as a function of their unique developmental histories and relational schemas" (Safran and Muran 2006).

Tasks and goals of the working alliance take shape within the framework of the bond which is established between therapist and patient. Bond formation is the core connecting the agreement on the therapeutic goals and the consensus on the therapeutic tasks. The last point which Bordin discusses in his seminal paper is the nature of the human relationship on which the therapeutic process is played out, which will in turn depend on the aims and engagement required by the kind of therapy adopted. Once again, Bordin's distinction is between the two extremes which he always seems to have in mind: the focused treatment of the cognitive-behavioral therapist, who gives the patient a form and asks him to make a daily record of his acts, and the treatment offered by the psychoanalyst, who directs the patient's attention toward the more protected recesses of inner experience. Although in both cases it is necessary for some basic level of trust to develop, the crucial point is that the kind of therapy adopted also determines what kind of relationship is established: "One bond may not necessarily be stronger than the other, but they do differ in kind. Another nuance in the therapeutic bond might be defined by the difference between a caretaker and a consultant. Some modes of psychotherapy have much more of a take-charge element built into them" (p. 254). Even the creation of the bond, either implicitly or explicitly, depends on the indications provided by the theoretical apparatus supporting the therapeutic intervention.

So what deeply informs Bordin's conceptualization of the working alliance is an even more fundamental factor which underlies all others without ever being thematized: the theoretical explanation of the elements of which the therapeutic alliance is a function. As we have already repeatedly stressed, however, the theoretical attitude, which is one of many possible modes of accessing experience, fails to grasp human life in its imminence and incompleteness—in its being constantly in the process of unfolding.

As we have seen, if we consider the psychic sphere not as an object of theoretical knowledge but in terms of the full enactment of the experience of living which always takes shape in life-contexture with others, factically realizing itself as concrete history, in order to access this sphere, we must resort to a kind of knowledge which in each case finds its determination in the actual situation at hand. In this respect, in our research, we have outlined the thematic areas of a form of psychology understood as an a-theoretical science, by taking as our starting point the general features of its peculiar field of investigation: ipseity. In line with the tradition of practical knowledge, the mode of knowing which distinguishes a-theoretical psychology is a "knowing how" that is grounded on general principles, but these general principles are continuously co-determined in the concrete situation in which the individual finds himself (Arciero 1989). Therefore, even the method which enables us to access another person's experience, the method of a psychotherapy based on this mode of knowledge, must take shape in a way that is consistent with the different life experiences it seeks to investigate.

7.8 Therapeutic *Subtilitas*: Different Strokes for Different Folks

How can we reformulate our conceptualization of the therapeutic alliance in the light of the above perspective? If it is through the full enactment of experience that each person finds his existential direction and an access to himself in every situation, suffering—like any other experience—presents itself to the patient in the unique situations of his life. The therapist must therefore examine those concrete situations in order to grasp their meaning, starting from the patient's perspective and adjusting his knowledge according to the sense required by those actual lived experiences. This, as previously noted, is the experience of the other's experience—something which evidently cannot be reduced to, or understood as, empathy.

For the therapist guided by practical knowledge, exercising this science means being capable of actualizing the knowledge in question in the concrete situation at hand. This is a very different form of knowledge (*techne*) from the one guiding the production of objects (*poiesis*), and according to which, the product—that is always self-identical—is already fully determined a priori. This is one of the areas in which psychotherapy falls within the domain of hermeneutics, as Frank had grasped. Indeed, the actualizing of the therapist's knowledge in accordance with the patient's experience corresponds to that *subtilitas applicandi* which, along with understanding (*subtilitas intelligendi*) and interpretation (*subtilitas explicandi*), constitutes the third moment of the act of interpretation taken as a whole: to understand and interpret for the therapist always means to apply his knowledge. However, the most interesting aspect, which—as we will see—may account for the fact that the person of the therapist is a critical factor in the success of therapy, has to do with what the term *subtilitas* describes.

With regard to this terminology and the ancient hermeneutic tradition from which it derives, Gadamer observes: "The process of understanding was regarded as consisting of these three elements. It is notable that all three are called *subtilitas*—i.e., they are considered less as methods that we have at our disposal than as talents requiring particular finesse of mind (*das besondere Feinheit des Geistes*)" (1975, p. 307).

The crucial point is that the therapist who develops this "finesse of mind" does not deal with the patient's experience and story from a third-person perspective but is involved in first person to varying degrees, remaining faithful to the patient's meaning in methodically regaining his lived experience. The distinctive sensitiveness to which the term *subtilitas* refers consists in knowing how to listen and see what is meaningful (*dejiudicatio*) in the particular situation at hand and how to keep to it.[13] It is precisely this difference in terms of finesse which may account, both for the fact that some therapists are more effective than others, regardless of the type of treatment (Lutz et al. 2007; Wampold and Brown 2005), and for the data indicating that the therapist is the most robust predictor of outcome of any factor ever studied (Wampold 2005).

There is evidence to suggest that much of the variability in outcomes in therapy is due to the therapist (Kim et al. 2006; Okiishi et al. 2003; Wampold and Brown 2005) but also that the characteristics or actions of the most effective therapists are not really known (Beutler et al. 2004). This variability is attributed to the better use made of the common factors by certain therapists (Hubble et al. 2010) or to the capacity of the more competent to establish better therapeutic alliances (Baldwin et al. 2007). On the other hand, the patients who felt that their therapy was successful attribute the positive outcome to the personal qualities of their therapist (Strupp et al. 1969; Luborsky et al. 1985). In an extensive review of more than 2000 process-outcome studies, Orlinsky et al. (1994) have identified, among the therapist variables that consistently had a positive impact on the treatment outcome, the ability to focus on the patient's problems and to direct the patient's attention to the patient's affective experience, along with the ability to engage with the patient.

From our perspective, the difference in terms of *subtilitas*—whereby what is established as the very heart of the method is the therapist's capacity to grasp in the actual situation at hand those aspects which enable the fulfillment of the course of treatment—might be the key to a comprehensive understanding of all the data just outlined.

The diversity in question is nonetheless to be interpreted in conjunction with another compelling indication provided by several studies on what clients say about therapy, which consistently show that patients emphasize feeling understood as one of the most significant helpful aspects of psychotherapy (Bohart and Tallman 2010, Ch. 4). Evidently, this also explains why patients primarily attribute the effectiveness of their treatment to the relationship with their therapists (Elliott and James 1989).

The most significant aspect to emerge from patients' narratives clearly points to an issue which has always been acknowledged in the field of psychotherapy, even though all attempts to account for it have proven fruitless: each treatment should be tailored to the uniqueness of the patient.

This long-standing question resounds in the title of the special issue of the *Journal of Clinical Psychology* published in February 2011: *Adapting Psychotherapy to the Individual Patient*. Norcross and Wampold, whose contribution opens this special issue, echo the aforementioned paper by Lazarus by repeating the same

[13] "In order to establish wheter the actualization is original or not, the proximity or distance of a sense context from the origin must be determined. This decision is called dijudication. The criterion for this decision lies in the origin itself" (Kisiel 1993, p. 129).

mantra: "different strokes for different folks."[14] However, whereas in 1993 this saying had given Lazarus the opportunity to develop a systematic critique of the various currents of psychotherapy, in the article published some 20 years later—no doubt the most revolutionary decades in the history of modern psychotherapy—its repetition comes across as the acknowledgment of a state of affairs which one must face up to. Thus although they emphasize the compelling need to find an answer to the problem, when confronted with the full extent of the issue they raise, Norcross and Wampold—just like Lazarus before them—fail to deal with its consequences in a suitably radical way.

The same impression may also be gathered by taking a broader look at a body of research which strongly suggests the need to be responsive to clients' requests and needs. This compelling indication translates into the ethical and practical imperative to privilege the client's experience, as suggested by J.C. As basic as it is to adjust therapy to the client, a complete commitment to "starting where the client is at" is critically important. "In practical terms, without preconceptions, this means organizing clinical services to clients: who they are, what they want, and what constitutes and influences the circumstances of their lives. It also means that assessments of the quality and the outcome must come from the client, the ultimate beneficiary of the service" (p. 36).

The question which must be addressed here, then, is: what method can we use to access the patient's living experience (i.e., where the client is at)? Faced with this obstacle, attempts are made to solve the problem by invoking the common factors and emphasizing the way in which the therapist delivers his treatment: better therapists use the common factors to achieve better outcomes.

Among the common factors, then, is explanation—i.e., the different explanations that the various forms of psychotherapy provide on the basis of their underlying theories—which acquires special relevance. Several studies attest to its importance. Data suggest that the lack of a theoretical structure orienting the treatment focus is a good predictor of a negative outcome (Lambert and Bergin 1994, Mohl 1995). In other words, the level of explanation of the patient's experience which the therapist provides must be perceived by the patient as being consistent and organized over time according to a framework which the therapist can govern. Not only that, but the depth of the conviction with which the therapist delivers treatment is regarded as a common factor that exists across all therapies. "Allegiance to a treatment approach—Wampold 2001 writes—refers to the degree to which the therapist delivering the treatment believes that the therapy is efficacious" (Dush et al. 1983; Hoag and Burlingame 1997; Luborsky et al. 1999, 2002; Paley and Shapiro 2002; Robinson et al. 1990; Shapiro and Shapiro 1982; Smith et al. 1980). In other words, while the strength of conviction is grasped as a common factor—and no doubt as a rhetorical element—it is attributed to the more or less consistent relation which the therapist has with the theory to which he subscribes and according to which he resignifies the

[14] "Ideally, therapists should select for each patient the therapy that accords, or can be brought to accord, with the patient's personal characteristics and view of the problem" (Frank and Frank (1991) p. xv).

patient's experience. In this respect, it makes no difference what kind of therapy or theory guides the treatment, since all treatments delivered by a competent therapist to a client motivated to engage in the process are equally effective. The relationship of collaboration between patient and therapist, the transaction between them, and the structure of the treatment that is offered thus acquire relevance across all forms of psychotherapy. This leads to the conclusion that what only matters is "that a treatment, conducted by a given therapist with particular client at a specific time and place, yielded positive results" (Norcross and Lambert 2005a, b).

But if the patient's life is accessed according to the therapist's theory—no matter what theory this may be—what remains of the ethical and practical imperative to make living experience the starting point for the treatment? If a therapist grasps the patient's lived experiences and understands them according to principles that are consistent with his own theoretical system of reference, is he not betraying the injunction to listen to the client's experiences?

To meet the patient where he is at is to grasp the meaning of his experience starting from the way in which he himself lives it—not to explain this experience to him in more or less persuasive terms, according to how the therapist understands it, so as to then turn this explanation into a common ground through which to establish the means and aims of the psychotherapy: the sense of an experience does not coincide with its explanation! On the other hand, one cannot make up for the lack of a method by resorting to a rhetoric that seeks to persuade the patient on the basis of a priori principles determining the possible explanations of the patient's experience—a practice akin to a sophist's attitude. A therapist of this sort is a far cry indeed from the Aristotelian view of the rhetor as someone who clearly grasps and mobilizes the possibilities of persuasion of the person he is addressing.

The pioneers just referred to may be credited with the great merit of having emphasized the common relationship factors, while at the same time questioning the effectiveness of established models, thereby ushering in a new era. However, one crucial aspect which apparently eluded them is that no psychotherapy founded upon a natural-scientific psychology can ever grasp the uniqueness of the patient's experience nor start analyzing it from where the client is at. And the reason for this is that, by its very nature, theoretical knowledge determines what is unique in the light of general principles, thereby erasing the personal meaning of every experience.

This underlying lack of clarity extends to the attempt to consider therapy a contextual phenomenon: a metamodel (Frank and Frank 1991; Wampold 2001a, b) which focuses not just on the common factors as the ground for the equivalence of diverse psychotherapeutic orientations in terms of effectiveness but also on the contextual aspects of the actualization of such factors. Critical contextual phenomena are seen to include the healing context and the meaning attributed to it by the participants (therapist and client). Wampold (2001a, b) sums it up as follows: "According to Frank and Frank, provision of new learning experiences, as an example, will not be therapeutic unless the client perceives the therapy to be taking place in a healing context in which he or she as well as the therapist believe in the rationale for the therapy; the therapist delivers therapeutic actions consistent with the rationale; the client is aroused and expects to improve; and a therapeutic relationship has been developed" (p. 26).

According to this model, all forms of psychotherapy share (1) an emotionally charged, confiding relationship with an empathic helping person, (2) a healing setting where the relationship is concretely implemented, (3) a conceptual framework (myth) through which to explain the patient's disorders and prescribe some remedies (rituals), and (4) means to carry out the treatment, based on collaboration between the patient and the therapist. Although the healing contextual phenomena acquire central importance, the web of elements at play is structured around the explanation, a common factor, which for all forms of psychotherapy is guided by a specific theoretical frame. In other words, what is contextualized is the way in which the theory/theories adopted is/are used to account for the experience under investigation: as Wampold (2001a, b) notes, "this position implies that ideally therapists should select for each patient the therapy that accords, or can be brought to accord, with the patient's personal characteristics and view of the problem. Also implied is that therapists should seek to learn as many approaches as they find congenial and convincing" (p. 25).

What is lost sight of here is a deeply paradoxical characteristic of this model: while, on the one hand, the contextual aspect of the therapeutic process is invoked, on the other, the therapist is shown how to decontextualize the experience under investigation according to the theory or theories which he is capable of adapting to it. Possibly misled by the soothing effect produced by the explanation provided, one fails to realize that the procedure in question ultimately betrays the ethical and practical imperative for the therapist to meet the patient where he is at—and this surely cannot be achieved through a theory that freezes the patient's living experience. Hence the role of rhetoric, understood as sophism, the special purpose of which is to persuade the patient by bringing the therapeutic explanation in line with the patient's meanings or personal perspective. The generative power of the patient's personal growth is therefore tamed according to the therapeutic orientation adopted and the goal consensus it implies.[15]

7.9 Empathy and Compassion

Promoting active involvement in the healing process by treating the patient with great respect and as an ally (Bohart and Greaves 2013) implies a use of rhetoric linked to the capacity to elicit listening. This solicitude is intended to restore the other's motility of life and freedom of choice, without any claim to take his place. Indeed, the purpose of rhetoric as such is not to persuade (*peisai*) "but to identify the appropriate means of persuasion in each case" (Reth. A 1, 1355 b 10).

In psychotherapy, then, this kind of rhetoric is used by mobilizing that powerful potential for change peculiar to each person and which may only be accessed via the patient's living experience. So if it is indeed the case that the therapist helps the patient to appropriate his own past starting from the present situation and in the light

[15] By sophism we mean the art of speaking and persuading by using fictitious arguments that only apparently apply to the thing under discussion.

of a possible future, contrary to what Frank argues (1987, p. 294), it is not enough for the resulting "narrative truth" to be persuasive to ensure its rhetorical appeal. Rather, it must have an evidential value consisting in the actual opening up of ipseity to new movements of understanding and therefore to the renewal of the patient's existential orientation through the re-enactment of the meaningful perspectives enclosed within his story.

Therapeutic discourse, which has the transformation of meaning as its aim, uses rhetoric in order to awaken within the patient the silent voice of his conscience, which speaks of the situatedness of his being in the world, and thus to reorient the opening up to oneself and hence the interpretation of one's own existence. This rhetoric is closely connected to hermeneutics: through the ordinary language of therapeutic dialogue, it fosters an understanding of one's being emotionally situated—i.e., knowledge of what it is all about—which the interpretation provided actualizes and preserves.

If we truly make the person in all of his or her uniqueness the center focus of the healing process, tailoring the therapeutic relationship without any need to transform the therapist into an authentic chameleon, only phenomenology—as a method faithfully directed toward the way in which living experience shows itself—can enable the science of healing to access the "who" of the other. Grasping the other's experience in its unfolding therefore implies a mode of comporting oneself—guided by formal indication—which keeps to the other's life situations and follows its traces so as to regain the prereflective experience and the sum of implications in which it is embedded.

"This form of knowledge directly translates into a practical attitude, an actual conduct and an involvement that takes shape in accordance with the unfolding circumstances, discoveries and developments in the patient's life, in line with the possibilities and interpretative tools afforded by such science" (p. 323–24). In order to reappropriate oneself, it is necessary first of all to take care of oneself!

This leads us to one of the most delicate aspects of relating to another person's suffering, one which all theoretically based forms of psychotherapy—guided by methodological requirements emphasizing the myth of interiority—have traced back to empathy. The same underlying assumptions today also inform the vigorous scientific interest in empathy and a range of related phenomena—emotional contagion, emotional sharing, motor mimicry, imaginative projection, and perspective taking—in the field of the neurosciences and of social cognition. Within this epistemological context, empathy plays a role similar to that played by perception within the framework of a theoretical construction. If the origin of objectivation lies in the fact that reality gives itself in what is perceived, understanding the other means accessing his mental life through the empathic grasping of expressive phenomena. From this perspective, it is impossible to realize that all sense data reach us from a totality which manifests itself as a horizon but whose presence is different from that of ordinary things subject to perception. Likewise, those who believe they can access other people's experience through empathy fail to realize that, in order to do so, they must already be able to detect some modes of being within the framework of a preliminary mutual understanding (Mazzola et al. 2010): within the framework, that is, of coexistence in a common world, understood as a shared experience, where empathy is only one of the possible ways of perceiving oneself in the actuality of the relationship with the other—certainly not a way in which a subject confined to an autonomous sphere (the therapist) meets and understands another subject who is equally isolated within his interiority (the patient). This means that, in order to detect the suffering of the other, it is not necessary to be empathic, whereas in order to be empathic an understanding of the other's suffering is required.

Unfortunately, from the days of Freud (1905) and Rogers (1957), down to contemporary developments (see Elliott et al. 2011), the prevailing tendency among therapists has been to consider empathy a higher-order category under which to nest the various modes in which an individual can understand other people's experience by putting himself in their shoes. All forms of psychotherapy, therefore, rest on the assumption that it is possible to grasp another person's state of mind directly (or indirectly) in the available expressive phenomena; with different emphases, they all use this assumption over the course of the treatment as a means to gain access to the patient's world.

But if the therapeutic understanding of the other amounts to an investigation which must keep to the unique life of the patient in order to re-enact its meaningful trajectories, encountering another person's experience, while respecting its original intention, entails a way of comporting oneself which only reaches fulfillment through the co-enactment of the original experience together with the patient. It is evident that, according to this perspective, expressive phenomena have a very different significance from the central one they are assigned by those who believe that, in order to understand the pain or joy of another person, the therapist must undergo the experience he perceives. In other words, the sharing of the patient's living experience does not automatically take place through a kind of merging with the therapist's emotions. On the other hand, it cannot be exercised instrumentally by the therapist as a chameleon-like attitude designed to put the patient at ease and make him feel understood or accepted nor can it be used as a means to resignify the patient's expressions, thereby reducing the other's experience to one's own, based on the assumption that in such a way it will be possible to get to know the patient better than he knows himself. Freud's *Moses* provides an indirect example of this.

As Scheler observes (Scheler 1970, p. 45), "In true unalloyed commiseration and rejoicing there is no state of sorrow or joy in oneself" (*Mitleiden und Mitfreuen*); rather, this sharing is always oriented toward—and guided by—the unique suffering and joy of the other. In other words, sharing the other's emotions means becoming involved in his way of feeling *qua* other, according to his peculiar individuality, without losing sight of oneself. Going along with the other's lived experience is a fundamental methodological attitude which is a constitutive part of the method of formal indication.

If we understand healing (*therapeuein*) as that form of investigation which *serves* a patient's story, oriented and guided by the patient's meaningful possibilities, then compassion, goodwill, solicitude, kindness, tolerance, and all the other forms of being respectfully involved in the unique experience of the other[16] within a therapeutic framework are strictly methodical, insofar as they stem from the practice

[16] As Ricoeur notes, only respect justifies sympathy (1954).

of going along with the other in the search for traces of the self. These qualities constitute the relational context focused on the unique existence of the other, which enables the therapist to co-enact lived experience together with the patient, as this is originally offered to him. To quote Scheler again, "True fellow-feeling betrays itself in the very fact that it includes the existence and character of the other person as an individual, as part of the object of commiseration or rejoicing. Can one rejoice more profoundly with a person than at his being the perfect, talented, unspoiled individual that he is? or commiserate more deeply than for his having to suffer as he does, being the sort of man he is?" (1970, p. 39).

This all-round sensitivity must therefore be nourished and practiced as a key element of the method, which in this respect shows an ethical significance that manifests itself precisely through the *subtilitas applicandi* which the therapist is capable of. This sensitivity also encompasses the "decisions regarding different relationship stances, including when and how to be directive, supportive, reflective, cold, warm, tepid, formal or informal" (Lazarus 1993, p. 404)—the decisions made by a well-meaning friend with a passion for healing, we might say. And it is only in the context of a method which has a concrete beginning and is always oriented by where the client is at that it is possible to newly take up—from a different perspective—the reflections made by the later Rogers on what is inappropriately described as the empathic process. These reflections thus acquire a clear and rigorous meaning, instead of being left at the mercy of authentic chameleons. Despite its length, the following quotation by Rogers is worth singling out:

> The way of being with another person which is termed empathic has several facets. It means entering the private perceptual world of the other and becoming thoroughly at home in it. It involves being sensitive, moment to moment, to the changing felt meanings which flow in this other person, to the fear or rage or tenderness or confusion or whatever, that he/she is experiencing. It means temporarily living in his/her life, moving about in it delicately without making judgments, sensing meanings of which he/she is scarcely aware, but not trying to uncover feelings of which the person is totally unaware, since this would be too threatening. It includes communicating your sensings of his/her world as you look with fresh and unfrightened eyes at elements of which the individual is fearful. It means frequently checking with him/her as to the accuracy of your sensings, and being guided by the responses you receive. You are a confident companion to the person in his/her inner world. By pointing to the possible meanings in the flow of his/her experiencing you help the person to focus on this useful type of referent, to experience the meanings more fully, and to move forward in the experiencing. To be with another in this way means that for the time being you lay aside the views and values you hold for yourself in order to enter another's world without prejudice. In some sense it means that you lay aside your self and this can only be done by a person who is secure enough in himself that he knows he will not get lost in what may turn out to be the strange or bizarre world of the other, and can comfortably return to his own world when he wishes (Rogers 1975).

Entering the private perceptual world of the patient, being sensitive moment to moment to his changing felt meanings, living in his life without making judgments, communicating your sensings of his world and checking with him as to the accuracy of your sensings, being guided by the responses you receive, and laying aside the views and values you hold for yourself, these precepts, which reveal Roger's remarkable sensitivity as a therapist, are destined to remain nothing but words unless they are made an integral part of a method which centers the treatment on the actual (and hence historical) existence of the person and unless the therapist's respect for the patient's individuality turns his finesse of mind into the heart of the therapeutic process and of psychotherapy, understood as a practical science.

In this respect, agreements on the therapeutic goals, just like the consensus on the therapeutic tasks, cannot be set in advance because they cannot be known a priori but must be redefined as the therapeutic process unfolds in its uniqueness. From their enactment it will be possible to gradually derive a roadmap for the treatment; conversely, on the basis of the trajectory of the treatment, it will be possible to formulate goals and tasks. This science sets out from a concrete situation in an attempt to understand what is expected of it, thus requiring the practitioner to enact it in an autonomous way. Unlike in the case of the theoretical sciences, the method is therefore defined by the "object" of enquiry in its uniqueness: an object whose character consists in being incomplete and therefore open to the possibility of being other than what it is and which directly engages the person who establishes a relationship with it, constantly becoming actualized in facts. The *subtilitas applicandi* is therefore unthinkable without *compassion*.

Phenomenological psychology and psychotherapy envisage a kind of knowledge and mode of treatment that are very different from those of medicine. The latter rests on an ontology which accesses life on the basis of that invariable logic which is grounded in the organs of the body, the mutual relations between them, and the way in which death interrupts their functions (Bichat 1799). This ontology—the ontology of death (1966), as Hans Jonas termed it—goes so far as to conceive of the organism from the perspective of the dissection table, which was established as the foundation first of physiology and later of psychology. This explains why the history of the development of psychotherapy is so closely entwined with medicine. This ontology also helps understand why the five components (complaint, explanation, mechanism of change, therapeutic ingredients, and specificity), which Walpert points to as the constitutive elements of a model common to both disciplines, endure as the medical model in both medicine and psychotherapy (Wampold et al. 2001; Imel and Wampold 2008).

But if the thematic field of psychology is rethought in the light of the motility of life, psychotherapy requires not only a method which can remain faithful to this essence but also—at the same time and inevitably—a new research dimension in the form of a new engagement with the life sciences and especially the neurosciences. This engagement must be developed in keeping with the tradition running from Helmholtz to Varela, a tradition which has sought to identify the mechanisms governing experience by examining the second-person accounts of the experiences under investigation. As we have seen, along this trajectory, an encounter has taken place between cognitive science and transcendental phenomenology, which has moved the former toward a perspective that is more consistent with the views of Husserl and Merleau-Ponty on intentionality, intersubjectivity, action, and embodiment (Gallagher 2012), yet has done so by taking the egological position as man's guiding idea. This perspective still resonates with the view of the living organism which led Johannes Müller to trace all forms of exteriority back to internal nervous impulses.

But if, as a positive science, psychology no longer shares the same assumptions as physiology, those which led Müller to state that *nemo psychologus nisi physiologus*, how are those life sciences based on the ontology of the dissection table affected by the fact that the incompleteness of man's being-there can hardly be reconciled with a view of the living organism as a closed system which remains the same down to its final end? What new possible relations can emerge from this ontological and methodological freeing of psychology from the natural sciences?

References

Alessandrelli M (2015) L'ordito Della Vita: Esperienza Del Divino Nella Poetica Di Aristotele. Lexicon Philosophicum: Int J Hist Texts Ideas 3

Anderson T, Ogles BM, Patterson CL, Lambert MJ, Vermeersch DA (2009) Therapist effects: facilitative interpersonal skills as a predictor of therapist success. J Clin Psychol 65:755–768

Arciero G (1989) From epistemology to ontology: a new age of cognition. Paper presented at the AAAS. S.Francisco

Arciero G (2006) Sulle tracce di sé. Bollati Boringhieri, Torino

Arciero G, Bondolfi G (2009) Selfhood, identity and personality styles. John Wiley & Sons, Chichester

Bachelor A (2013) Clients' and therapists' views of the therapeutic alliance: similarities, differences and relationship to therapy outcome. Clin Psychol Psychother 20(2):118–135

Baldwin SA, Wampold BE, Imel ZE (2007) Untangling the alliance-outcome correlation: exploring the relative importance of therapist and patient variability in the alliance. J Consult Clin Psychol 75:842–852

Bandura A (1997) Self-efficacy: the exercise of self-control. Macmillan, New York

Beutler LE, Malik M, Alimohamed S, Harwood TM, Talebi H, Noble S et al (2004) Therapist variables. In: Lambert MJ (ed) Handbook of psychotherapy and behavior change, 5th edn. Wiley, New York

Bichat, Marie - François - Xavier (1799) Traité des membranes en général et de diverses membranes en particulier. Richard, Caille et Ravier, Paris

Bohart AC, Greaves AL (2013) The client in psychoterapy. In: Lambert M (ed) Bergin and Garfield's Handbook of psychotherapy and behavior change, 6th edn. Wiley, New York, pp 219–257

Bohart, A. C., & Tallman, K. (2010). Clients: the neglected common factor in psychotherapy. In B. Duncan, S. Miller, B. Wampold y M. Hubble (Eds), The heart and soul of change: delivering what works in therapy (2nd ed., pp. 83-111). Washington: American Psychological Association.

Bordin ES (1979) The generalizability of the psychoanalytic concept of the working alliance. Psychotherapy 16(3):252

Bromberg PM (1998) Standing in the space. Analytic Press, Hillsdale

Burke K (1969) A rhetoric of motives, vol 111. University of California Press, Berkeley

Carroll KM, Nich C, Rounsaville BJ (1997) Contribution of the therapeutic alliance to outcome in active versus control psychotherapies. J Consult Clin Psychol 65(3):510

Dush DM, Hirt ML, Schroeder HE (1983) Self-statement modification with adults: a meta-analysis. Psychol Bull 94(3):408

Elliott R, James E (1989) Varieties of client experience in psychotherapy: an analysis of the literature. Clin Psychol Rev 9(4):443–467

Elliott R, Bohart AC, Watson JC, Greenberg LS (2011) Empathy. In: Norcross JC (ed) Psychotherapy relationships that work: evidence-based responsiveness. Oxford University Press, Oxford

Foucault M (2001) L'herméneutique du sujet. Cours au Collège de France 1981–1982. Gallimard-Seuil, Paris

Frank JD (1961) Persuasion and healing. Johns Hopkins University Press. Garner, Baltimore

Frank JD (1987) Psychotherapy, rhetoric, and hermeneutics: implications for practice and research. Psychotherapy 24(3):293
Frank JD (1995) Psychotherapy as rhetoric: some implications. Clin Psychol Sci Pract 2(1):90–93
Frank JD, Frank JB (1991) Persuasion and healing: a comparison study of psychotherapy. JHU Press, Baltimore
Freud S (1905) Jokes and their relation to the unconscious. F. Deuticke, Leipzig
Gadamer HG (1975) Truth and method (1960). Crossroad, New York
Gallagher S (2012) In defense of phenomenological approaches to social cognition: interacting with the critics. Rev Philos Psychol 3(2):187–212
Garfield SL (1995) Psychotherapy: an eclectic-integrative approach. John Wiley & Sons, New York
Greenberg LS, Johnson SM (1986) Affect in marital therapy. J Marital Fam Ther 12(1):1–10
Greenberg LS, Safran JD (1989) Emotion in psychotherapy. Am Psychol 44(1):19
Greenberg RP, Constantino MJ, Bruce N (2006) Are patient expectations still relevant for psychotherapy process and outcome? Clin Psychol Rev 26(6):657–678
Greisch J (1994) Le phénomène de la chair: un «ratage» de Sein und Zeit. In: Floviral G (ed) Dimensions de l'exister. Etudes de anthropologie philosophique, Bibliothèque Philosophique de Louvain. Peeters, Leuven
Grencavage LM, Norcross JC (1990) Where are the commonalities among the therapeutic common factors? Prof Psychol Res Pract 21(5):372
Guidano VF (1987) Complexity of the self: a developmental approach to psychopathology and therapy. Guilford Press, New York
Heidegger, M. (1926). On the essence of truth (Pentecost Monday) (TJ Kisiel, Trans.). Becoming Heidegger: on the trail of his early occasional writings, 1927, pp 275–288.
Heidegger GA 2 1977
Heidegger GA 58 1992
Heidegger GA 9 (1976)
Hoag MJ, Burlingame GM (1997) Evaluating the effectiveness of child and adolescent group treatment: a meta-analytic review. J Clin Child Psychol 26(3):234–246
Horvath AO, Bedi RP (2002) The alliance. In: Norcross JC (ed) Psychotherapy relationships that work: therapist contributions and responsiveness to patients. Oxford University Press, Oxford
Horvath AO, Greenberg LS (1994) The working alliance: theory, research, and practice, vol 173. John Wiley & Sons, New York
Horvath AO, Luborsky L (1993) The role of the therapeutic alliance in psychotherapy. J Consult Clin Psychol 61(4):561
Horvath AO, Symonds BD (1991) Relation between working alliance and outcome in psychotherapy: a meta-analysis. J Couns Psychol 38(2):139
Horvath AO, Del Re AC, Flückiger C, Symonds D (2011) Alliance in individual psychotherapy. Psychotherapy 48(1):9–16
Hubble MA, Duncan BL, Miller SD (1999) The heart and soul of change: what works in therapy. American Psychological Association, Washington
Hubble MA et al (2010) Introduction. In: Duncan BL, Miller SD, Wampold BE, Hubble MA (eds) The heart and soul of change: delivering what works in therapy. American Psychological Association, Washington
Husserl E (2006) Logische Untersuchungen. Ergänzungsband. Zweiter Teil.: Texte für die Neufassung der VI. Untersuchung. Zur Phänomenologie des Ausdrucks und der Erkenntnis (1893/94-1921), vol 20. Springer, Berlin
Imel Z, Wampold B (2008) The importance of treatment and the science of common factors in psychotherapy. In: Brown SD, Lent RW (eds) Handbook of counseling psychology. Wiley, New York, pp 249–266
Johansson H, Jansson JÅ (2010) Therapeutic alliance and outcome in routine psychiatric out-patient treatment: patient factors and outcome. Psychol Psychother Theory Res Pract 83(2):193–206
Kim DM, Wampold BE, Bolt DM (2006) Therapist effects in psychotherapy: a random-effects modeling of the National Institute of Mental Health Treatment of Depression Collaborative Research Program data. Psychother Res 16(02):161–172
Kirsch I (1985) Response expectancy as a determinant of experience and behavior. Am Psychol 40(11):1189

Kisiel T (1993) The genesis of Heidegger's being and time. University of California Press, Berkeley
Krupnick JL, Elkin I, Collins J, Simmens S, Sotsky SM, Pilkonis PA, Watkins JT (1994) Therapeutic alliance and clinical outcome in the NIMH Treatment of Depression Collaborative Research Program: preliminary findings. Psychotherapy 31(1):28
Krupnick JL, Sotsky SM, Simmens S, Moyer J, Elkin I, Watkins J, Pilkonis PA (1996) The role of the therapeutic alliance in psychotherapy and pharmacotherapy outcome: findings in the National Institute of Mental Health Treatment of Depression Collaborative Research Program. J Consult Clin Psychol 64(3):532
Lambert MJ (1986) Some implications of psychotherapy outcome research for eclectic psychotherapy. J Integr Elect Psychother 5(1):16–45
Lambert MJ (1992) Implications of outcome research for psychotherapy integration. In: Norcross JC, Goldstein MR (eds) Handbook of psychotherapy integration. Basic Books, New York, pp 94–129
Lambert M, Bergin AE (1994) The effectiveness of psychotherapy. In: Bergin A, Garfield S (eds) Handbook of psychotherapy and behaviour change, 4th edn. Wiley, New York, pp 143–189
Lambert JJ, Ogles BM (2004) The efficacy and effectiveness of psychotherapy. In: Lambert MJ (ed) Bergin and Garfield's handbook of psychotherapy and behavior change, 5th edn. Wiley, New York, pp 139–193
Lazarus AA (1993) Tailoring the therapeutic relationship, or being an authentic chameleon. Psychotherapy 30(3):404
Luborsky L, McLellan AT, Woody GE, O'Brien CP, Auerbach A (1985) Therapist success and its determinants. Arch Gen Psychiatry 42(6):602–611
Luborsky L, Diguer L, Seligman DA, Rosenthal R, Krause ED, Johnson S et al (1999) The researcher's own therapy allegiances: a "wild card" in comparisons of treatment efficacy. Clin Psychol Sci Pract 6(1):95–106
Luborsky L, Rosenthal R, Diguer L, Andrusyna TP, Berman JS, Levitt JT et al (2002) The dodo bird verdict is alive and well—mostly. Clin Psychol Sci Pract 9(1):2–12
Lutz W, Leon SC, Martinovich Z, Lyons JS, Stiles WB (2007) Therapist effects in outpatient psychotherapy: a three-level growth curve approach. J Couns Psychol 54(1):32
MacIntyre A (1984) After virtue: a study in moral theory. University of Notre Dame Press, Notre Dame, IN
Mahoney MJ (1984) Integrating cognition, affect, and action: a comment. Cogn Ther Res 8(6):585–589
Martin DJ, Garske JP, Davis MK (2000) Relation of the therapeutic alliance with outcome and other variables: a meta-analytic review. J Consult Clin Psychol 68(3):438
Mazzola V, Latorre V, Petito A, Gentili N, Fazio L, Popolizio T et al (2010) Affective response to a loved one's pain: insula activity as a function of individual differences. PLoS One 5(12):e15268
Mohr DC (1995) Negative outcome in psychotherapy: a critical review. Clin Psychol Sci Pract 2(1):1–27
Norcross JC (2010) The therapeutic relationship. In: Duncan BL, Miller SD, Wampold BE, Hubble MA (eds) The heart and soul of change: delivering what works in therapy. American Psychological Association, Washington
Norcross JC, Lambert MJ (2005a) What should be validated? The therapy relationship. In: Norcross J, Levant R, Beutler L (eds) Evidence-based practices in mental health: debate and dialogue on the fundamental questions. American Psychological Association Press, Washington, pp 209–218
Norcross JC, Lambert MJ (2005b) The therapy relationship. In: Norcross JC, Beutler LE, Levant RF (eds) Evidence-based practices in mental health: debate and dialogue on the fundamental questions. American Psychological Association, Washington
Okiishi J, Lambert MJ, Nielsen SL, Ogles BM (2003) Waiting for supershrink: an empirical analysis of therapist effects. Clin Psychol Psychother 10(6):361–373
Orlinsky D, Grave D, Parks B (1994) Process and outcome in psychotherapy. In: Bergin A, Garfield S (eds) Handbook of psychotherapy and behaviour change, 4th edn. Wiley, New York
Paley G, Shapiro DA (2002) Lessons from psychotherapy research for psychological interventions for people with schizophrenia. Psychol Psychother Theory Res Pract 75(1):5–17

Priebe S, Richardson M, Cooney M, Adedeji O, McCabe R (2010) Does the therapeutic relationship predict outcomes of psychiatric treatment in patients with psychosis? A systematic review. Psychother Psychosom 80(2):70–77
Putnam FW (1988) The switch process in multiple personality disorder and other state-change disorders. Dissociation 1(1):24–32
Rachman S (1981) The primacy of affect: some theoretical implications. Behav Res Ther 19(4):279–290
Ricoeur P (1954) Sympathie et respect: phénoménologie et éthique de la seconde personne. Revue de Métaphysique et de Morale 59(4):380–397
Ricoeur P (1992) Oneself as another. University of Chicago Press, Chicago
Ricœur P (2001) Le juste 2. Ed. Esprit, Paris, p 9
Robinson LA, Berman JS, Neimeyer RA (1990) Psychotherapy for the treatment of depression: a comprehensive review of controlled outcome research. Psychol Bull 108(1):30
Rogers CR (1957) The necessary and sufficient conditions of therapeutic personality change. J Consult Psychol 21(2):95
Rogers CR (1975) Empathic: an unappreciated way of being. Counsel Psychol 5(2):2–10
Rosenzweig S (1936) Some implicit common factors in diverse methods of psychotherapy. Am J Orthopsychiatry 6(3):412
Safran JD, Christopher Muran J (1996) The resolution of ruptures in the therapeutic alliance. J Consult Clin Psychol 64(3):447
Safran JD, Muran JC (2006) Has the concept of the therapeutic alliance outlived its usefulness? Psychotherapy 43(3):286
Safran JD, Muran JC, Eubanks-Carter C (2011) Repairing alliance ruptures. Psychotherapy 48(1):80
Scheler MF (1970) The nature of sympathy. Transaction Publishers, London
Shapiro DA, Shapiro D (1982) Meta-analysis of comparative therapy outcome studies: a replication and refinement. Psychol Bull 92(3):581
Sloane RB, Staples FR, Cristol AH, Yorkston NJ, Whipple K (1975) Psychotherapy vs. behavior therapy. Harvard University Press, Cambridge
Smith ML, Glass GV, Miller TI (1980) The benefits of psychotherapy. Johns Hopkins University Press, Baltimore
Strupp HH, Fox RE, Lessler K (1969) Patients view their psychotherapy. Johns Hopkins Press, Oxford
Wampold BE (2001a) Contextualizing psychotherapy as a healing practice: culture, history, and methods. Appl Prevent Psychol 10(2):69–86
Wampold BE (2001b) The great psychotherapy debate: model, methods, and findings. Lawrence Erlbaum Associates, Mahwah
Wampold BE (2005) Establishing specificity in psychotherapy scientifically: design and evidence issues. Clin Psychol Sci Pract 12(2):194–197
Wampold BE (2007) Psychotherapy: the humanistic (and effective) treatment. Am Psychol 62(8):857
Wampold BE (2010) The basics of psychotherapy: an introduction to theory and practice. American Psychological Association, Washington, DC
Wampold BE (2011) Qualities and actions of effective therapists. American Psychological Association, Washington
Wampold BE, Bhati KS (2004) Attending to the omissions: a historical examination of evidence-based practice movements. Prof Psychol Res Pract 35(6):563
Wampold BE, Brown GSJ (2005) Estimating variability in outcomes attributable to therapists: a naturalistic study of outcomes in managed care. J Consult Clin Psychol 73(5):914
Wampold BE, Ahn H-n, Coleman HLK (2001) Medical model as metaphor: old habits die hard. J Counsel Psychol 48(3):268–273
Wampold BE, Imel ZE, Bhati KS, Johnson-Jennings MD (2006) Insight as a common factor. In: Castonguay LG, Hill CE (eds) Insight in psychotherapy. American Psychological Association, Washington, pp 119–139

Part III

The Renewed Pact: Corporeality, Experimentation, and the Care of Self

Corporeality and Ipseity

8

8.1 The Body, the Corporeal, and Phenomenology

The architecture of the present work has revolved around the phenomenon of the motility of life: a motility which over the course of our investigation we have described in a variety of ways. At this stage, however, when confronted with the positive sciences of life, it is necessary to examine motility from a hitherto unexplored perspective. The motility of life constantly takes shape in the flesh of a body living in an environment which, in turn, is formed within this relation.

What we have, then, is a unique phenomenon: while corporeality, which is to say carnality, is understood as a body whose limits are in each case enacted according to a way of living which intertwines with the world it encounters, the body—whose limits coincide with one's own skin—withdraws from the domain of experience. The body of physiology, in other words, participates in human existence at the price of disappearing—this is the parameter for normality.

As Canguilhem (1966) puts it, quoting René Leriche, "health is life lived in the silence of the organs" (p. 65). What this means is that unawareness of one's own body coincides with the condition of health, whereas the coming of the body to our awareness marks a limiting condition, thereby threatening one's well-being.

I am sitting at my desk when I realize that I have a train to catch in 15 min but have not bought a ticket yet. I have a work appointment I cannot miss. I thought I had plenty of time but discover that I am late. I head to the station. I start walking faster. I am afraid I will miss the train, so I start running. I dash up the stairs leading to the platform. A train has just pulled in. It is the train I need to catch.

In this sequence, the body—viewed from the outside as a running, physical body—is the precondition, so to speak, for my catching the train on time. And yet, the body never revealed itself during my journey. It "enabled" every stage of my rush to the station without ever showing itself.

If, as we have seen, the structure of a phenomenon is given by the sense of relation, the sense of content, and the sense of enactment, to what extent can phenomenology regard the body as a phenomenon? In my rush to the station, my body is not

G. Arciero et al., *The Foundations of Phenomenological Psychotherapy*,
https://doi.org/10.1007/978-3-319-78087-0_8

just my physical body but rather takes part in my relationship with the world: in connection to the need to make it to my appointment on time, the (withdrawn) body manifests itself in the speed which enables me not to miss my train. Can the body therefore be said to manifest itself as a phenomenon in my making haste?

The answer is certainly no, if we conceive the flesh starting from the notion of presence. The body, that is, does not appear as a phenomenon in my heading off to the station or my appointment, if we grasp it as a body which has increased its heart rate, muscular tension, or perspiration in making haste. The body is only accessible as a phenomenon if we grasp the corporeal starting from its enactment as an opening up to what we encounter. The gaze which sees the body as an object on hand cannot account for the phenomenon of carnality: for this is not reducible according to any form of egology, including the organismic one. Each person is corporeal in relation to what he occupies himself with or worries about—i.e., in relation to what concerns him. The corporeal only appears within the framework of my relationship with the world I encounter and in connection to my being together with other people. Therefore, grasping the body as a thing among other things, or as a phenomenon, presupposes a way of conceiving the self and, more generally, man's mode of relating to his environment.

In living the world in view of this or that thing, human corporeality occurs as a bodying forth (*Leiben*) based on modes of being that emerge in accordance with what we address, shaping our life at any given moment. In other words, in being constantly open to what meaningfully meets him, each human being extends the habitability of his own world, by perceiving and understanding the meaningfulness of whatever addresses him. The limits of the corporeal thus coincide with those of our opening up to the world and constantly change with it.

We have already seen how the situations we encounter over the course of our life are always grasped according to an emotional position, which provides an orientation for us in the environing world (*Umwelt*) with respect to what is "to be done" and, at the same time, a sense of our condition (be it pleasant or unpleasant, light or burdensome, etc.). It is always in accordance with this being emotionally situated that the domains in which we "get things done" are disclosed. Hence, movement, defined as an actual mode of conducting oneself in the world, is emotionally determined as an orienting of oneself, as the factical finding of a position within the domain disclosed by one's encounter with a situation, and limited by the bodying forth (*Leiben*) of one's body in relation to the current event. In this respect, movement is related to spatiality. Quoting Heidegger, we will refer to this phenomenon of conjunction between being-in-the-world and corporeal situatedness—encapsulated by the fact of comporting oneself in such and such a way within a space disclosed by the encounter—as the corporeal phenomenon (*Leibphaenomen*) (GA 89 n.d.).

From this perspective, corporeality has an eccentric meaning, insofar as it accompanies our being-in-the-world in all of its manifestations. The corporeal is not the center which marks the point zero of our orientation, but rather what anchors the possibilities on the basis of which our direction concretely takes shape in each circumstance. It originates from the relationship (*Bezugssinn*) with those meaningful contexts (*Gehaltssinn*) which it encounters and which in each case determine its concrete enactment (*Vollzugssinn*). Hence, a phenomenology of the living body must rest on the foundation of actual living.

The crucial guiding element here is the fact that our corporeal nature—the living body—takes shape through our being-in-the-world. Hence, no phenomenology of perception can gain access to a phenomenology of the flesh: for a phenomenon is a mode of being, not a presence.[1] Hence, it is by being oriented toward what meets us in each particular situation, which is to say toward the meaningfulness we encounter, that the body gets a feel of things, so to speak, and contributes to enacting existential possibilities. The phenomenon of spatiality, the topic of the famous §23 of the first section of *Being and Time*, reveals this crucial aspect: the corporeal is determined precisely by its orientation with respect to what man engages with, by the meaningful directions on the basis of which the places, territories, and itineraries of our everyday transactions and, more generally, the spatiality of our environing world take shape.

8.2 Spatiality and Its Pathology

In a short popular essay written to take a stance on the controversy surrounding Spinozism (*Pantheismusstreit*), which had fired the souls of German intellectuals in the second half of the eighteenth century (creating a rift between those who invoked a leap of faith as the only means to approach the Absolute and those who instead put their trust in reason), Kant (1786) endeavors to show how we can orient ourselves in the darkness of thought when metaphysical research is no longer capable of providing any guidelines; hence the title of the essay "What Does It Mean to Orient Oneself in Thinking." The interesting thing about this essay—and what explicitly links it to §23 of *Being and Time*—is the fact that, in order to establish an analogy with orienting oneself in thought, Kant sets out from a geographical and, later, mathematical conceptualization of this notion.

Leaving aside the idea of finding one's bearings in thought, which is the first term of the analogy and which constitutes a reference to the dispute in which the philosopher is required to take sides (this being the context for the writing of the essay), it is interesting to focus on the second term of the analogy: orienting oneself in a geographical and mathematical sense.

After a measured introduction, Kant opens his argument by emphasizing that, although orienting oneself literally means identifying all other points on the basis of a cardinal one, the east, the reference to the sun and stars alone is not enough. Unless I refer to the difference I perceive between right hand and left hand—i.e., the fact of being able to determine a difference in positioning a priori—I will not be able to tell whether the west is to be situated to the right or to the left of a horizon marking the south. Kant writes, therefore, that "even with all the objective data of

[1] The structure of perception, in other words, does not take shape through hyletic data which are subsequently to be assigned meaning.

the sky, I orient myself geographically only through a subjective ground of differentiation" (p. 8).[2]

In the following passage, the geographical concept of orientation is extended to include space in general. In his discussion of this second point, Kant repeats the argument he had formulated for the first. In this case too, the objective guidelines present in any given space are not enough to provide an orientation, as their position can only be determined on the basis of a subjective criterion. For example, I could orient myself in the darkness, in a room familiar to me, by touching an object whose position I recall, even if someone had played the trick on me of moving all the objects on the right to the left and vice-versa, while preserving their order. In this case too, I could determine their position "through the mere feeling of a difference between my two sides, the right and left" (p. 9). It goes without saying that this orientation, based on the structure of a subject without any reference to the world he finds himself in, has shaped the course of the study of spatiality in the field of the natural sciences.

The issue we are interested in here, however, is a different one: if our starting point is not the subject, but ipseity, corporality—which in each case is constituted through one's relationship with the world—necessarily determines and circumscribes the directions of one's orientation, not because the process of orienting oneself is centered on the body but rather because the opposite is the case: if we are always already oriented toward the meaningfulness we encounter, it is through this relationship that corporeality takes shape, in the sense that in each case it emerges with an orientation toward those facts that are being disclosed. Indeed, this is the only way in which we can distinguish between right and left, close and far, and high and low. In discussing the example of the room which Kant uses to establish the subjective criterion of orientation, Franco Volpi (1996) rightly notes: "In order to orient myself, 'the mere feeling of a difference' between my two sides is of no use to me unless I recognize a given object, unless I 'have impressed its location upon my memory', as Kant states in passing. But what all this means is that I necessarily orient myself according to a being-involved-with a world 'familiar' to me and on the basis of this" (p. 35).

The phenomenon of spatialization, then, reveals one essential fact: the corporeal has its foundation in the relationship I find myself in with this or that thing. Hence, corporeality is oriented insofar as my relationship with the world is originally directed toward something which concerns me—which addresses me in its meaningfulness. The characteristics of this relatedness cannot be located in the physical body situated at a given point in space, but are defined by my opening up and being receptive to what I encounter. This relation is always a spatial one too, and it is this spatiality of existence which provides the possibilities for the body to orient itself. Hence, Heidegger can argue that corporeality "basically is *not* inanimate matter but

[2]Also orientiere ich mich geographisch bei allen objektiven Datis am Himmel doch nur durch einen subjektiven Unterscheidungsgrund p 353 *Immanuel Kants Werke. Band IV. Schriften von 1783–1788. Herausgegeben von Dr. Artur Buchenau und Dr. Ernst Cassirer. Berlin: Bruno Cassirer 1913.*

a domain of that non-objectifiable, optically invisible capacity to receive-perceive the meaningfulness of what it encounters, which constitutes the whole being-there" (GA 89 n.d., p. 293). It is therefore necessarily co-determined by that relatedness which is accomplished through the various modes of being-there.

Corporeality, which bodies forth according to the various modes of conducting oneself in the world, in each case enters into a close or distant relationship with what occurs, mobilizing and rearranging with respect to the event the sphere of one's motility and receptiveness—seeing, hearing, smelling, tasting, and touching. As Greisch (1999) puts it, evoking the Prologue of John, inhabiting is the only measure and limit of corporeality.[3]

Inhabiting the world always implies an orienting of oneself which takes the concrete form of a mode of being involved with others and things by emphasizing certain aspects and according to specific sensory and motor combinations. Clearly, groping about in a dark room—as in Kant's example—is different from being captivated by the voices of a female choir streaming out of a church at the end of an alleyway in Rome, gazing raptly at the stars (what earned Thales a snide remark from a Thracian maidservant), being carried away by the taste of a *madeleine* (as in the famous passage from *Du côté de chez Swann*) or sensing the familiar scent of one's home. Each of these modes of being constitutes—together with all others and in view of what meaningfully becomes present—an overall configuration which cannot be reduced to an organismic whole, as suggested by *Gestaltpsychologie*, but which rather represents a mode of openness that grounds the possibilities for relatedness of ipseity. This openness is at the same time a mode of finding oneself and being emotionally disposed in a situation—in a world.

In its supremacy, each of these sensory modes carries with it a way of being involved with and a related feeling of situatedness. Touch, for example, implies an awareness of proximity, which by *fully* anchoring corporeality to what is touched erases the hand as a separate organ—a very different perception from the distant view of those stars which prevented Thales from noticing the well he was about to fall into! It was this distance—which caused the eyes to disappear and, with them, all perception of the body—that provided that sense of detachment which the ancients cultivated as a prerequisite for acquiring stable knowledge.

The different extension and use of touch, sight, and all other sensory modes show that each form of receptiveness is rooted in our position—be it close or distant—with respect to what we encounter. From this perspective, corporeality can only come about because it is rooted in the spatial relation of human being-there in the world, which is always emotionally situated.[4] This is particularly evident in the case of agoraphobia, a disorder characterized—according to the description of it provided by Westphal (1871)—by the fear of open spaces.[5] "For several years—Westphal writes—patients have repeatedly come to me with the peculiar complaint

[3] "And the Word became flesh and took up residence among us" (John 1:14).

[4] All this could easily be put to the test by demonstrating the variation of peripersonal space in relation to the emotional openness of a person.

[5] For a review of diagnostic problems and the DSM categorization, see Asmundson et al. (2014).

that it is impossible for them to go into open spaces and down certain streets, and because of this fear their freedom of movement is disturbed." Indeed, for various different reasons, starting at a given moment of instability in their lives, agoraphobic patients begin to thematize certain living spaces and the possibility of moving within them, anticipating their incapacity to cope through a suitable motor-emotional response to the situations which those spaces might engender (Arciero and Bondolfi 2009). The crucial defining feature of agoraphobia, then, is a vicious circle which, starting from an initial anxiety, anticipates the current and hypothetical impossibility of suitably moving within certain spaces, thereby causing even greater anxiety and creating a vicious circle.

Westphal describes the symptomatology of one of his patients as follows: "If he wishes to walk across an open space and he is seized by the feeling of anxiety, then it seems to him that the square is so wide that he will never be able to cross it, and his steps become shorter and shorter so that he does not move forward at all." Westphal identifies the very core of this disorder and sheds light on the phenomenon of spatialization, albeit without grasping its genesis. The crucial fact, which is borne out by his description, is that the anxious disclosure of given spaces (perceived in terms of the uncontrollability of one's being-involved-with) co-determines corporeality, generating a redefinition of the sense of distance and proximity and of the related capacity to move and hence orient oneself with respect to the space thus disclosed. People's emotional position with respect to the events marking their life journeys, therefore, always also coincides with an integral way of finding oneself in the world together with others and of being disposed toward the ongoing situation. This situatedness discloses a space—which is also a meaningful domain—and orients one's movement, directing one's actions.

Corporeality, then, finds "its spaces" based on our relationship with this or that thing—not the other way round. From a broader perspective, corporeality presupposes the fact of being oriented according to a meaningful direction, while it "essentially and intimately" participates in this orientation itself by regulating our movement according to the meaningfulness we encounter. Within the context of this existential intimacy, "material corporeality"—the flesh—in each case becomes such, and our own, on the basis of those possibilities of relating (i.e., those meaningful possibilities) according to which ipseity finds a position in the world and comes to itself.

Patočka writes: "The personal body is not a thing in objective space: it is a life which *spatially* exists in itself, *produces* its own localization, and makes itself spatial" (1995, p. 59). Even the simplest hand gesture, grasping, can only fully be understood if it is viewed not merely as a mechanical gesture, but in relation to a meaningful context which is always determined by how the person is bringing forth his active relationships with the world (Mazzola et al. 2013): "Like everything else bodily, the human hand belongs so directly to human being-in-the-world that it is determined in everything it is and does by how its owner is relating to what he is at that moment encountering" (Boss 1977, p. 104). As an element of existence, the flesh thus participates in ipseity, in each case occurring together with it.

8.3 Corporeality and Presentification

Corporeality is what enables ipseity to find itself in the world and feel emotionally situated within it and hence always exposed to the facts, things, and people it encounters, as well as all the entities by which it is governed. The way in which I perceive myself—the emotional tones according to which I constantly experience myself as being anchored to my corporeality—corresponds to the way in which in each case I am involved with others and with things, to how I am factically engaged in the world. The most evident consequence of this, and also the most disturbing for theoretical-scientific forms of knowledge, is that, if the way in which I perceive myself coincides with the way in which I am always my body, carnality always lies beyond my corporeal presence and hence is never fully accessible through any organicist reduction.

My corporeality is always "animated" within the framework of its relation to the existence which gives it its eccentric character; therefore, it is always open to the other and the situation it relates to. Anything corporeal, then, is not the expression of an interiority, but rather a mode of comportment which in all cases is disclosed by the ongoing situation and the other with whom I establish a relationship. Dispelling what has become a common assumption in contemporary studies on embodiment, it is crucial to understand that ipseity does not stand in contrast to otherness according to the bipolar relationship which lies at the heart of the metaphysical metaphor of intersubjectivity, but that being with the other in each case corresponds to a determination, to a position of one's own being-there. If the phenomenon of the flesh is conceived as the enactment (*Vollzugssinn*) of a way of relating (*Bezugssinn*) to things and other people, based on the opening up to a common world (*Gehaltssinn*) of which the corporeal is a constitutive element, it is once again easy to grasp why the understanding of the other is only made possible by the "interpersonal" structure of corporeality.

It is only because corporeality always already participates in my relationship with the other person—even when the other is neither present nor perceived—that I can directly grasp the other's experience and situate myself with respect to him without the need to bring any form of empathic mediation into play. "Dasein in itself is essentially Being-with," Heidegger writes in a famous passage of *Being and Time* (GA 2 1977, p. 117). In other words, the human being does not become Being-with *after* discovering its world or *after* realizing that it exists in the presence of another entity that shares its disclosing nature. Like every mammal which comes into the world, man is always constitutively connected to otherness, in the flesh, through a co-originary relationship: through the gestation of the other in the same, to quote Levinas (1979).[6] If we examine from this perspective the theories about social cognition that are now dominant within the neurosciences—action understanding and mind reading—and which seek to account for the mechanisms underpinning the understanding of the other, we can see that

[6] It is from this perspective that we have understood and reinterpreted Bowlby's studies on attachment within the framework of psychology (Arciero 1994).

they operate within the context of an egological view based on the idea of the body as a center, a centering, and the point zero of sense perception. The view of embodiment developed by these theories, therefore, shares an inherited picture of man as a self-contained and self-supporting entity, a picture which has been in vogue since Descartes.

On the other hand, in all its possible forms, my finding myself together with the other in a world where the two of us coexist is shaped by the defining experience of the corporeal: the flesh freed from the dissection table, which extends its limits beyond the body and regains life's distinctive motility, cannot be reduced to a pre-given, self-subsistent body but is rather associated with a form of corporeality that is constantly in the process of being actualized—of bodying forth. Hence, the body of the natural sciences—the body grasped in the present, according to that peculiar approach to time inherited from ancient thought—cannot account for all those corporeal phenomena which instead take shape within the framework of man's existence and hence through "what happens or may happen to someone."

The open quality of the flesh, which exceeds mere presence and surpasses the present moment of observation on the part of the anatomopathologist, points to a different mode of access to corporeality: corporeality, understood as a phenomenon, can only be grasped in the light of that movement toward the world through which each person's existence is realized at every moment and in which the flesh is always involved. In view of this eccentricity of the corporeal, the analysis of presentification becomes the cornerstone allowing Heidegger (GA 89 n.d.) "to approach the phenomena of the flesh" (p. 105).

In order to grasp this point we must return to the previous example. I am sitting at my desk when I realize that I have a train to catch in 15 min. The moment in which I realize that I am late, I visualize the station and the route I must take to reach it. In other words, while I am still sitting in my studio, in accomplishing the presentification of the station I already find myself involved with it, although factually I am still at my desk. Hence, I engage with the things that are before my eyes in my studio, while also engaging with the station through the mode of presentification.

Heidegger concludes that being-involved-with—whether it is being involved with that which is perceptively present in flesh and bone or being involved with what I presentify without really perceiving it—coincides with the open quality of man's being, whereby man is always solicited (*Beanspruchung*) by objects. What presentification implies is precisely the station before everyone's eyes, as it appears to me when I must catch the train to Basel—not a representation or fanciful image of it: presentification points to what is present within it. My being-involved-with something, therefore, is not restricted by the fact that I am sitting in my studio, since my being here can take many different forms: I check my email before leaving, the suitcase to make sure I have everything I need, the appointments scheduled for tomorrow, the quickest route to the station, and so on. All these possibilities of openness which I can enact through a change of attention do not suppress my being here but only change the way in which I am here. So while human being-there is always here on the basis of its peculiar receptivity, at the same time it is always elsewhere: it is always in a direct relation with something.

This, then, is the conclusion which our analysis of presentification leads to. Corporeality and its boundaries, which are co-extensive "with the reach of my sojourn" (p. 113), are qualitatively different from the boundaries of the body, as marked by the skin. To quote Greisch (1994), "Carnal phenomena such as turning red, grasping things, the corporeal extension of a backache, and laughing and crying all point to a clear-cut distinction between the boundaries of the body and those of the flesh. But what is even more important is to note the specific qualitative difference between the two types of boundaries" (p. 173).

Unlike the body, corporeality is not an entity present in a world, a thing among other things; rather, as we have seen, by participating in existence and being exposed with it to the events of life, to occurrences, corporeality constantly alters its limits and horizons. Heidegger (GA 89 n.d.) states: "The bodying forth of the body is determined by the way of my being. The bodying forth of the body is therefore one of my modes of being there" (p. 153, 113).[7] So while presentification enables the person who performs it to be involved with that which he visualizes, extending the domain of perception to something that is not physically present and thus sense-perceptible, as well as orienting existence toward the future (e.g., toward the station when I need to catch a train), it also enables the person to visualize something experienced in the past, as we have seen in relation to Valery's rereading of Stendhal's *Lucien Leuwen*. In this respect, corporeality implies a peculiar relation between spatiality and temporality.

The ontological significance of this interpretation of corporeal phenomena has been emphasized, with different nuances, by different researchers, who have focused on the relation between embodiment and ipseity (Greisch 1994, 1999; Benoist 1992; Gros-Azorin 1998). Indeed, the relation with time is the underlying theme of the *Zollikon Seminars*: Heidegger explores the topic by grasping corporeality as the constant enactment (*Vollzugssinn*) of the body in view of its potentiality to be, in relation (*Bezugssinn*) to what it encounters (*Gehaltssinn*). Hence, grasping the body as a phenomenon in each case is not only the way to access a phenomenology of the flesh—which is ontological from top to bottom—but is also what enables Heidegger to clearly define the various boundaries of the corporeal against the body which ends with the skin, in the light of a certain way of conceiving the temporal nature of the body. On the one hand stands the body (*Koerper*), a self-subsistent entity which submits every "not yet" to the stable present (organization); on the other stands the corporeal (*Die Leiblichkeit*), which on the basis of occurrences bodies forth again and again, thus newly disclosing on every successive occasion the embodied traces of time. In this sense, the corporeal, which becomes such through factual happening, constantly renews its relationship with that which it encloses.

[7] "Das Leiben des Leibes bestimmt sich aus der Weise meines Seins. Das Leiben des Leibes ist somit eine Weise des Daseins" (GA 89 n.d., p. 113).

8.4 Corporeality and Therapeutic Discourse

In his last conversation with Medard Boss (GA 89 n.d.), after once again stressing that the occurrence of corporeality is fully part of existence, Heidegger states: "This bodily [nature] develops in such a way that it can be used in dealing with the inanimate and animate 'material things' which are encountered. Yet, in contrast to a tool, the bodily spheres of existing are not set free *[entlassen]* from being-human. They cannot be cared for in a toolbox. Rather, they remain in the sway of being human, held in it, and belonging to it so long as the human being lives" (p. 293).

The body, therefore, contributes to the concrete realization of existence, by embodying it, and preserves its traces: those traces which being-a-man, as being-in-the-world, in each case interprets in the light of that which it encounters. Hence, with the enactment of facticity which the flesh thus contributes to bring about, and through which life finds meaning, corporeality at the same time discloses the historicity embedded within it.

The analysis of presentification conducted as part of our research on therapeutic care proves particularly interesting in another crucial respect: if psychotherapy is a word-based therapy which takes shape through words uttered and listened to, where is the body in speaking and listening? Two extracts from the Zollikon Seminars provide some general guidelines with regard to this point. The first extract is an enigmatic statement made on 6 July 1965, within the framework of a discussion on the issue of method in the natural sciences and in phenomenology. Heidegger states: "The problem of the body is primarily a problem of method" (p. 2). The same claim was repeated on 8 July, followed this time by an indication and specification of the method of phenomenology. After stressing the fact that being involved with something in "body and soul"—for example, being absorbed in a topic of discussion or being involved with the station—implies the participation of bodying forth in the phenomenon, Heidegger rhetorically asks: what method allows us to disclose the function of bodying forth in these examples? What does it mean for each person to be involved with that which he or she encounters? The philosopher continues as follows: "You had to free yourselves from the common notion of a merely subjective representation of things inside your head, and you had to engage in the way of existing [*Weise des Existierens*] in which you already exist. It was necessary to perform specifically this 'engagement' [*Sicheinlassen*] in the mode of being in which you already exist" (p. 109).

In order to grasp being-involved-with, it is necessary to specifically perform (*Vollziehen*) the "engagement" in the mode of being in which man's relating to what he encounters takes shape. Only in such a way can we grasp that the body, as a phenomenon, consists in being engaged with something. Forty years after Heidegger's first courses in Freiburg, the phenomenological method still guides the research and enables him to shed light on the problem of the corporeal. Once again, we can detect the influence of Husserl's teaching—the teaching of Heidegger's esteemed master (*mein verehrter Lehrer*. p. 171)—as expressed in the first and sixth Logical Investigation, and especially of the method of formal indication.

The movement of a hand putting a watch away or wiping one's brow, blushing, speaking, and listening are all gestures, which is to say movements conceived as a way in which man comports himself—solicited by an entity—as a being-in-the-world determined by corporeality. Within this context, dialogue too—in our case, therapeutic discourse—shows itself in its carnal dimension.

If therapeutic questioning aims to newly mobilize those affective dispositions that orient the patient's sphere of intelligibility, the act of speaking and listening on the part of the therapist coincides with the fact of being involved with those experiences which are elicited by the therapist's questions and recounted by the patient: it coincides with a bodying forth related to what is understood and co-enacted with the patient. On the other hand, the patient who is solicited by the therapist's questions expresses those meaningful complexes embedded within the flesh (as possible openings), which—once activated by therapeutic questioning—draw him to himself and become the foundation of speech: a speech which encompasses the body and directs the therapist by showing him how to re-enact the meaning of what has been said (formal indication).

In this sense, the patient who is summoned by the voice of the healing other is drawn to himself, to those significations which the flesh discloses, and is led to take charge of the new possibilities it ensures in view of the reorienting of his existential horizons. This is precisely the foundation of word-based therapy: the flesh which co-determines man's being as being-in-the-world—that flesh rooted in existence and constantly co-defined by it which had bound the patient's existence to a symptom—is awakened to new possibilities. Effective treatment therefore consists in restoring life's possibility of openness through the overall appropriation of its sense. By following this trajectory, we can understand the following question raised by Patočka: "In principle, is our body too not to be understood in these terms, as an organized whole, which coincides with the sense-enactments taking place within it, which are made possible by it and in turn make it possible? Should feelings, emotions, perceptions and actions not be understood as significative complexes which, being inwardly conditioned by the body, organize corporeality and in turn condition it?" (Patočka 1995, p. 22).

8.5 The Normal and the Pathological

Alongside this therapeutic perspective, another, complementary one emerges: the perspective on sickness, on the alteration which curtails existential possibilities—that suffering for which treatment is sought. In his *Zollikon Seminars*, Heidegger speaks of sickness as a privation, as deficient health, and hence as a quantitative deficiency of health. This is, for instance, the case with schizophrenia: as the basic distinguishing feature of being-human is the fact of being constantly open to the things which concern us, Heidegger speaks of a privative alteration of this disclosure to account for the dearth of contact characterizing schizophrenia.

This conception of sickness is echoed by Medard Boss in his valuable medical treatise (1977), where he repeatedly stresses that sickness is invariably nothing more than deficient healthiness. Both Heidegger and Boss therefore share the same conception of pathological phenomena in terms of quantitative (hypo- or hyper-) variations of the corresponding physiological phenomena. This way of envisaging sickness and health is paradoxically similar to that dogma which both Comte and Claude Bernard enunciated in very different conditions, and with very different intentions, compared to Heidegger and Boss (Canguilhem 1966).

While grasping sickness as a deficiency has the advantage of placing it on the same ontological level as health, it is not sufficient to account for the phenomenology of the sick, symptomatic body. "The content of the pathological state—Canguilhem (1966) argues—cannot be deduced, save for a difference in format, from the content of health; disease is not a variation on the dimension of health; it is a new dimension of life. … Disease is both deprivation and change" (p. 108).

The most evident feature of this change, in all of its forms, is the fact that the flesh to some extent withdraws from existence and becomes "visible" as the hindrance, obstacle, limitation, burden, limb, organ, or presence toward which the sick person develops a peculiar attitude. What returns to the fore with this awareness of the body is the problem of limits raised by Greisch (1999) and of the articulations of the phenomenon of carnality, which changes in accordance with our way of inhabiting space. From this perspective, another key moment emerges: the body which sickness has made visible inhabits corporeality, imposing itself upon it.

Grasping this point from the angle of general phenomenological pathology, Boss makes the following observation with regard to the sick person's experience of his own sickness: "His report reveals his attitude toward his condition, and the physician may learn whether the patient has accepted the illness as a challenge, revolted against it, tried to deny its existence, or surrendered to it" (Boss 1977, p. 197). What we have here is the relation between the sick person and sickness: between ipseity and self-estrangement. It is within this framework that the problem emerges of the limit represented by factual living with a disorder.

In sickness, the corporeal, which ceases to be absorbed by this or that thing, shows and imposes itself as a phenomenon by withdrawing from existence and hence from invisibility to the point that it literally reenters its own skin. We have seen this in relation to Giovanna's heart disorders: with the change of certain life conditions, a small alteration of the stability of her life was enough to trigger a mechanism leading to visceral symptoms. Much the same applies to the previously described case of the agoraphobic patients. The narrowing down of the horizon of these patients' being-in-the-world goes hand in hand with an alteration of the sense of their own position within given spaces, an alteration which they perceive in their flesh. Clearly, what is limited here is the accessibility of the world.

So whereas in a condition of health carnality vanishes, merging with the sense according to which it is oriented, in the condition of sickness it ceases to be the invisible part of one's relationship with the world and makes its presence—its being present—felt: carnality *bodies forth*. Thus another unique phenomenon occurs which Heidegger does not—and cannot—see: in sickness the limits of corporeality (quantitatively) withdraw within the limits of the body. Heidegger cannot see this phenomenon because of his ontologizing perspective, whereby he grasps the body in the present merely as a *Koerper*, as a thing among other things, as an entity which is grasped in the light of the present and never as a phenomenon, as corporeality superimposed upon the limits of the physical body and extending even beyond it, down to its "inner spatiality."

Even when Heidegger focuses on the way in which one's feeling alive is anchored to the physiological states of the body, his analysis is always subordinated to a way of envisaging the flesh exclusively as an element of one's being-in-the-world, without ever addressing the problem that being-in-the-world itself may take shape in view of the body. In his work on Nietzsche, Heidegger writes: "But because feeling, as feeling oneself to be, always just as essentially has a feeling for beings as a whole, every bodily state involves some way in which the things around us and the people with us lay a claim on us or do not do so. When our stomachs are 'out of sorts' they can cast a pall over all things" (Heidegger GA 6.1 1996, p. 99). However, this casting of a pall is never thematized.

Heidegger, in other words, fails to see that the contraction of corporeality (*Leiblichkeit*), which may even lead it to coincide with the body (*Koerper*) or to inhabit the latter "from within," emerges within existence itself through that alteration of one's openness which is determined by sickness but is not merely a deficiency, a matter of quantity: for it is also a matter of quality. It is "the replacement of an order one is used to and loves with another order which one must enter in contact with or suffer," to quote Canguilhem (1966, p. 158).

It is impossible to understand sickness by conceiving it merely as a (hypo- or hyper-)privation, as the Positivists did; sickness also coincides with the imposition of a new way of feeling alive which the sick person now endures—a mode of being which he is forced to adopt and which reduces his existential possibilities by binding them to the presence of the symptom, to the present of the body. The new element, the qualitative difference imposed by sickness, is precisely this: the body in the present—not because sickness is an ontologically different entity, nor because the body is grasped as a thing among other things according to an ontology of presence, but rather because its being in the present stems from being sick as a particular mode of being in the world. This being in the present imposed by sickness could permanently be fixed in one's organs by death.

In the fact of existing in view of the world, whereby the life of each person constantly realizes itself, the body no longer disappears, but rather makes its presence felt, becoming a limit. In not disappearing, it becomes an obstacle and sometimes a hindrance to relating with that meaningfulness which one encounters in each situation. The body becomes other (*heteros*) at the heart of the ipseity which it supports. To phrase it in different terms, the sick body becomes present within my existential horizon as a difference within ipseity, as a body which I can no longer govern and which therefore reduces my potential ways of being; in this respect, sickness is an alteration of the flesh.

What appears with sickness, therefore, is not just a privative alteration of the "I can" factor, but also—in qualitative terms—an alteration of my power of being in the world. This is a crucial point which Ricoeur sums up most effectively in his own distinctive fashion: "With the decrease of the power of acting, experienced as a decrease of the effort of existing, the reign of suffering, properly speaking, commences" (1992, p. 320). There is no distinction between physical and psychological suffering here![8]

This experience of suffering unto death, which accompanies sickness and in relation to which the therapeutic alliance is established in all its forms, is precisely what enables us to grasp carnality as a body—or the self as other, as Ricoeur puts it. In making my body different, in turning it into something other than its corporeality, pathology freezes it in the actuality of suffering. The experience of suffering, through which carnality reveals itself as a body in the present, therefore enables us to see why my sick body also appears to me as a body among other bodies. We can therefore understand the "passivité-altérité" syntagm which Ricoeur repeatedly uses to indicate how the otherness of the flesh—otherness understood as something almost synonymous with passiveness—takes shape as a difference to oneself within ipseity. "Otherness is not added on to ipseity from outside, but belongs instead to the tenor of meaning and to the ontological constitution of selfhood" (p. 317), Ricoeur writes, reinterpreting from the perspective of the above syntagm the whole sphere of the passiveness intrinsic to the flesh, as defined by the great French scholars—that is, from the soul's passions, through mood embodiment and tones, down to self-affection.

Ricoeur's fundamental gesture consists precisely in having set at the heart of ipseity this intimate passiveness of the flesh, the "primary otherness" that is proper to it—and which differs from the otherness of the other. The experience of sickness is what best reveals this passiveness, by engendering a difference to oneself within ipseity. By identifying the flesh as the realm of the proper and of alteration, thus tracing the difference to oneself back to the flesh, Ricoeur opens up a new perspective which fosters an assimilation and engagement with the "great genera" of philosophical discourse.[9]

Although, as emphasized by Greisch (1999), the understanding of the body as other first took shape within the theoretical context of Husserl's *Fifth Cartesian Meditation*, where this distinction is developed within the framework of the question of the construction of the other as foreign subjectivity, Ricoeur attempts to detach the problem of the flesh from this original context by grasping the flesh as a paradigm, not for the otherness of the other, but for otherness to oneself. It is precisely through this mode of access that Ricoeur opens up a dialogue with ancient philosophy. The relation between ipseity and otherness, which rests on the flesh, is framed within the context of the meta-category of otherness with an explicit reference to Plato's metaphysical dialogues and their intertwining with Book III of Aristotle's *Metaphysics* (Ricoeur 2001, p. 69). In the light of all this, what really stands out is the title of Ricoeur's greatest work: *Oneself as Other*.

[8] Ricoeur thus appropriates that intimate sense which according to de Biran corresponds to the "inner" experience of effort: to how it feels to exist within one's own flesh.

[9] Being and nonbeing, rest and movement, the self and the other, the one and the many.

8.6 Corporeality and Its Limits

We have already encountered the meta-category of otherness at the beginning of this work, in relation to what Aristotle argues about the art of healing. To illustrate the concept of potency, along with that of architect, the philosopher makes the example of the physician and of the art of healing as a principle of change. The physician—Aristotle claims—can heal not just his patients but also himself when he falls ill. Aristotle therefore stresses the fact that the potency to heal which enables the physician to treat his patients also allows him to treat himself, when sick: the physician treats himself as other.[10]

Following the path of the great genera, Aristotle's example leads us back to sickness as difference to oneself. We have repeatedly spoken of the presence of sickness as an alteration of one's power to be in the world. Goldstein wrote: "Pathological phenomena are the expression of the fact that the normal relationships between organism and environment have been changed through a change of the organism and that thereby many things that had been adequate for the normal organism are no longer adequate for the modified organism" (1939, p. 328 Eng). The person who is medically sick, in other words, restricts his existential possibilities and narrows down his life context to a limited environment which differs from his previous life condition. In this respect, sickness brings about a reduction of the possibilities to access a shared world and hence a limitation of one's freedom.

If, on the one hand, therefore, sickness translates into a reduction of the openness to the meaningfulness which addresses us, and hence of that openness which distinguishes human being-in-the-world, on the other hand, what makes its appearance onto the stage of life, of the experience of living, is the body in the present—the Stone Guest that manifests, or indeed imposes, its presence where it was least expected.

A few examples may help to clarify this point. If I *am all ears* when listening to a concert, the fact of being completely absorbed in the music is precisely what enables me to pay attention to the unfolding of the performance; "being all ears" means that the corporeal vanishes behind listening. However, a gallbladder spasm—an inner organ painfully revealing itself—is all it takes to immediately interrupt this vanishing of the flesh within music: pain makes the body appear, while the music may become a mere acoustic phenomenon, a sense perception, so to speak. Pain refocuses my whole existential sphere on the affected organ.

Something similar also occurs with psychological disorders. If a fortuitous contact with the person seated next to me at the concert fills me with the dread of having been infected with some disease and with an irrepressible desire to go and wash myself, what ceases to be, once again, is the disappearance of the flesh in the

[10] This distinction clearly echoes the one which Aristotle draws between oneself and the other by defining the notion of otherness through the more particular one of difference (*diaphora*): "We call different (*diaphora*) those things which though other (*heteros*) are in some respect the same (*to auto*)" (Δ 9 1018a 12–13). On the *heteros/allos* distinction, see too I 3 1054 b 23–26 (Alessandrelli com. pers.).

presence of the music, while my "dirty" hands, tense body, and unrestrainable sequence of ruminations—that fuel the anxiety from which they originate—make their presence felt, spoiling, so to speak, my capacity to listen to the music. The indecipherability of the fortuitous contact (and of physical proximity more generally) narrows down the openness of existence to a present centered on the epidermis, on my "dirty" skin.

Let us consider another case which seems to combine the two previous ones. The lights in the concert room go out and the orchestra begins its performance. The first movement of Mahler's Sixth Symphony rumbles in with kettledrums, trumpets, and oboes setting the powerful rhythm that marks the opening of the symphony. Then, gentle, elusive and melancholy, comes Alma's theme. At that moment I sense my heartbeat, which grows faster and faster. I turn my eyes to the exit, but it is far away and I cannot stand up in the middle of the concert—I am trapped. My heart is besieging me: I can hear it in my ears and I fear that something might happen to me. The marching drums of the Sixth Symphony are only perceived as an unbearable clangor! Here the limit of carnality has withdrawn to the heartbeat and its rhythm, while the music has become mere noise.

These different examples make it quite evident that, in order for listening to occur, it is necessary for carnality to be involved with the music: for in being involved with the music, the flesh withdraws and disappears. Thus the limit of corporeality becomes that of the concert hall. When suffering appears in any of its various forms, what is interrupted is precisely this vanishing of the flesh through my being involved with the music. Suffering brings the limits of carnality back within the body.

In each of the cases just described, the contraction of the ability to engage freely in music listening points to a different limit from that of the concert hall: a limit which prevents the disappearance of the flesh because it forces us to be involved with our own body—our bowels, tense muscles, bones, epidermis, and so on. The more corporeality is forced to withdraw in various ways down to the skin and into the skin and into the inner organs, limbs, or head, thereby limiting our possibility of living the world, the more the body which had vanished as corporeality reaffirms its presence by bursting forth into our existence.

The imposing itself of the body brings to the fore the experience of the body in the present and existence as a whole is re-attuned to the corporeal sphere. Suffering, in other words, captures the motility of life, arresting it and contracting its possibilities of realization, as all resources are mobilized with respect to it. Sickness, in all its various forms, thus brings out a difference to oneself, a deficiency, which is what distinguishes the fact of feeling sick or at any rate *altered.*[11]

---

[11] The alteration of our openness to the world which accompanies suffering—and which has to do with our focusing on the affected area—monopolizes our attention, motivating behaviors aimed at ending, reducing or escaping exposure to the source of the noxious stimulation. This monopolization is particularly evident in the "sorrow of the soul" which characterizes depression (Arciero and Bondolfi 2009).

It is in the light of this "objective" presence—with respect to which one's conscience and will prove powerless—that the nature of the human body has been envisaged as a substantial reality and hence one that may be measured and worked out. This body forever in the present may be regarded as the experiential foundation of the seminal discovery made by Bichat and that ontology of death on which medicine rests, namely, that in the innermost tissues of a dead body traces survive of the cause underlying the clinical manifestation of an illness.

Based on these assumptions, the natural sciences have deemed it legitimate to extend their investigative criteria to human existence, purporting to work out its meaning by reducing it to the causal relations between organic structures. Following the lead of von Helmholtz's research, up until the 1930s, this tradition has engaged with a causality commensurate with experience to explain the functioning and organization of life. Then, from McCulloch to the present day, it has come to be guided by mathematical modeling. This mode of operating, which is so evident in the neurosciences, is what currently distinguishes the practice of a multitude of researchers (and philosophers) who, faithful to the assumption that certainty and truth are one and the same thing, strive to carry out experiments, create models, and develop theories to explain the various psychiatric pathologies without ever having met a sick person! Besides, the lack of a methodology enabling clinicians to exchange views with those researchers who practice the natural sciences in their laboratories makes the latter—quite understandably—distrustful of observations and suggestions coming from the clinical field.

8.7 Corporeality and Suffering

A new question must now be addressed: if the way in which I inhabit the world is the setting of corporeality (which is an integral part of it) and what determines its limits, how does the emergence of sickness affect my being-in-the-world?

The underlying theme of Heidegger's *In-der-Welt-sein* has surfaced again and again in the pages of the present study: we have understood man's receptive openness to the meaningfulness he encounters as a being-in-the-world within which human corporeality (*Leiblichkeit*) occurs as a bodying forth (*Leiben*). This relating (to entities by making them appear) ceases to be as soon as it is enacted: it disappears as one becomes involved with the station, music or—like Thales—the stars; so much so that even the limits of existence and of the corporeality which takes part in it constantly change. Hence, in his *Zollikon Seminars*, Heidegger states that "the limit of bodying forth … is the horizon of being within which I sojourn" (p. 113).

As we have seen in relation to sickness, however, with the impairment of definite abilities to enact our own being-there, corporeality ceases to be an opening up or relating to things, and no longer disappears by referring to something else; rather, it points to the body which supports it. All forms of sickness, therefore, are characterized by an impairment of potential ways of relating to the world within the specific spheres they affect. Carnality, in other words, makes its presence felt and strains our being-in-the-world. This is, for example, clear even from a painful spasm: the difference between the limits of carnality and those of the body becomes evident because our openness to the world, and hence the horizon of our corporeality, are re-attuned to the aching organ, which imposes itself upon that openness as otherness to oneself. Carnality and the body, which may be said to coincide in a present, subsistent body, combine in the aching part of the body: life loses health and, with it, the silence of the organs.[12]

The emergence of this otherness at the very heart of ipseity seems even more striking in the case of the existential suffering due to an unexpected loss, or an unwanted and unanticipated separation, for instance. These alterations of human being-in-the-world, which entail the shattering of our familiar way of inhabiting the world (associated with a loss), engender a mode of bodying forth characterized by two aspects that are closely and mutually integrated in the experiencing of sustained sadness: inaction and suffering (Arciero and Bondolfi 2009; Bondolfi et al. 2015). Different languages employ the same expressions used for physical pain (hurt, injured, harmed, broken bones, heartbreak, etc.) to describe painful experiences that exceed physical pain, such as rejection or the loss of a loved one, and all of them point to the appearance of the body as something totally damaged, something wholly suffering. In grieving, otherness to oneself is perceived as a global suffering in being alive, as the enduring of oneself, and at the same time as a powerlessness to change the circumstances which bring this condition about; hence the sense of immobility, of standstill, of being incapable of avoiding the presence of the loss which accompanies such otherness (Arciero 2006).

As is widely known, the work of mourning consists in reintegrating this otherness within oneself by understanding the loss one has experienced within a new existential horizon. On the other hand, when an individual is incapable of engendering new possibilities of opening up to the world and others, the ongoing engagement in painful feelings becomes the focus of ipseity, creating a downward spiral (between sadness and themes of derogation, unacceptableness, unworthiness, defectiveness, and unamiability) which, if stabilized, leads to a depressive state (Bondolfi et al. 2015). Therefore, the melancholic disposition—like, on the opposite side of the spectrum, the maniac one—does not merely correspond to a deficient existential attunement in terms of a lack or excess, as Medard Boss suggests. The contraction of receptive openness has to do with an otherness which affects the depressed person by wholly binding him to the flesh (sadness), while freeing the maniac individual from all bonds (euphoria). Hence the different phenomenologies distinguishing these two conditions, which vary in terms of the quality of the experience and cannot be reduced to mood elevation or lowering: the former is invariably inwardly oriented, the other is affectively incontinent and unstably attuned to even the slightest variation of the ongoing context.

The inverted signs given by the flesh—whereby corporeality ceases to be an opening up to things pointing to the body which supports it instead—are of crucial interest in psychopathology. We have already mentioned this in relation to the obsessive disorder, the phobic disorder, and the hypochondriac one in Giovanna's

[12]This reflexiveness of the flesh concerns all organs except the brain.

case. It emerges with even greater prominence in relation to eating disorders and hysteria (Arciero and Bondolfi 2009).

The otherness of the flesh, which becomes detached from the existence within which it took shape, elicits a shift of attention onto the suffering it engenders, focusing being-in-the-world on it. This leads to a self-referential circle—between prereflective states (the contexts of intelligibility in the background), which have been stripped of any meaningful connection and are therefore suffered, and the ongoing experience. This downward spiral is only exacerbated by the sick person's *inappropriate* attempts to makes sense of his situation, whether cognitively, narratively, or in any other way. The otherness to oneself—as we have seen in Giovanna's case—only increases, turning from being an effect to being a cause according to a circular pattern. This vicious circle, whereby the otherness of the flesh increases the altering conditions, constitutes the mechanism at the basis of symptoms in psychopathology.

The freezing of the motility of life in the repetition of a symptom limits the possibility of disclosure of the world, as it binds every form of initiative to a presence engendered by the mechanism over and over again. The symptom, in other words, confines the motility of life to a self-referential present dominated by a present, subsistent body; and in sickness the body, insofar as it is present, coincides with corporeality—or, rather, with its self-referential circle.

Sickness thus reveals something more fundamental: if it entails a decrease of the power to act which corresponds to a repetitive halting of the motility of life, this loss of possibilities and hence (in) capacity precedes me, so to speak—I do not have any power over it, but am rather dominated by it. The suffering over which I have no power, and which therefore detaches my will from the flesh which supports it, reveals the passive dimension of the motility of life, making it appear to be independent of all power.

In this respect, Ricoeur grasps the flesh as a paradigm of otherness: "I, as this man, is the foremost otherness of the flesh with respect to all initiative. Otherness here signifies primordiality with respect to any *design*. Starting from this otherness, I can *reign over*. Primordiality, however, is not a reign. Ontologically, the flesh precedes the distinction between the voluntary and the involuntary. … In short, it is the origin of all 'alternation of ownness'" (p. 324). As we have seen, the passiveness-otherness dimension is what explains the distinct tendency in the natural sciences to hypostasize the motility of life as a substantial reality, and hence as a present, subsistent body (*Koerper*).

Indeed, whereas in its passive dimension motility appears as that material body which dominates me, in its free engagement with the world it shows itself as a model of realization of existence, as an activity. Sickness thus brings the motility of life to light as the original sphere that combines the apparent dimension of subsistence, through which the flesh manifests itself as the body in the present, with the dimension of receptive openness to what one encounters that is a defining feature of existence. In each case, this motility of human life concretely marks the point of articulation of activeness and passiveness where this distinction appears, becomes blurred, or vanishes and hence the limit between the body and corporeality.

8.8 Corporeality, Violence, and Trauma

What forces corporeality to "lay bare" the flesh on which it rests, and which it engenders even more so than sickness, is violence. But whereas sickness makes the body appear as "other" vis-à-vis its corporeality, thereby freezing corporeality in the present suffering "from within," i.e., from the flesh itself, violence reveals the otherness of the body "from without": what engenders it is the other.

An act of violence intentionally perpetrated upon my existence by others brings out that primordiality of the flesh on which I can no longer *reign* that foremost otherness on which the oppressors are now exercising their design and domain. Given that it would be quite unthinkable to find oneself in the world without a body, since it is only through a body that being-there can in each case be situated and become what it is, the violence of man against man consists in curtailing this active motility through which individual existence is realized and which rests on—and is ensured by—the body. This connection of violence with freedom, understood as a specific feature of the opening up of existence to the world, a connection that it violates, makes violence a distinctly human phenomenon.[13] In a famous essay outlining her political thought, Hannah Arendt (1970) stresses the distinctly human aspect of violence—torture providing particularly striking confirmation of this. In doing so, she envisages violence within the framework of the much broader dimension we have just described: "What makes man a political being—Arendt writes—is his faculty of action; it enables him to get together with his peers, to act in concert, and to reach out for goals and enterprises that would never enter his mind, let alone the desires of his heart, had he not been given this gift to embark on something new. Philosophically speaking, to act is the human answer to the condition of natality" (p. 82).

As to act is the human answer to the condition of natality, the violence of man against man, which curtails the capacity to act, denies the human condition precisely in terms of the fundamental freedom granted by the fact of being born, of having come into the world as a human being. In other words, if being alive corresponds to an open possibility of meeting the meaningfulness which elicits our attention, and if what lies at the basis of man's humanness is precisely this freedom to open up to meaning in view of the world as a whole (as we will see in the following chapter), then the violence of man against man consists in denying this fundamental freedom which Arendt interprets as the freedom of initiative.[14] For Arendt, this freedom is the answer to that *initium* which marks the beginning of the miracle of life in its uniqueness, and which allows man to introduce the element of innovation into

---

[13] Unlike aggression, violence cannot be conceived according to the categories of ethology (Arendt 1970; Ricoeur 1992; Tinbergen 1968; Lorenz 1966): "Neither violence nor power is a natural phenomenon, that is, a manifestation of the life process; they belong to the political realm of human affairs whose essentially human quality is guaranteed by man's faculty of action, the ability to begin something new" (Arendt 1970, p. 82).

[14] Ricoeur observes: "*Initiative:* this is the key word on the phenomenal-anthropological level, to which there corresponds, on the political level, *consent* with regard to living-together" (Ricoeur 2010, p. 25).

the course of history; hence, to deny the freedom to act is to deny the humanness of man and, more generally, to wound humanity as a whole.

If the ontological basis of this actual finding oneself within the world is corporeality as the possibility to open up to the other and to each situation, violence can erase corporeality—as is the case with torture—by reducing the motility of life to the mere passiveness of the flesh.[15] Thus, by making corporeality coincide with the flesh, violence interrupts existence, curtailing—down to the finest tissues—the constitutive motility through which it realizes itself.

Far from regarding a trauma as a disruption of the "protective dam" a person has erected through early experiences to protect the inner space containing his mental states (which thus overflow), we are led to envisage each trauma as an experience which is not only irreconcilable with the individual's previous existential horizon but which utterly erases his sense of belonging in the world, the accessibility of meaning. A trauma is a wound inflicted upon existence by the tormentor's power over his victim: to exercise this domination is to reign over the primordiality of the flesh of the other, which is forced to emerge in all of its passiveness under the aegis and weight of the threat of death, which may be implemented by the oppressor at any moment.[16] In this act, the corporeality of the victim withdraws within the flesh because it is erased—because it is violated. This *violence* leaves only the impossibility of meaning.

The incommensurability of a traumatic experience compared to the ordinary progression of existence is thus rooted in the very structure of trauma: as an inflicted experience, it is disarticulated the very moment in which it is felt. In other words, insofar as this experience is imposed by the tormentor, it may be said that to some extent it does not belong to the person who has it: for while this person stands at the origin of the experience, he or she suffers it. This dispossession of the *initium* also coincides with the reduction of the victim to a material body which, under the torturer's blows, has complete control over the person. Torture annihilates the victim within suffering!

As a consequence of this imposition of the passiveness of the flesh, the temporal existence of the body is deprived of its openness to the future, of the sense of what may be, and is confined to the dimension of the subsisting present: the present that is being inflicted. The victim lives in a more or less prolonged frozen present, absorbed in the suffering of his or her own flesh, which reveals its primordial otherness under the tormentor's blows.[17]

---

[15] The same aspect is also to be found with some differences—and certainly in view of different aims—in the case of chemotherapy patients.

[16] The act of enduring this torment, thereby bearing witness to the fact that one is ready to forego one's flesh, engenders martyrdom.

[17] Examples include concentration camp experiences, torture, disasters, prolonged exposure to life-threatening circumstances (e.g., hostage situations—prolonged captivity with an imminent possibility of being killed) and, more generally, all those conditions capable of bringing about what ICD 10 identifies as "enduring personality change after catastrophic experience" (F.62). Enduring personality change may follow the experience of catastrophic stress. The stress must be so extreme that it is unnecessary to consider personal vulnerability in order to explain its profound effect on the personality.

As eloquently described by a patient of ours, one's own material body becomes a sort of enemy which forces one to live, even when the suffering endured becomes unbearable:

> "My despair turned to rage and then I got it into my head to devour my own skin.
> I was to pierce the vessels carrying my blood, which was hostage to my veins—accomplices to this murderous body.
> A warrior strength seized me. Suddenly I was sheltering a beast within my womb.
> A monster with a ravenous appetite. Insatiable. Enraged.
> Moved by this violent urge, I attempted to grasp the smallest bit of skin which showed itself, vulnerable, between the ropes.
> I was to drive my ivory nails deeper and rip my veins open!
> My frenzied bites sunk into the delicate skin of my wrist but despite my 'cannibalistic' hunger, only some snippets of skin came off.
> I could feel the bloody consistency of my flesh tear under my teeth and soak my lips with a sweet taste of impotence.
> The suffering of my skin, torn by my self-inflicted bites, seemed indecently soothing compared to other screaming wounds.
> My veins defied me. They refused to yield their blood, overflowing with life within a body that had long been dead.
> I wallowed in my oppressive impotence.
> My urges were suppressed by my impossibility to die, or even to fall asleep…"

This feeling of self-expropriation, which in torture coincides with an escalation of suffering and fear as the victim awaits the final moment, constitutes the prereflective foundation of the difficulty of expressing a trauma in words. The experience is paralyzed in the flesh through a circle engendered by a crescendo of pain and fearful anticipation of even more intense pain, which freezes the body in a continuous present. To quote our patient again:

> Time was suspended. Heavy. Intense. Oppressive!
> It was as thick as the sludge of excrement I was covered in.
> Silence too no longer had the same texture.
> I was suffering and sunk into complete inner loneliness.
> Here stood a subterranean world made of screams, gasps and moans: my own…
> I was at war with myself—with this suffering body which had become my worst enemy.
> It filled the whole space of this flesh which had become an executioner, which only conveyed new and growing pains to me. Thousands upon thousands of fragments of this living torture is all that I was made of.
> My last, desperate prayer was: May I not survive this night!
> I was to entice death so that it would finally take interest in me—so that it would desire me and seize me completely, so that it would erase me.

So, on the one hand, the traumatic experience compromises the ordinary progression of life, expectations, hopes, projects, memories, and initiatives, because it suspends the sense of existence upon a permanent present; on the other hand, it cannot be verbally expressed in the first person, because it affects the very structure of experiencing: the victim is deprived of his reign over the primordiality of his own flesh. The ipseity estranged from its flesh—which is no longer capable of heeding the call of what opens up before it—is confined by its tortured flesh within a subsisting present; hence, it can no longer be grasped according to the ordinary categories of life—i.e., according to those meaningful perspectives which, as Ricoeur puts it, configure temporality, enabling language to convey it through a narrative and anchoring this narrative to temporality (Ricoeur 1983). Already at the prereflective level, then, a traumatic experience resists the "natural" underlying cohesion of lived experiences which constitutes the integral unity of experience. This is the basis of the traumatic amnesia which has consistently been noted in a wide variety of traumatized patients. In this respect, a trauma amounts to the fracturing of oneself.

The unspeakability of the traumatic experience, which prevents it from being articulated and appropriated, is thus due to its peculiar prereflective conformation: it remains cut off from the flow of the temporality of life because it does not share its structure. This basic difference accounts not only for the failure to integrate the experience within the web of one's experiences but also for the symptomatology which may spring from the trauma. Indeed, the posttraumatic experience of all patients is characterized by the fact that they are pervaded by intrusive sensory elements of the trauma, which are incorrectly referred to as flashbacks. These can occur in a variety of modalities (auditory, visual, olfactory, kinesthetic, and affective), usually in a fragmentary way, and can intrude upon everyday life, to the point of taking the form of genuine dissociative states. Dissociation, therefore, is not so much a defense mechanism activated by the victim in an attempt to maintain mental control during the trauma—since physical control is lost—as the marker of a violation of the flesh, of a defilement of the innermost structure of human experience. Clearly, this kind of dissociation is very different in nature from the dissociative forms related to the range of clinical conditions that Janet brought together under the classical term hysteria, on the basis of their sharing of the fundamental pathogenic mechanism of dissociation. On the other hand, the "occupation" of the domain of the proper by the experience of trauma brings about an alteration of the general attitude of the person, who lives in a chronic condition of anticipation of the possibility of imminent danger and of vigilance, seeking to grasp in his dealings with the world a series of signs that might help him avoid repeating the experience of suffering.

The patient is usually incapable of accounting for this condition.[18] If, however, we re-enact the emergence of the dissociative experience with the patient, it becomes quite clear that what we are dealing with is not a memory at all. Given a series of conditions, a trivial perceptual stimulus, which is to say a certain meaningful thing in the world, such as an object of some sort, may trigger the dissociative experience. Here too, as in the previously discussed case of parapraxes, there is no need to invoke the unconscious. If we consider the fact that the traumatic experience has

[18] The role of past traumatic experiences on the current lives of people has significantly influenced the development of contemporary psychiatry. As Von der Kolk has rightly emphasized, Charcot, Janet, and Freud all noted that fragmented memories of traumatic events dominated the mental life of many of their patients and built their theories about the nature and treatment of psychopathology on this recognition (1989).

never reached the level of speech and has, therefore, been frozen in the body in the present from the very moment in which the trauma occurred and if we further consider the fact that this mode of being is among the possibilities for relatedness characterizing the sphere of the proper, then an event which strikes the person is not so much a memory of this experience as something which reawakens it. A particular sensory circumstance here becomes a way for the person to access a sphere of his experience which he has never been capable of verbally expressing because it is unspeakable.[19]

Over the course of a difficult pregnancy that had her bedridden for 5 months, a patient of ours started experiencing episodes of irrational fear, triggered by everyday situations and objects—e.g., the deep green of a cactus plant, a key, a bright electric light, and particular sounds or noises. Acoustic or visual perception was enough to trigger a symptomatology characterized by deep fear, with a dynamic akin to that of a panic attack. Although this was the diagnosis, the phenomenal characteristic of these states was a dissociative condition lasting a few minutes, in which the patient would be absorbed in an intense fear accompanied by pain in different parts of the body. This marked the beginning of a symptomatology which was to last 15 years, before it was finally identified as the formal indication of a trauma, a step which made it possible to trace the experience back to its origin and to develop a narrative of it. What was ultimately brought to light from behind these "bizarre" symptoms was a detailed account of the abduction, torture, and group violence the woman had been subjected to for several days before being left lying in a coma on the steps of a hospital at the age of 17.

This freeze-frame present is what provided the current sphere of intelligibility required for the patient to understand and appropriate her "irrational" fears, bizarre reactions of avoidance or flight and dissociative states, as well as—more generally—to reestablish within the motility of life the experience which her tormentors had frozen within her flesh: a meaningful context which was regained through a slow and progressive narrative appropriation.

But why does the mere fact of narratively expressing the traumatic experience already have such an evident therapeutic effect? Is the account itself cathartic? In a way, it is.[20] The account enables the patient to regain—via therapeutic dialogue—his or her first-person perspective on the lived experience. As Storolow notes from a different point of view, dialogue provides a relational home for the patient's painful emotional states (2008). Indeed, linguistic articulation makes it possible to appropriate the traumatic experience, by restoring the sense that it belongs to oneself and framing it within a renewed hermeneutic of the self. Its reintegration within

[19] Once human existence is seen as bearing the spatiotemporal worldwide realm of openness as, itself, a perceptive and actively responsive openness, the problem of location takes on a wholly different aspect. It is self-evident, then, that the possibilities for relatedness belong fundamentally and immediately to the whole realm of perceptive openness.

[20] It is important to note, however, that the mere fact of expressing a traumatic experience in words does not resolve the dissociative symptomatology, particularly in the case of long-running pathological conditions.

the dynamics of one's life represents precisely the crucial variable on which the successful or unsuccessful resolution of the symptomatology depends.

The points just made may shed light on what Elie Wiesel notes when attempting to explain the many suicides which occurred in the Jewish community in the aftermath of World War II: "the writers who were part of the shrinking community of Holocaust survivors endured severe hardship. Despairing of the written word's power, some chose silence. The silence of death" (1996, p. 471). In the face of so much horror, the unspeakability of the trauma becomes the impossibility of speaking about the atrocity of evil; and it becomes even more terrible for those who have better means to describe things than others, but no longer have the words to do so.

References

Arciero G (1994) I and Thou: notes on the phenomenology of reciprocity. Paper presented at the meeting "Ethology and Knowledge" -Nov. 1994 University of Siena-Published in "Conoscenza e Etologia" (1996) Melusina ed

Arciero G (2006) Sulle tracce di sé. Bollati Boringhieri, Torino

Arciero G, Bondolfi G (2009) Selfhood, identity and personality styles. John Wiley & Sons, Chichester

Arendt H (1970) On violence. Houghton Mifflin Harcourt, New York

Asmundson GJ, Taylor S, AJ Smits J (2014) Panic disorder and agoraphobia: an overview and commentary on DSM-5 changes. Depress Anxiety 31(6):480–486

Benoist J (1992) Chair et corps dans les séminaires de Zollikon: la différence et le reste. In: Courtine J-F (ed) Figures de la subjectivité. Éditions du CNRS, Paris

Bondolfi G, Mazzola V, Arciero G (2015) In between ordinary sadness and clinical depression. Emot Rev 7(3):216–222

Boss M (1977) Existential foundations of medicine & psychology. Jason Aronson, Northvale

G. Canguilhem (1966) On the normal and the pathological. D. Reidel Publishing Company, Dordrecht, Boston and London. Il normale ed il patologico (1994) trad. it Einaudi

Goldstein K (1939) The organism: a holistic approach to biology derived from pathological data in man. American Book Company, New York

Greisch J (1994) Le phénomène de la chair: un «ratage» de Sein und Zeit. In: Floviral G (ed) Dimensions de l'exister. Etudes de anthropologie philosophique, Bibliothèque Philosophique de Louvain. Peeters, Leuven

Greisch J (1999) Les limites de la chair. In: Olivetti MM (ed) Incarnation. CEDAM, Padua

Gros-Azorin C (1998) Le phénomène du corps (leib): Une entente participative. Les Études Philosophiques, pp 465–477

Heidegger GA 2 1977

Heidegger GA 6.1 1996

Heidegger GA 89 (n.d.)

Kant I (1786). What does it mean to orient oneself in thinking? In Allen W. Wood (transl and ed) Immanuel Kant, religion and rational theology. Cambridge University Press, Cambridge, 1996

Levinas E (1979) Totality and infinity: an essay on exteriority, vol 1. Springer Science & Business Media, Dordrecht

Lorenz K (1966) On aggression, trans Marjorie Latzke. Methuen, London

Mazzola V, Vuilleumier P, Latorre V, Petito A, Gallese V, Popolizio T et al (2013) Effects of emotional contexts on cerebello-thalamo-cortical activity during action observation. PLoS One 8(9):e75912

Patočka J (1995) Papiers phénoménologiques. Editions Jérôme Millon, Grenoble

Ricoeur P (1983) Narrative time. In: Mitchell WT (ed) On narrative. University of Chicago Press, Chicago
Ricoeur P (1992) Oneself as another. University of Chicago Press, Chicago
Ricœur P (2001) Le juste 2. Ed. Esprit, Paris, p 9
Ricoeur P (2010) Power and Violence. Theory, Culture & Society 27(5):18–36
Stolorow RD (2008) The contextuality and existentiality of emotional trauma. Psychoanal Dialogues 18(1):113–123
Tinbergen N (1968) On war and peace in animals and man. Science. 160(3835):1411–1418
Van der Kolk BA (1989) The compulsion to repeat the trauma. Psychiatr Clin North Am 12(2):389–411
Volpi F. (1996) Introduzione. In Kant I (1996). Che cosa significa orientarsi nel pensare? Adelphi
Westphal CFO (1871) Die Agoraphobie, eine neuropathische Erscheinung. Archiv für Psychiatrie und Nervenkrankheiten 3:138–161, and 219–221
Wiesel E (1996) Mémoires 2, Paris. Éditions du Seuil, Paris

9 Corporeality and Organisms

9.1 Corporeality and Existence

There is no doubt that Heidegger grasped the otherness of the flesh, the "thrown" otherness of which being-there in each case takes charge (see Greisch 1994); and it is interesting that in his work the relation with this otherness is precisely what marks the difference between man and animal. As we will see, this otherness is the key to understanding the kind of privative zoology (Dastur 1995) which can only be defined by thematizing the proximity between man and animal.

To describe the phenomenon of thrownness (the fact of finding oneself in the world without being aware of one's origin and destination)—the condition constituting the foundation of Arendt's theme of natality[1]—Heidegger uses the German term *Geworfenheit* (thrownness), an expression echoing the verb *werfen*, which in everyday language is applied to animal birth giving, to describe the expelling (*werfen*) of a litter (*Wurf*). Wolfgang Marx notes: "Cats have a litter – we speak of a litter of cats, dogs or pigs, and say: 'They have been released into the world by the dams'. Things are no different in the case of human beings … This is also the distinctly human facticity: thrownness, which is to say facticity" (p. 85, Marx 1998).

The condition just described may better be understood in the light of our emotional, existential situatedness, which combines self-disclosure with openness to the world, affection with the understanding of what we encounter, and the passive bearing of the weight of existence with the duty of bearing this burden.[2] Existence, therefore, is not grasped as an object of knowledge by means of reflection, but rather relates to oneself through its mode of actualization—by relating to the other (and the world).

[1] For a more detailed analysis, see Durst (2004).

[2] From this perspective it is easy to appreciate the difference between the conception of the flesh as a motility which precedes me, so to speak, and the notion of the experiencing of the body as an effort, as a motor impulse (see, e.g., Maine de Biran).

G. Arciero et al., *The Foundations of Phenomenological Psychotherapy*,
https://doi.org/10.1007/978-3-319-78087-0_9

However, Heidegger never developed a conceptual apparatus enabling him to consider corporeality and the body as two aspects of the motility of life itself, of the power of corporeal motility—a motility which is not to be confused with, or reduced to, that of the motor system, since perception too is a form of motility.

It is against this problematic background that Patočka develops his criticism of Heidegger's analytics, accusing him of making the ontology of existence too formal (1988, p. 93). As ipseity always amounts to the establishment of a relationship, which in each case coincides with the appearance of a world and the manifestation of oneself, every possible disclosure of a field of view is correlated with the fulfillment of certain existential possibilities rather than others. In other words, through its enactment, ipseity makes things appear and, in turn, these shed light on ipseity. Therefore, the mode of being of that entity which each of us is, the existence of each one of us, differs from that of other entities precisely in the following respect: in enacting its meaning, it relates to other entities, bringing them to manifestation; it is entirely its own praxis, the execution of motility.

In order to be actually fulfilled, the meaningful possibilities with respect to the situations we encounter over the course of our existence must be concretely implemented. This means that factically engaging with the world necessarily implies the activity of a corporeal being-there. This point constitutes precisely the focus of Patočka's criticism of analytics of existence: if it is only by virtue of corporeality that existential possibilities—whatever they may be—are *practically* actualized, existential analytics does not provide any explanation for how this enactment occurs.

According to Patočka, factical engagement with this or that situation is based on an original possibility, on which all other possibilities rest: the possibility of acting and suffering, the motility of life. Thus corporeality is grasped as the primordial possibility, as an ontological condition which, as such, co-determines all existential possibilities. So by grasping existence in terms of its enactment, Patočka envisages the body from the perspective of actual life: in terms of that motility which lies at the basis of all possibilities. The characterization of existence as that motility of life, whereby a body becomes such (*leiben*) in every situation it encounters over the course of its life, marks a fundamental difference between corporeality and the other entities in the world.

As we have seen in the previous chapter, the ontological status of corporeality is necessarily overlooked or, rather, circumscribed by Heidegger, insofar as he considers the body in the present as a subsistent thing and never as an existential possibility. Hence, he cannot posit that the body, as motility, is the possibility of all possibilities and that, as such, it has an ontological position, in the sense that it may be understood as one of the existentials of being. But if the body in the present and corporeality, as the openness to this or that thing, are two manifestations of the motility of life itself—as passiveness and activeness—then motility as a whole must be understood in terms of the body. "Our existence – Patočka writes – is essentially corporeal, embodied, and our own body – as a living body, capable of movement, over which we have power – is the foundation of every living of experience. Power over the body is that being which, as a whole, coincides with the understanding of its most fundamental possibilities, without which the living of experience, which is itself self-understanding, would not be possible. For this reason, it is necessary to state once more: our existence is such that not only movement intrinsically belongs to it, but itself, its whole nature, is movement" (p. 107, 1995).

Patočka thus stresses a key point which binds together the body, existence, and motility: the fact that each person's body, as actual motility, coincides with that fundamental dimension which enables existence to find a place in the world through the factical fulfillment of the person's distinctive possibilities. In the context of this enactment, every possibility is at the same time a way of relating to oneself—every understanding, a being affected by something. Patočka writes: "In existential terms, the body is the sum of possibilities which we do not choose but within which we find a place, the possibilities with respect to which we are not free but which we *must be* ... it is only on the basis of them that 'free' possibilities are disclosed" (p. 94, 1988). In this respect, existence is necessarily embodied, and corporeality, as a concrete enactment of life, is existentially determined.

The dimension of corporeality which realizes itself through motility, de facto binding existence to the world, reveals the interworldly quality of existence, while at the same time showing how the "I am" differs from other entities. Unlike things, which are ever-identical subsistent entities, the motility characterizing being-there is never self-coincident; existence is always incomplete. It is always intertwined with a deficiency, lack, or want, with a "not yet," with an enduring difference to oneself. This difference between the modes of being of entities, this gulf between the "who" and the "what," helps account for how, in enacting his existential possibilities, each person relates to other entities by making them appear, while at the same time receiving himself. In other words, each person enters into a relationship with himself by relating to something else, and this relationship implies corporeality as situatedness.

Corporeality, then, manifests itself as the moment of articulation of our belonging to the world (through the fulfillment of our own distinctive possibilities), of the appearance of the world (in our relation with the entities we encounter), and of our being with others (as entities that, like ourselves, have a world). If existence in its worldly dimension is grasped as corporeality, it may be seen to ground the possibility of meeting the other as an entity, which in turn is bound to the world and reveals itself in the things it does and says—in its own encounter with the circumstances of life. The fact of finding ourselves in a world in which the other coexists with us, in each case situating ourselves with respect to him, is thus based on this fundamental ontological trait: a mandatory starting point for anyone wishing to explore the phenomenology of love.

9.2 The Horizon

Following a different path, we once again come across the analyses of the motility of life which we outlined at the beginning of the present work. We have seen how in Aristotle movement—which is synonymous with change—is understood according to the paradigm of the production of objects, as the transition from something which is no longer present to something else which comes into presence while its substrate (*ousia*) endures in its selfsameness. Heidegger's great intuition was precisely to bring to light the fact that this ontological horizon, centered on a view of being as a thing which is present and available (*ousia*), endures as the foundation of modern ontology. We have traced the development of this metaphysics of presence down to the very heart of the contemporary natural sciences and neuroscience. Heidegger was to "destroy" this perspective by remaining faithful to the incompleteness of man's being-there, without ever betraying its nature: he deconstructed the *ousia*, envisaging history in terms of change, starting from an ontological interpretation of time.

The significant contribution which Patočka made to Heidegger's perspective, and which makes it possible to overcome the impasse created by the ontologizing approach of the *Zollikon Seminars*, is the understanding of motility as corporeality starting from existence. Besides, Medard Boss too in his *Existential Foundations of Medicine and Psychology*—a text which "evolved under Heidegger's watchful eye" (p. xxiv) and may therefore be regarded as an outgrowth of the *Zollikon Seminars*—establishes corporeality as an existential (p. 166). Boss provides the following outline: "Man's bodyhood occurs exclusively as the bodying forth of his existential dwelling amid the beings that address themselves at any given time to his perception and require from him an appropriate response" (p. 103).

If being alive means being open to that which we encounter, according to given possibilities of enactment, in each case being-there becomes what it is through movement. The motility of life (*kinesis tou biou*), in other words, does not correspond—as in Aristotle—to an entering into the presence of a pre-given potentiality resting within a substrate which endures unchanged. Movement, the coming to oneself, the process of occurring, is rather what leads being-there to itself, what enables it to access itself by generating the substrate anew on every occasion. Thus the body is constituted and reconstituted at each instant in and through the motility of existence.

"On the foundation of corporeality, our activity is always a movement *from... toward...*, it always has a point of departure and a goal. On this foundation, our existence is always burdened, in regards to its activity, with the weight of need, of repetition, of restoration and of the prolonging of corporeality itself. The circle of existence (existing in view of oneself, in view of the mode of one's being) always contains in a certain way the circle of life that accomplishes vital functions in order to return in itself and to come back to itself – in such a way that life is the goal of all its particular functions" (1988, p. 105). Motility, as a movement which changes things, as a movement which "is capable of doing," and hence as a primordial movement which qualifies life, is defined as a centrifugal energy which directs life toward things: a "seeing force" which orients the individual's encounter with things against the background of a horizon that is constantly receding and enveloping the environment which the living being discovers and through which he establishes himself as such.

Corporeality emerges here as the absolute zero object—"one so near that it is not thematized" (Patočka 1998, p. 27)—and the origin of that seeing force which only makes itself visible via refraction, by reflecting oneself from things, just as a beam of light is refracted on the surface it encounters; it is present as a dynamism which we experience but which does not appear before our eyes, whereas it is things that manifest themselves and the contexts which they point to within the framework of a given horizon. Patočka notes: "The living of experience is like a weft stretched between two horizons: one is my self, the other the world" (p. 63, 1995).

Along this trajectory, Patočka's view of the horizon, with its spatial emphasis, converges with Koselleck's analysis, which stresses its historical dimension. It may be argued that both are summed up by the dictionary definition quoted by Koselleck: "Horizon: an apparent line which separates sky and earth, and which recedes when approached" (Drozdzynski in Koselleck p. 261). For Koselleck the horizon of expectations refers to what one expects from the future and which is evidently restricted by one's previous experiences. By crossing the horizon line established on the basis of previous experiences, every new experience shifts its boundaries. This unique interweaving of experiences and expectations, of past and future, is a distinguishing feature of historical time and of its transformability.

On the other hand, according to Patočka, the creation of a horizon points to that centrifugal energy which dominates our life and binds it to the world according to a range of vanishing points—directing it further and further into the distance—while from the world we return to ourselves, situating ourselves within it. The world "is sometimes defined as the horizon of all horizons: the horizon of reality as a whole, within which each partial horizon, each closed connection of meaning and understanding finds its place" (p. 64, 1995). The motility of life directed toward the world, therefore, engenders—on the basis of the situatedness ensured by the body—meaningful possibilities which at each moment combine within the weft stretched between two horizons, the horizon of myself and that of the world, between myself as a seeing force and the possible meaningful horizons according to which individual reality takes shape.[3]

9.3 Corporeality and Incompleteness

Patočka repeatedly stresses the orienting function of motility, which consists of two elements: a prereflective force which sheds light on things as an "I can," "I do," and "I find myself" and the horizons that make up the world as an overall context of meaningfulness. "As a seeing force, it must contain something like a light, a light by which it illuminates its path. This path is a movement towards and penetration of the world, the very possibility of which is implicated in the force that takes on the title of myself" (p. 66, 1995).

[3] What clearly emerges here between the lines is Heidegger's notion of *Bewandtnis*: each object has its own state, its own meaning, and its own "involvement" in the light of the sum of references which situate it within a context in view of which it acquires a given meaning.

This perspective leads to an existential analysis of sensibility founded on emotional situatedness (p. 47, 1998): our emotional position with respect to the events we encounter co-determines the disclosure of sensory fields as possible meaningful horizons and the coordinating of the various "sensible horizons," in each case engendering a certain mode of relating to the present situation.[4]

As a life context which is articulated and stabilized in different ways as life unfolds, changing in accordance with its transformations, the world binds life to itself, acting as its content; as such, the world can seduce, entice, enchant, and enchain life—it can drag it and put it to sleep (*Hang*). This is what *letting oneself live* means! On the other hand, we have the gravity of life, the tendency which leaves one no choice, the urge (*Drang*) springing from this irresistible tension, the blinding need which imposes itself and no longer allows the world to appear, undermining all meaningfulness. In the domain of psychology, we have defined these two dimensions of man's existential relatedness, which are ontologically rooted in life, as outwardness (*Hang*) and inwardness (*Drang*) (Arciero and Guidano 2000; Arciero 2002, 2006; Arciero et al. 2004; Bertolino et al. 2005; Rubino et al. 2007; Arciero and Bondolfi 2009; Mazzola et al. 2010, 2014, 2018). These are the two complementary aspects that merge in life, as it takes care of things—the two unavoidable dispositions whose combination in each case determines our openness to what we encounter.[5]

To the extent that existence entails a given sense with respect to the meaningfulness of every event (actualized in the way in which the seeing force flows into things), sensory fields open up and, with them, motor possibilities, which enable the engagement of certain forms of sensibility (starting from one's own emotional situatedness). From this perspective, not only does motor activity also imply perception, but the latter is itself the enactment of motility.[6] Consequently, if those fields of sensations that accompany our experience are oriented according to a more primordial motility, any attempt to understand sensibility in physiological terms must address the following question: what is the point of studying the visual phenomenon of the perception of the color brown, for example, by isolating the individual function and the cerebral loci related to it, if phenomenology has shown that the various sensory fields converge in the unity of an object, in the light of a receding horizon? Sensory fields become actualized with the appearance of things against an invisibly meaningful background (Mazzola et al. 2018).

If corporeality lies at the center of these fields, as the axis on which they are oriented, it gives and knows itself, on every occasion, with respect to the situations which occur. Hence, Patočka (1995) can argue that the body scheme is "a prereflective awareness of one's own dynamic and its development, the *creation* of a

[4] Patočka's analyses on the seeing force, which echo his research on "negative Platonism," here overlap with those on *Sorge* in Chap. 6 § 41 of GA 2 1977 (see too Patočka 1998 lecture xvii).

[5] The distinction just made is clearly indebted both to § 41, Chap. 6, Section 1 of *Being and Time*—where care (*die Sorge*) is understood as the precondition for these two dimensions—and to Scheler's reflections on sympathy (1970).

[6] Patočka repeatedly evokes the inseparability of *kinesis* and *aesthesis*, a notion dear to Aristotle.

position, possessing a non-thematic inner clarity – an action which, without perceiving itself, is nonetheless sensitive to what hinders or supports it: it is as though my action within the world drew its own force lines" (p. 67–68). The sensory field here coincides with the perspective and possible horizon according to which we meet other people and things, which in turn are laden with references. Dreams, the past and the future, the various spheres of the imagination, abstraction and schematization, nature, history, society, everyday life, work, eroticism, loneliness, familiarity, and unfamiliarity: these are only some of the many possible horizons within which the events we encounter may take shape.

We can situate ourselves within each of these horizons, accessing a reality which addresses us and draws us in. It is within the framework of these horizons that we act and accomplish our tasks. Hence, unlike animals, which depend on the context they find themselves in, for human beings what is directly present is not limited to what is immediate. The horizon always leads man to be ahead of himself: it provides a range of possibilities on the basis of which a course of action is capable of undermining reality. This capacity to be ahead of oneself is precisely what becomes altered in the attention deficit hyperactivity disorder (ADHD), where the transition from one actuality to the next is detached from all connections. Indeed, with the vanishing of the horizon, the present coincides with the immediate.

The range of possible horizons constituting the world makes every human moment pregnant and makes it exceed the present in terms of possibilities. This excess dominates and surpasses us, as the seeing force which leads us toward the world and which coincides with the generation of horizons, orients us while situating us, or situates us while orienting us. On the one hand, the emergence of horizons in the direction of the world ensures that we will position ourselves within its domains; on the other hand, self-refraction—how we experience our return to ourselves—occurs as a situating of ourselves within the world by embracing its possibilities, so as to be integrated within it.[7]

This centrifugal energy which Patočka defines as "myself" is the corporeal. "My" corporeality is the corporeality of a force. In this sense, Patočka understands corporeality as the motility which stands at the basis of all possibilities and at the origin of all possible horizons. "The personal body is not a being in the way that a thing is, but as a relationship, or rather as a relating to oneself that is the subjective relationship that it is only by making the detour through an extraneous being" (p. 59, 1995). As a relationship, motility implies a constitutive difference, a "from... to...," an incompleteness which is always also a form of precariousness. This gap, this non-present, this coming-into-being ever anew, is always in search of an actualization, a "whither," a meaning; and as certain existential possibilities are fulfilled, the incompleteness of the movement is renewed, the "whither" is disclosed, and the horizon recedes.

We have repeatedly stressed the fact that understanding change starting from what persists through it—a substrate which remains ever the same—does not allow

[7] This situating ourselves may solidify into a sort of selfsameness which we experience as facts taken for granted.

us to grasp the movement of human existence, which is distinguished by a constitutive impermanence: a line which draws back as we approach it. The fatal mistake made by natural-scientific psychology has been precisely to unquestionably adopt this perspective, in the belief that it could account for the life of people by referring to unchangeable traits. Consequently, this has also been the starting point of natural-scientific forms of psychotherapy, which are all oriented according to the Aristotelian principle of healing, modeled on the art of crafting things.

If motility implies a shift from one thing to another that is always accomplished through the enactment of a possibility, and is therefore constantly renewed, and if the movement of life always coincides with the fulfillment of possibilities with respect to things, followed by the discovery that this movement is once again unaccomplished, then it is motility itself which achieves the synthesis of "whence" and "whither," by orienting our life toward the world which on each occasion refracts it, making it revert to itself. From this perspective, change does not mean transformation with respect to that which subsists, but rather the realization "of something that is not yet present and that can take the given into itself and forge it into a unified meaning" (Patočka 1998 p. 147).

Doing, here, means concretely finding our bearings with respect to that which, in approaching us, enables the enactment of given existential possibilities through the appropriation of the possibilities we are offered to situate ourselves within the world; "and it is only through this dialectical path, through self-reification, through bodying-forth, that we develop and come to perceive ourselves as a free, historical and incomplete being which passes beyond each of its stages while preserving it" (p. 93, 1995). As a movement which, in refracting itself in things, reverts to its origin, corporeality fulfills its distinct possibilities through this orienting of oneself in the world and in each case bodies forth, renewing its incompleteness.

Patočka thus contributed to radicalizing that kind of research on human existence which, in remaining faithful to the precariousness of life, had turned the phenomenology of intentionality into the phenomenology of living (Barbaras 2007). But whereas Heidegger had grasped embodiment (*Leiben*) as "the immediate emergence of all of our so-called material, bodily nature of the possibilities of the self" (GA 89 n.d. p. 296)—starting from the fundamental constitution of human existence (as the disclosure of a world)—Patočka goes one step further: he accounts for the belonging to the world of being-there and hence of that ontological trait that anchors the possibility of meeting the other and all things as entities that are themselves intraworldly.

9.4 Living Beings and Life

Corporeality, therefore, is not the substrate of existence but a bodying forth which, in becoming such again and again, renews its horizons and regenerates its history. The understanding of corporeality as the principle of acting, and the precondition for the enactment of the possibilities of being in the world—where the latter is understood as the "whither" that orients our effectuations—enables us to define life and existence as two moments of motility: the "whence" and "whither" of the movement. Starting from motility, we can account for that tension between the existential determination of the body and the embodiment of existence which constitutes the essence of motility. The motility of life implies a constant regeneration of this shift and hence the creation of new possibilities. Echoing one of Heidegger's fundamental theses, Patočka argues that "life is existence, but existence is not life through and through" (in Barbaras 2007, p. 123).[8] Living is both a mode of belonging to the world and the precondition for its appearance: the "how" of motility. Human life, in other words, coincides with this ongoing establishment of oneself within the world and, at the same time, with one's awareness of it.

If we conceive of man's humanness starting from the fact that, by enacting itself (*Vollzugssinn*) as a relationship (*Bezugssinn*) with things and others (*Gehaltssinn*), life lends them meaning—and as this relationship becomes established, living and the world in which we live in each case take shape—then this "fundamental belonging of humanity to life" (Barbaras 2007, p. 129) raises some fresh questions. First of all, given that man shares the sphere of life with all other living beings, how can we account for its unity and underlying differences? This problem conceals an even more crucial one: how can we account for the mode of being of life, since it is precisely this mode of being which stands at the basis of the unity of all living creatures and the differences between them?

In the first part of this work, we retraced the research trajectories in physiology which led from Müller's intuitions, through the methodology inaugurated by Helmholtz and the mathematical model later developed by McCulloch and other researchers, to the developments in second-order cybernetics. Within this framework, life—in its differences and unity—was grasped starting from the conditions determining the emergence of the distinguishing features of every living organism: biological autonomy and self-organization (Dumouchel and Dupuy 1983; Varela 1979; Atlan 1979).

In accordance with its biological makeup, every living organism selects—among countless perturbations—those environmental stimuli to which it can react without losing its identity qua organism. The effects of the world upon the organism are due to the translation of external stimuli into nervous impulses, sensory-motor structures, and modifications of the inner dynamics of the living creature. Among the countless perturbations from the external world, those which may turn into stimuli do not provide any information on their own nature, but only on the organism which recognizes them as such by referring them to its own inner structure (self-reference).

[8] Heidegger states: "Bodying forth as such belongs to being-in-the-world. But being-in-the-world is not exhausted in bodying forth. For instance, the understanding of being also belongs to being-in-the-world. [This understanding of being includes] understanding the fact that I am standing in the clearing of being, and [it also includes] the particular understanding of being, that is, of how being is determined in [this] understanding. This limitation [being-in-the-world] is the horizon of the understanding of being. Bodying forth does not occur here (*Hierbei geschieht kein Leiben*)" (GA 2 1977, p. 266–267).

What emerges between the lines here is Müller's old principle, according to which stimulability and the capacity to modify one's activity by responding to a stimulus constitute the distinguishing traits of living beings. Hence, while the unity of life is to be sought in the common mechanism—i.e., organizational closure (as theorized by Varela 1979) or self-organization more generally (Atlan 1974, 1979; Zeleny 1981)—which underlies biological autonomy and enables its realization and preservation, the diversity among living beings has to do with the different ranges of stimuli to which each individual organism is capable of reacting. Different organisms promote the conditions for their survival bringing forth different living worlds in relation to their own biological makeup, which remains the same over the course of their entire life cycle.

As has been stressed, analyses of this sort still rest on the paradigm of production: that unthought Kantian debt, whereby on the basis of ancient ontology, organisms are grasped in the light of a conception of being tied to the notion of permanence and endurance. The relation of each organism with its environment, understood as a cognitive relationship, is determined by its organization: by preserving its stability over time through modifications of its inner dynamics, the organization of the organism defines those stable properties according to which environmental transformations are to be dealt with. What resurfaces here, in a strictly biological form, is the very core of the Kantian revolution, according to which it is objects that must conform to our cognition—and not the other way round. Following the path paved by Fichte, who grasped the circularity between doing and knowing as a primordial and absolute experience (see p. 50), second-order cybernetics extends the Kantian principle to all living beings. Maturana writes: *Living systems are cognitive systems, and living as a process is a process of cognition* (p. 12, 1970).

Therefore, not just man but every living system, with or without a nervous system, in acting within the world in a way that is relevant to its own adaptation, distinguishes features which, being relative to the organism that has detected them, are peculiar to the organism and, more generally, to its species. In this respect, every organism is reflectively closed upon itself in a recursive fashion through its openness to the world that it upholds. As a cognitive biologist might put it, *every act of knowing brings forth a world.* Hence, for every living being (and not just for the I, as Schiller suggested), the world is like a ball that is thrown and then caught in reflection; and each living being throws and catches this ball on the basis of that physiological a priori—the body understood as a sensory-motor whole—in the light of which, as Rudolph Magnus argued, the reality of the world becomes graspable.[9]

In the first part of this study, setting out from the difficulties raised by the emergence of neurophenomenology, we deconstructed the perspective which extends from Kant to Fichte, only to resurface—over a 100 years later—in second-order cybernetics. We have therefore conducted our analysis by focusing on neuroscientific developments from Helmholtz to Varela. Although in this reconstruction our emphasis has been on the ontology which has supported and determined this

[9] This perspective further clarifies the key maxim of constructivism, namely that objects do not exist independently of the subjects who sense them.

development and which has never really been thematized by those endorsing it, we have still followed an "institutional" path, so to speak. After all, the path in question had already been partially traced by McCulloch when, in investigating the genealogy of his own view, he had drawn a link between the idea of a priori synthetic judgment and the work of his own teacher, Magnus' pupil J.G. Dusser de Barenne.

In accessing the whole issue according to this retrospective investigation, however, we have failed to grasp the contribution provided by one of the least known yet most significant figures associated with the developments in question: Jakob von Uexküll. Von Uexküll's importance is not limited to the influence he exerted on his friend Magnus, and hence on cybernetics, but also extends to the direct engagement of Heideggerian phenomenology with his thought and, finally, to the new horizons which his insights opened up in the evolutionary field, down to the contemporary prospects of a new synthesis.

9.5 The Many Living Worlds: Nothing but Soap Bubbles!

Von Uexküll's influence on Magnus is eloquently described by Sir Henry H. Dale in an introductory note to the Lane lectures (1930). After recalling Magnus' admiration for and indebtedness to Wilhelm Kühne (Helmholtz's successor at Heidelberg), with whom Uexküll too was working in those years, Dale writes: "Magnus' lifelong friendship with Otto Conheim and with Jakob von Uexküll also began in Heidelberg. All who knew him later with any degree of intimacy will remember the respect, spiced with the humor of true affection, with which Magnus spoke of Von Uexküll's philosophical outlook and its deep influence on his own intellectual development." The two not only shared the same keen interest in Goethe—in his role as naturalist and author of the *Theory of Colors*—so much so that they repeated his experiments using the same tools (Otto Magnus 2002), but in the marine station in Naples, they also conducted some research on the marine worm *Sipunculus* (Uexkull 1903) and on the hydromechanical mechanism exhibited by it. According to Dale, this work constituted one of the foundations of Magnus' later studies on the factors controlling the changes of animal posture in relation to gravity.

The full extent of the influence which Dale speaks of is revealed precisely by the aforementioned last lecture of Magnus', which offers a new perspective. The underlying topic of the lecture, which is explicitly explored through the examples of physiological a priori provided by Magnus—i.e., reflection, the regulation of the position of the eyes, "righting reflexes," etc.—has to do with the way in which different animals are tied to the earth, which is to say with the (species-dependent) animal posture with respect to gravity and the muscular tone by which such posture is maintained. This posture is the "ordinary position" to which human beings or animals always revert after the arbitrary movements evoked by their cerebral cortex and is what enables all the senses of the body to regain their precise relation to the outer world.

This a priori relationship between an organism and its environment is determined by a group of reflexes that sets the body segments' collective orientation against external disturbances, the precondition for all perception, and the starting point for all action. This fundamental relationship, which varies from species to species and which Magnus explores in his study of posture (1926), is merely one aspect of the more general principle that the world within which an organism is structured entails that only certain stimuli may act upon the organism.

Von Uexküll uses the expression "environing world" (*Umwelt*) to describe this peculiar relationship which every animal establishes with the world, by carving out for itself a living space tailored to its own needs, and hence on the basis of its distinguishing traits as an organism: "Every animal is a subject – von Uexküll writes – which, in virtue of the structure peculiar to it, selects stimuli from the general influences of the outer world, and to these it responds in a certain way. These responses, in their turn, consist of certain effects on the outer world, and these again influence the stimuli. In this way there arises a self-contained periodic cycle, which we may call the functional circle of the animal" (1926, p. 126).

In each animal's world, then, there are things that belong exclusively to that specific animal. An object belonging to a world peculiar to a given organism activates the latter's sensibility by affecting its organs of perception, which only pick up those signals which the animal is predisposed to detect.[10] This activation is transmitted to the nervous centers constituting the perceptual web (*Merknetz*) in which the objects responsible for the activation are recomposed. The reception of the stimuli then sets the effector organs and the corresponding activity web (*Wirknetz*) in motion. The connecting structure, with its activity webs linking receptor and effector organs, constitutes the animal's inner world, and it is precisely this constant sensory-motor interaction which shapes the animal's receptiveness to the environment. Uexküll states: "The functional circuits of animals start with perceptual markers – the properties of objects – and extend through the inner world of the body and turn back to the object with the effectors" (p. 218, 1921).

The functional circle ties each animal to its environment by various instinctual modes (attack defense, reproduction, feeding, and *milieu*), which connect the inner world to the external environment through a closed circularity. Long before second-order cybernetics extended the Kantian perspective to every living system, reconceptualizing life in terms of cognition, Uexküll—following in Helmholtz's Kantian footsteps—made this perspective his own by attributing to every organism the capacity to build a world in accordance with its own experience. The natural world—which Darwin had regarded as a single whole and entrusted to natural selection—multiplies into countless functional circles, the *Umwelten*, associated with an endless variety of living beings, each as perfect and well-adapted as the rest. Individual animals uphold as many worlds in complete isolation by preserving an impenetrable autonomy. This is how Uexküll sees these worlds: as soap bubbles.

[10] It is worth recalling here that Uexküll provided the foundations for one of the most famous experiments in cognitive biology (i.e., Lettvin et al. 1959). To illustrate how things act as sources of stimuli for an animal on the basis of its species-specific organismic traits, he mentions, among other examples, that of the frog: "The frog zeroes in on the fly and launches her tongue out toward it, in the space occupied by the fly" (p. 127, von Uexküll 1946–2014).

The *Umwelt*, in other words, constitutes a kind of virtual perimeter around the organism, marking both an "inside" within which certain things are significant and meaningful and an "outside" in which all other things are nonexistent—"hidden in infinity" (1926 p. 42). These multiple worlds intersect and combine, as though in a grand compositional harmony.

The acknowledgment of this new position of the animal with respect to things poses a radical challenge to the anthropocentric perspective in life sciences. In the preface to *Theoretical Biology*, Uexküll schematically yet accurately redefines the task of biology, suggesting two possible ways of further developing Kant's research: "(1) by considering the part played by our body, and especially by our sense-organs and central nervous system, and (2) by studying the relations of other subjects (animals) to objects" (p. XV).

This new understanding of each animal and its world—which, as Agamben (2002) notes, coincides with the radical dehumanization of the image of nature—goes hand in hand with a repositioning of man and in particular of the biologist: like all other animals, man too creates his own environing world, and biology is indeed such a world, which completely cuts him off from the universe. But if the biologist can distinguish the *Umwelt* of an animal by studying those meaningful elements—perceptual markers—through which the animal orients itself, he can only do so because he approaches this *Umwelt* as a single whole set against a background, which Uexküll calls the surrounding environment, *Umgebung*. In fact, the surrounding environment which serves as a background for the animal's *Umwelt* coincides with our environment, the biologist's *Umwelt* (Canguilhem 1952).

This intertwining, or rather overlapping, of different worlds, which constitutes the study domain of biology, raises a new crucial problem: that of the relation between the observer, who in order to study living organisms must use his sense organs and central nervous system, and the world of the organism that is being observed.[11] Indeed, if the observer only has access to his own world and not that of the animal as well, all that he can phenomenally perceive of the latter's external world are those features which his own perceptual apparatus enables him to detect, whereas those belonging to the subjectivity of the animal are bound to elude him. Uexküll returned to the issue of this unbridgeable distance again and again over the course of his career, developing a methodology which may summed up in a sentence that retains a strong Helmholtzian flavor: "Insofar as it represents an established connection, the whole life of each animal can and must be conceived mechanically" (1921 p. 47), if only in the form of a self-organizing machine.

Thus, in an attempt to grasp the features perceived by an animal—and which coincide with his own—the observer, placing himself within the functional circle, must proceed from the organs selecting the stimuli to the underlying organization, reconstructing those rules that govern both perception and all conformity to the organizational plan of the organism. This confirms Helmholtz's methodological

[11] The problem is expressed in the very formulation of point 1 of Uexküll's *Theoretical Biology*, namely, the assertion that brings the lecture on Magnus' physiological a priori to a close (see p. 39) and which was destined to endure as one of the key issues of cognitive biology down to Varela.

maxim, which accounts for the behavior observed by invoking the mechanism that engenders it. In a text from 1928, Cassirer observes: "The key principle of this biology consists in the fact that access to the different worlds of individual life forms can only be ensured by studying its organization" (p. 51, 1996–2003). However, the study of the building plan implies a constitutive ambiguity between its current existence relative to the various organisms and the modelization of those organisms implemented by the observer (Brentari 2015).[12]

9.6 Freiburg, the Winter of 1929–1930

The differences and similarities between human beings and animals are the subject of one of the most interesting works of Heidegger, *The Fundamental Concepts of Metaphysics* (GA 29-30 1983), which brings together the lectures he delivered in Freiburg in the winter of 1929–1930. In the epigraph, Heidegger informs us that Eugen Fink, the dedicatee of the volume, had repeatedly expressed the desire for this lecture course to be published before all others and tells us the reason for this. Probably, Heidegger suggests, in following this course "with thoughtful reticence," at the time (40 years before) Fink had already "experienced something unthought of his own that determined his path." What is Heidegger referring to? What is this "something unthought of his own" on the basis of which Fink, Husserl's favorite pupil, who had followed Heidegger's lectures, considered this text a privileged access to Heidegger's *Gesamtausgabe*? We will have to wait until the end of the following chapter to answer this question.

There is then another undoubtedly significant circumstance which is bound to shape our understanding of this course, because it allows us to identify its place in relation to Heidegger's broader research trajectory. In the summer term of 1929, which is to say, the one just before the course in question, the main topic of Heidegger's lectures was German idealism and its bearing on the philosophical situation of his own day. The course, which set out from the relation between anthropology and metaphysics—two areas which Kant had been the first to bring together by disputing the very possibility of metaphysics—actually led to a new engagement with Fichte's thought.[13]

The main point of interest for Heidegger is the return—understood as a repetition (*Wiederholung*)—to Fichte's notion of facticity, which had already been prominently featured in his habilitation work at Freiburg (1915–1916). In other words, Heidegger sought to return to that Fichte who had discovered, beyond the "I think"

[12] The problem of the observer—a central concern of Uexküll's throughout his career (see Uexküll 1926 p. 139; see von Uexküll 1921 p. 217)—is openly investigated in the last chapter of *Umwelt und Innenwelt der Tiere*, entitled *Der Beobachter* ("The Observer"). At the end of this chapter, to which substantial changes were made for the second edition, published over 10 years after the first one of 1909, in the light of this problem, Uexküll outlines 21 principles that the biologist "ought to bear in mind."

[13] While repeatedly discussing German idealism in his extensive oeuvre, Heidegger was never to return to Fichte.

(conceived as the irreducible and unsurpassable limit of reflection), the absolute position of the I as an active subject: the unavoidable facticity (*Factizitat*) of life that cannot be questioned any further. In his habilitation thesis, Heidegger had especially been influenced by the work of Emil Lask, who had introduced him to the concept of facticity, expressed in terms of "factualities" at the time. Lask influenced Heidegger "more than [Heidegger] himself is conscious of," as his thesis supervisor, Heinrich Rickert, observed (see Kisiel 2008, p. 45). The impact which the logical investigations conducted by the Fichtean Lask had on the first steps, which the young Heidegger was taking toward his formally indicative hermeneutics of facticity, had to do with the facticity of meaning and hence with the intentional relation, as Kisiel clearly notes.[14] Kisiel sums up the whole issue as follows: "The form accommodates ("tailors": Lask II, 59) itself to a particular matter such that it is itself particularized in meaning" (p. 51).

Over the course of the years and in its many different formulations, this way of conceiving meaning, and hence intentionality, remained the core both of the method of formal indication and of the structure of human experience (Sheehan 2014): the sense of relation, sense of content, and sense of enactment are all integrated within an overall temporal sense.[15] However, is the motility of life—variously described by Heidegger over the years—not what this formal scheme outlines?

The course held in the summer term of 1929, then, marked a fresh engagement with Fichte. The lectures centered on the first three principles of the 1794 *Doctrine of Science* and hence of that idealistic *Wissenschaftslehre* which, according to Lask, foreshadowed the Positivist Fichte (Kisiel 2008). More generally, however, the lecture course provided a new opportunity for Heidegger to engage with neo-Kantianism—an engagement which had already acquired a polemical twist by the spring of the same year through Davos' dispute with Cassirer. From a broader perspective, we should consider Heidegger's engagement with Scheler—as highlighted by Claude Romano (2009)—who in a work published in those years (1928) reopened the Kantian question of the essence of man within the framework of anthropology by introducing the issue of the comparison between human beings and animals.

On the other hand, in the first part of this study, we already noted that the interpretation of Kant developed by Fichte in his Positivist years shaped the only half-acknowledged conceptuality which guided Helmholtz's research, down to the contemporary developments in neuroscience and in the cognitive sciences. This conceptuality has spawned a natural-scientific epistemology and method but especially—and most crucially—a technology capable of mirroring neural functions by means of mechanical or electronic devices. In its inevitable intertwining with

[14] In his dissertation *Fichte's Idealism and History*, Lask reveals that the term *Factizitat* was coined by Fichte himself in the middle of his Positivist period, around 1800. For an outline of the notion of facticity from Fichte, through Lask, to the early Heidegger and for an exhaustive discussion of Lask's influence on Heidegger, we will refer here to Kisiel's studies on the topic (Kisiel 1993, 2000, 2002, 2008).

[15] This way of conceiving meaning is reflected in the use of the middle voice verb form which bears the mark of intentionality: "being directed toward"/"directing itself toward" (see Kisiel 1999).

existence, this technology in the image and likeness of man has turned man himself, and not just the earth, into an object of technological domination.[16]

In our view, all these elements make up the underlying fabric of Heidegger's engagement with Uexküll, the neo-Kantian responsible for introducing in biology the concept of an animal's connection to its environment as "environing world": the very environment (*Umgebung*) whose specificity as the meaningful context for things—and hence as a signifying element—the neo-Kantian Lask had been the first to intuit, grasping it as a pre-objective, pre-theoretical world. In 1919 Heidegger wrote: "Lask discovered in the ought and in value, as in an experienced ultimate, *the world,* which was non-thinglike, non-sensorily metaphysical, as well as not unthinglike, not extravagantly speculative, but rather was factic" (quoted in Kiesel p. 252, 2000).[17]

9.7 Formal Indication and Factical Life

The real, deep debate to which the 1929–1930 volume bears witness is the debate with the neo-Kantianism guiding the natural sciences and, in particular, Uexküll's biology. In the framework of a theoretical biology oriented in this direction, the question newly emerges of the essence of man and how he differs from animals. Alongside this, we have the comparison with a stone, already drawn in the course on Fichte as an example of a subsistent entity, which is to say one with no possibility of relating to its own being (GA 28 1997 p. 67, p. 287).

Thus the topic that really shapes the course is the issue of the essence of life vis-à-vis the lifeless, of the vitality of the living being—be it the animality of animals or the humanness of man—vis-à-vis a stone, "which has no possibility of dying." This difference between the living and the lifeless revolves around the issue of one's relationship with the world, as shown by Uexküll, who made this relationship the foundation of his theoretical biology. The world manifests itself to each individual organism as a subjective and self-referential reality, the *Umwelt*, whereas inanimate matter is worldless. Hence, what distinguishes being alive is the fact of having a world which takes shape in a different way for each subject. Here biology does not at all discriminate between an amoeba and Einstein, as Popper liked to believe!

In order to continue living, each organism constantly reshapes both its *Umwelt* and its inner activity—what today we would call the connectivity between the receptive function and the active one—organizing itself according to different levels of complexity. So what every animal, including man, perceives of the world is what its organs, integrated within a single organism, are capable of detecting, according to its distinct sensory-motor organization. The difference between human beings

[16] McCulloch's computational dream was to recreate a brain, yet the running of Nazi death camps was equally technological!

[17] In *Being and Time*, the term *Bewandtnis* is articulated in terms of its prepositional nexus "with/in/for/in/order to/for the sake of," which articulates and defines the meaningful whole that we call the environing world.

and animals, then, does not only reflect a peculiar way of relating to the world (understood as the sum of all accessible entities): for it also concerns the degree of extension of this relationship, which coincides with the range of possible perturbations that the organism is capable of decodifying .[18]

The conceptualization of life just outlined traces everything back to the level of construction, to a stable form of organization that produces and renews itself while self-referentially distinguishing the world in which it realizes itself. What this perspective fails to grasp is the fact that the being of an animal manifests itself in a peculiar motility of life which determines it as such. The crucial point, as Lask first noted, is that the world is not an entity transmitting stimuli which a receiving apparatus distinguishes and arranges according to its own inner rules: the difference between human beings and animals—or, more generally, between living beings—rather concerns the way in which they relate to, bind, and deal with the world.

This centrality of the motility of life, as the ongoing establishment of a relationship, corresponds to the emergence of the world as an accessible entity and of the organism as a being open to encounters. Hence, whereas according to Uexküll and the life sciences, the unitary totality of an organism coincides with the corporeal surface, according to phenomenology an organism is a fundamental mode of being of life which realizes itself by generating itself in its relationship with the world, again and again; every organism finds its world by enacting its capacities with respect to what it encounters as something meaningful.

"All life is not simply organism but is just as essentially process, thus formally speaking motion," Heidegger states. He continues: "But in what sense? As a sequence of unfolding events? Not at all, although it is always possible to treat the processes of life in this way. Even in our everyday experience we are familiar with the birth, growth, maturing, aging, and death of animals. But all this reveals to us a motility (*Bewegtheit*) of a peculiar kind, for here the organism as we now understand it does not simply happen to get caught up as it were in this motility. Rather, this motility determines the being of the animal as such" (GA 29 p. 385).

In this sense, factical life for Heidegger is ontologically more fundamental than "life" (*Leben*) as studied by biology (§10 GA 2 1977). As we have seen, the essence of factical life is motility, understood as a mode of becoming in each case rooted in the world and, at the same time, as a condition for the appearance of the world, whose structure is logically grasped through a formal schematism. The method of formal indication, therefore, enables us to access human life and hence to grasp the other's experience through the co-enactment of tendencies and motives that are faithful to the situation under scrutiny and which are implicitly there in the factual life of the entity we go along with in such a way. It is important to stress once more

[18] In the first version of *Umwelt und Innenwelt der Tiere*, published in 1909, Uexküll writes on the problem of the observer: "The lower animal, together with its environment, presents itself to the observer as a closed unit, whereas the unity of the higher animal with its environment can never be grasped by the lower animal. This conception of the animal kingdoms creates the notion of ever greater circles each enclosing the next smaller one" (von Uexküll 1909: 252). What we have, in other words, is a series of concentric circles with the human observer at their summit, who encloses them all within an environment encompassing all others: the environment of science.

that this does not mean reducing ourselves to the other's experience, as though we were the other, nor reducing the other to ourselves by identifying with his lived experiences; rather, it means going along with the other while remaining fully ourselves.[19]

The possibility of co-enacting the experience of another human by becoming attuned to it rests on the fact that ipseity is always a relation, that existence is always a shared experience which structures itself as a way of relating to another, even in solitude. In other words, we can either go along with the enactment of another person's experience, contributing to his access to the world, or be completely indifferent to him, or even incapable of going along with him, precisely on the basis of this original sharing of existence which distinguishes ipseity. But can we also go along with an animal's experience? Can we co-enact the relation which the animal has with its world? And if so, on what basis?

Investigating the difference between human beings and animals, then, raises a methodological question, in the etymological sense of the word, insofar as it is the method adopted which must provide a route to gain access to the animal as such. The step which Heidegger takes is to extend the method of formal indication to living beings in general, thereby stripping them of their human quality.[20] The formal schematism is therefore reformulated as:

(a) An animal's capacity to relate to (*Bezug*) the surrounding environment by moving ahead of itself along "this instinctual reference toward" (p. 240 § 59). Each animal has some instinctive modes of being (attack defense, reproduction, feeding, and *milieu*) which are proper to it—behaviors corresponding to its specific capabilities. In this being ahead of itself, in this instinctive drive toward things, the organism holds itself (*Einbehalt*) and thus acquires its specific unity. Using a language more attuned to contemporary biological research, we might speak of "canalization".[21]
(b) What is capable of disinhibiting those instincts (*Gehalt*)—which Uexküll refers to as *Merkmaltrager*—in such a way that the corresponding behavior will be enacted but which never manifests itself to the animal as this or that entity (*Gehalt*). Indeed, an animal only grasps an entity in accordance with its instinctiveness; hence, in disinhibiting the animal's instinct as it approaches it, it erases itself as an entity and sets the behavior in motion. We may refer here to Uexküll's famous example of the tick (*Ixodes rhitinis*), which never perceives the mammal on which it falls as an animal, but only as a sum of elements: the smell of

[19] We have seen how this going along with the other essentially leads therapeutic treatment to foster the return of the other to himself.

[20] In such a way, animals too are no longer subjected to the objectifying approach of the theoretical perspective.

[21] This term coined by Waddington (1957) refers to the directness of development that, while reducing developmental plasticity, makes the development of an adaptive phenotype resistant to environmental and genetic perturbation that would produce deviations from optimal form (West-Eberhard 2003; Siegal and Bergman 2002).

butyric acid (sweat), a temperature of 37 °C, and a hairless place enabling it to penetrate into the mammal's skin. All these elements in succession disinhibit the attack instinct and the feeding instinct, which in turn enables the disinhibiting of the reproductive instinct.

(c) The enactment (*Vollzug*) of this instinctive mode—varying from species to species—with respect to that which is the *raison d'être* of the capacity.

The question of method thus emerges at the very core of the much discussed issue of privative zoology (Dastur 1995).

References

Agamben G (2002) The open. Animal and man. Stanford University Press. (Italian Edition) L'aperto: L'uomo e l'animale, vol 118. Bollati Boringhieri, Torino

Arciero G (2002) Studi e dialoghi sull'identità personale: riflessioni sull'esperienza umana. Bollati Boringhieri, Torino

Arciero G (2006) Sulle tracce di sé. Bollati Boringhieri, Torino

Arciero G, Bondolfi G (2009) Selfhood, identity and personality styles. John Wiley & Sons, Chichester

Arciero G, Guidano VF (2000) Experience, explanation, and the quest for coherence. In: Neimeyer RA, Raskin JD (eds) Constructions of disorder: meaning-making frameworks for psychotherapy. American Psychological Association, Washington

Arciero G et al (2004) Identity, personality and emotional regulation. In: Freeman A, Mahoney MJ, Devito P (eds) Cognition and psychotherapy: second edition. Springer Publishing Company, Berlin

Atlan H (1974) On a formal definition of organization. J Theor Biol 45(2):295–304

Atlan H (1979) Entre le cristal et la fumée: essai sur l'organisation du vivant. Éditions du Seuil, Paris

Barbaras R (2007) Le mouvement de l'existence: études sur la phénoménologie de Jan Patočka. Éditions de la Transparence, Chatou

Bertolino A, Arciero G, Rubino V, Latorre V, De Candia M, Mazzola V et al (2005) Variation of human amygdala response during threatening stimuli as a function of 5′ HTTLPR genotype and personality style. Biol Psychiatry 57(12):1517–1525

Brentari C (2015) Jakob Von Uexküll: the discovery of the umwelt between biosemiotics and theoretical biology, vol 9. Springer, Berlin

Canguilhem G (1952) Le vivant et son milieu. In idem. La connaissance de la vie. Vrin, Paris

Cassirer E (1996) The metaphysics of symbolic forms. (Italian Edition) Metafisica delle forme simboliche (2003). Sansoni

Dastur F (1995) Pour une zoologie privative. In: Alter, Revue de Phénoménologie, n° 3. Editions Alter, Paris, pp 281–318

Dumouchel P, Dupuy JP (1983) L'auto-organisation: de la physique au politique. Éditions du Seuil, Paris

Durst M (2004) Birth and Natality in Hannah Arendt. In: Husserliana A (ed) Does the world exist? vol 79. Springer, Dordrecht, pp 777–797

Greisch J (1994) Le phénomène de la chair: un «ratage» de Sein und Zeit. In: Floviral G (ed) Dimensions de l'exister. Etudes de anthropologie philosophique, Bibliothèque Philosophique de Louvain. Peeters, Leuven

Heidegger GA 2 1977

Heidegger GA 20 1988

Heidegger GA 28 1997

Heidegger GA 29-30 (1983)
Heidegger GA 89 (n.d.)
Kisiel T (1993) The genesis of Heidegger's being and time. University of California Press, Berkeley
Kisiel, T. (1999). L'indication formelle de la facticité. Etudes phénoménologiques, 15(29/30), 107-126.
Kisiel T (2000) Heidegger-Lask-Fichte. In: Rockmore T (ed) Heidegger, German idealism, and neo-Kantianism. Humanity, New York
Kisiel T (2002) Heidegger's way of thought: critical and interpretive signposts. A&C Black, London
Kisiel T (2008) On the genesis of Heidegger's formally indicative hermeneutics of facticity. In: Raffoul F, Nelson ES (eds) Rethinking Facticity. State University of New York Press, New York
Lettvin JY, Maturana HR, McCulloch WS, Pitts WH (1959) What the frog's eye tells the frog's brain. Proc IRE 47(11):1940–1951
Magnus R (1926) The physiology of posture: Cameron lectures. Lancet 211(53):1–536
Magnus R. (1930) The physiological a priori, Lane lectures on experimental pharmacology and medicine, Stanford University Publications, University Series, Medical Sciences 2(3) pp. 97–103, Stanford: Stanford University Press
Magnus O (2002) Rudolf Magnus: physiologist and pharmacologist (1873-1927). Springer Science & Business Media, Dordrecht
Marx W (1998) Uber das Märchen vom Ende der Philosophie. Eine Streitschrift für systematische Rationalität. Königshausen und Neumann, Würzburg
Maturana HR (1970) Biology of cognition. Biological Computer Laboratory Research Report BCL 9.0. University of Illinois, Urbana
Mazzola V, Latorre V, Petito A, Gentili N, Fazio L, Popolizio T et al (2010) Affective response to a loved one's pain: insula activity as a function of individual differences. PLoS One 5(12):e15268
Mazzola V, Marano G, Biganzoli EM, Boracchi P, Lanciano T, Arciero G, Bondolfi G (2014) The In-Out dispositional affective style questionnaire (IN-OUT DASQ): an exploratory factorial analysis. Front Psych 5:1005
Mazzola V Arciero G, Fazio L, Lanciano L, Gelao B, Popolizio T, Bertolino A, Bondolfi G (2018) An association between emotion-body link disposition and salience—action brain network underlies individual differences in people's anger response
Patočka J (1988) Le Monde naturel et le mouvement de l'existence humaine, traduit par E. Abrams. Kluwer, Dordrecht
Patočka J (1995) Papiers phénoménologiques. Editions Jérôme Millon, Grenoble
Patočka J (1998) Platone e l'Europa. Vita e Pensiero, Milan
Romano, Claude (2009). Le monde animal: Heidegger et von Uexküll. In S. Jollivet et C. Romano (éd.), Heidegger en dialogue 1912-1930. Vrin Paris
Rubino V, Blasi G, Latorre V, Fazio L, d'Errico I, Mazzola V et al (2007) Activity in medial prefrontal cortex during cognitive evaluation of threatening stimuli as a function of personality style. Brain Res Bull 74(4):250–257
Scheler MF (1970) The nature of sympathy. Transaction Publishers, London
Sheehan T (2014) Making sense of Heidegger: a paradigm shift. Rowman & Littlefield International, London
Siegal ML, Bergman A (2002) Waddington's canalization revisited: developmental stability and evolution. Proc Natl Acad Sci 99(16):10528–10532
Uexkull JV (1903) Der biologische Bauplan des Sipunculus. Zeitschr f Biol 44:269–344
Uexküll JV (1909) Umwelt und innenwelt der tiere. Springer, Berlin Heidegger GA 29–30
Uexküll JV (1926) Theoretical biology, tr. DL Mackinnon. Kegan Paul, Trench, Trubner, London
Varela FJ (1979) Principles of biological autonomy. North-Holland/Elsevier, New York
von Uexküll J (1921) Umwelt und Innenwelt der Tiere, 2nd edn. Springer, Berlin

Jakob von Uexküll (1946-2014) Der unsterbliche Geist in der Natur (Italian Edition) L'immortale spirito nella natura (I Timoni)
Waddington CH (1957) The strategy of the genes. A discussion of some aspects of theoretical biology. With an appendix by H. Kacser. Allen & Unwin, Sydney
West-Eberhard MJ (2003) Developmental plasticity and evolution. Oxford University Press, Oxford
Zeleny M (1981) Autopoiesis, a theory of living organizations. North-Holland/Elsevier, New York

10 Organisms and Freedom

10.1 The Coming Forth of the Animal's Organism

In its relationship with its environment, an animal can clearly actualize its specific capabilities through its encounter with things, i.e., according to its pre-existing phenotypical structure—a structure which preserves a history of genomic and environmental influences. On the other hand, it is precisely the actual enactment of the organism over the course of its life cycle that helps to explain how physical development influences the generation of variation. Indeed, from the perspective of evolutionary developmental biology (Müller 2007), the environmental responsiveness of organisms not only influences their individual development but can also stand at the origin of a selectable phenotypic variation that fuels adaptive evolution. Epigenetic studies have shown that the same genotype can produce different phenotypes in response to diverse external conditions, so much so that adaptive variation initially introduced in response to environmental pressures may promote the colonization of new environments and facilitate speciation (West-Eberhard 2003; Pfennig et al. 2010). In this sense, genes can follow, not lead, evolutionary changes (West-Eberhard 2003).

Following Uexküll, Heidegger speaks of an environmental circle of disinhibition, understood as the *Umwelt* which surrounds the animal and in which the animal actually lives. From an evolutionary perspective, it corresponds to the sphere according to which adaptive variations occur, channeling the generation of phenotypic variants and hence partly determining the variants subject to selection—thereby biasing the direction and rate of evolution (Laland et al. 2014a). Over the course of its life span, the animal will fight for this environmental circle within which it is absorbed—and captivated, as Heidegger puts it—by articulating the manifold disinhibitions and bringing forth a species-specific field of possible stimulability in its actual life: evolutionary biologists call it "niche construction" (Odling-Smee et al. 1996, 2003).

This circle shared with an integrated community of conspecifics who live, develop, and evolve together constantly takes shape through a web of enduring

G. Arciero et al., *The Foundations of Phenomenological Psychotherapy*,
https://doi.org/10.1007/978-3-319-78087-0_10

behaviors that constitute the actual mode of life of the animal—a mode of life which can be stabilized and/or adaptively modified and which may evolve across successive generations. The animal modifies the environment and is in turn modified by it. This disinhibiting circle may be seen as the context for the possibilities of transferring habits, skills, and preferences from one generation to the next that Jablonka and Lamb (2005) refer to as behavioral inheritance systems (BISs). According to these researchers, an animal "transmits" behavioral information only in the sense that through its behavior other animals acquire that information (p. 158). Jablonka and Lamb therefore distinguish three types of BISs:

1. The transfer of behavior-influencing substances, such as, for instance, young rabbits' juniper food preferences, which are transmitted through the placenta, milk, and feces of their juniper-eating mothers.
2. Inheritance through nonimitative social learning. This kind of socially mediated learning is characterized by the observation on the part of conspecifics of the conditions in which the behavior of experienced individuals is taking place, as well as of the consequences of such behavior, and by the use of this information to reconstruct a similar behavior. The best-known example comes from 1940s England: here the habit spread among birds of opening the milk bottles left by the milkmen on the doorstep of people's homes. The birds did not learn by imitation: they did not copy the actions of the bottle top removers, since some opened bottles one way and some another way.
3. Imitative learning, as in the case of vocal imitation, which occurs in some birds, dolphins, and whales. There is some evidence that the mirror mechanism is also involved in imitation learning (Rizzolatti et al. 2014).

The cross-generational transmission ensured by these learning modes implies the preservation over time of certain acquired behaviors, on the basis of which new skills can be discovered and additional habits can be established. In the light of this, we also understand the concept of a "culture" of animals, conceived as an independent agent of evolution, to quote Jablonka et al. (2014). In other words, modes of relation acquired in the past are preserved and become the foundation on which additional habits can be established. One famous example here is the development of a new lifestyle on the part of the macaques of the Island of Kōjima. In the 1950s, Japanese primatologists left sweet potatoes on the beach in order to lure the macaques out of the forest. The successful stratagem produced an unforeseen consequence: one young female, Imo, started washing the potatoes in a nearby stream to remove the mud from them. This behavioral innovation spread out to other macaques, who soon started washing the potatoes in the salty water of the sea. Within a short time, they learned to bite the potatoes before dipping them in the salt water to clean them, in such a way as to season them. Something similar occurred a few years later, when seeds were left on the beach for the macaques. To separate the seeds from the sand, Imo used the same procedure again, noticing that when the seeds were thrown into the water, they remained afloat. This innovation too spread out to other monkeys and started to be transmitted from mother to child. But the most interesting aspect is the chain of variations which this behavioral innovation produced in the macaques' lifestyle.

The sea became a new setting for a whole range of behaviors. For instance, the infants started playing and bathing in it, jumping and diving into the water became popular, older males began eating fish that the fishermen had discarded, and so on. This new web of behaviors acquired by social learning endures to this day among the macaques on the Island of Kōjima, over 50 years and six generations after Imo's behavioral innovation.

What has happened since the scientists first started feeding the macaques on the Island of Kōjima is that the formation of new habits and of a new tradition has left its imprint on the dynamics that bring forth the organismic unity of the macaques. Just as in living the world human corporeality in each case realizes itself by becoming actualized as a bodying forth (*Leiben*), the organismic unity of an animal in each case takes shape precisely on the basis of the animal's relationship with the environment that it thereby contributes to modifying.

10.2 The Awakening of Dormant Genes

The process just described is particularly evident in the case of the taming of the silver fox, which Jablonka and Lamb (2005) regard as an exhaustive example of a change which, arising in response to altered environmental conditions (i.e., epigenetic conditions), can influence evolution through the genetic system. In the late 1950s Belyaev, a convinced Mendelian geneticist, started a long-term experiment which, within twenty generations, led to the selection of a population of docile silver foxes, which is to say, of tamed animals, anxious to please their human handlers and compete for their attention.

The most interesting aspect is that this behavioral transformation brought about by selection for tameness was matched by both hormonal and physical changes: not only had the reproductive season of the females become longer and the time of molting changed, along with the levels of stress and sex hormones, but the ears of some foxes now drooped and the way some carried their tail was different; some had white spots on their fur; and a few (only about 1%) had shorter legs or tails or a different skull shape. In addition to these heritable phenotypic changes, the foxes showed changes in their chromosomes.

Belyaev interpreted these data by attributing the appearance of new phenotypes to the activation of "dormant" genes, of which animals have a large reservoir. Belyaev suggested that in stressful situations, such as during domestication, the selection for domesticated behavior altered the foxes' hormonal state, which in turn affected chromatin structure and thus activated many normally silent genes in both the soma and germ line, causing these inactive genes to become heritably active. The body of an animal therefore becomes such by unmasking the possibilities it holds (already existing genetic variation) in the light of stressful situations (domestication) which are imposed upon it in the context of its environmental circle. To put it in the language of evolutionary developmental biology, the epigenetic changes caused by environmental stresses reveal hidden genetic variations that produce new phenotypes (Jablonka et al. 2014).[1] The unmasked phenotypes can be genetically assimilated through selection, in such a way as to be retained, independently of any stimuli. In this sense, the range of conditions and resource frequencies that the organism can tolerate are confined within its fundamental niche (Turner 2000a, 2004a; Laland et al. 2014b).

According to Heidegger, this essence of the organism, conceived as a closed environmental circle, is the reason behind the much-reviled poverty of the animal world (GA 29-30 1983): unlike man, no animal can be free for beings as such. "Man's being open is a being held toward ..., whereas the animal's being open is a being taken by ... and thereby a being absorbed in its encircling ring" (GA 29-30 1983, p. 498).

10.3 The Humanness of Man

We can now catch a first and generic glimpse of the difference between man and animal, starting from the motility of life and how man has access to it. Whereas an entity will never reveal itself as such to an animal, because it vanishes—so to speak—with the disinhibition of the animal's instinct, in the case of man, an entity will reveal itself in all its meaningfulness, requiring a relationship in order to manifest itself in one way or another and hence become accessible. In this sense, man is world-forming: he brings wholeness as such—the world in its general intelligibility—to manifestation prereflectively. This formal indication occurs again and again in the second part of GA 29-30 1983.[2] As we have already seen in relation to Heidegger's Freiburg lectures, the table (*Katheder*) meets me starting from the range of references that situate it within a context in view of which it acquires a different meaning for the cleaning lady, the postman, the electrician, the student, myself, and someone who has no familiarity with it at all.

The crucial point—associated with Lask's overturning of the neo-Kantian tradition—is that what we grasp of an entity is not merely a characteristic it possesses in relation to us, i.e., a quality of the table which we refer back to ourselves, but rather a quality proper to the specific object as such. The table grasped in such and such a way thus becomes manifest starting from a totality, from a meaningfulness which we do not explicitly grasp and which changes with our being-there. Those contexts of meaning which encompass all objects are like categories within which we prereflectively live and which environ, horizon, or border things themselves, enabling us to constantly situate ourselves within the world and to directly grasp its sense. In

[1] Evidently, past selection has made the networks of interactions underlying development indifferent to minor changes.

[2] Formally indicating philosophical concepts are defined as follows: "the meaningful content of these concepts does not directly signify or express what it refers to, but only provides an indication, a reference to the fact that the conceptual connection exhorts the person who understands it to implement a transformation of himself in the direction of being-there" (p. 430 GA 56/57 1999).

other words, the accessibility of the world and, therefore, the sense according to which ipseity in each case opens up to what it encounters rest upon the meaningfulness of the world as such.

"Living in an environment (*Umwelt*) – Heidegger states in his famous *Kriegsnotsemester 1919* – it signifies to me everywhere and always, everything has the character of world, it *becomes world* [*es weltet*]" (GA 56/57 1999, p. 73). The world, therefore, takes shape within the framework of a relationship through which ipseity—as we have already repeatedly stressed—experiences itself by *finding itself* situated within it. This openness is ensured by the emotional tones which enable one to find a place within the world, while making the latter appear. Evidently, corporeality—which Heidegger never focuses on—plays a crucial role here.[3]

For a human being, considering something intelligible—i.e., understanding the entity he encounters in its meaningfulness—is a way of constructing his own existence. Therefore, this structure of intelligibility which precedes speech and constitutes its foundation is crucial for the formation of a world and for bringing it to manifestation (Volpi 1996b). We thus grasp the unity of understanding, of situating oneself, and of the situation as the preverbal precondition for language.

This keeping oneself united with something else, this unique relationship of oneself with something as such and such a thing, which at the same time constitutes a mutual revelation, coincides with meaning. And it is the very essence of man: man "understands and gives shape to intelligibility".[4] It is through this reciprocal agreement that words allow us to understand what may be understood.

Out of meanings, sounds arise: the human voice is very different from that of an animal, which lacks meaning.[5] For man, the capacity to understand what he is dealing with as this or that (unifying apprehension) is the precondition for agreeing through sound on that which he is keeping himself united with, that which he understands: the symbolic dimension.

In a symbol, man keeps himself united with something else (the world in which he lives) and, at the same time, reaches (or fails to reach) an agreement with other men with regard to how that symbol should be interpreted. This openness of man to that which he encounters, and which is pre-predicatively manifested in its totality, represents the precondition for discourse (*adequatio*), something that is denied to animals, insofar as openness for them coincides with captivation. Every assertion, therefore, is made starting from a totality to which man keeps in ever-different ways

---

[3] Clearly, this grasping is an enactment, and hence intelligibility—understanding a thing as this or that (*etwas als etwas*)—constitutes an indication starting from one's own situatedness. As "being is not a genus," it is not accessible by generalization but only by formalization. It is necessary, then, to distinguish between a universal mathesis which always deals with an object, and phenomenological formality, which primarily deals with a relation: intentionality (p. 209 Kisiel 1996).

[4] "From the very beginning our essence is such as to understand and form the possibility of understanding" (GA 29-30 1983 p. 307).

[5] For an exposition of the psychology of linguistic development from this perspective, see Arciero Bondolfi 2009, Ch. 2.

and according to different possibilities: indifference, absorption, detached observation, engagement, hesitation, marvel, etc..[6]

The crucial difference between man and animal, and hence the possibility of a method, revolves around this difference in how the world is accessed. While the world is that accessible entity to which both human beings and animals are open, human beings and animals are open to the world in an essentially different way: the former freely relate to the meaningful things they encounter by letting themselves be bound to them, whereas the latter do not have this possibility—animals are bound to their environment and captivated in the enactment of this relationship. Heidegger claims that man is world-forming, whereas animals are "poor in world."

Poverty here means that an animal does not access an entity qua entity: "it goes without it," even though it has access to the world. It is on the basis of this, then, that man can enter the environmental circle of an animal, albeit without being able to go along with the dealings which the animal has with its world, since he cannot have "direct" access to it. Heidegger sums it up as follows: "The animal intrinsically displays a sphere of transposability, and does so in such a way that man, to whose being-there belongs a being transposed, already in a certain way finds himself transposed into the animal" (GA 29-30 1983, p. 309).

The engagement with Uexküll's neo-Kantianism thus reaches its culmination. Whereas according to Uexküll the observer penetrates the external world of the animal because he phenomenally perceives those qualities of it which his perceptual apparatus is capable of detecting, thereby establishing a sort of coincidence of worlds, Heidegger ontologically bases the possibility of studying an animal on the transposability of man into the organismic unity through a deconstructive analysis (*Betrachtung*). This transposing of oneself—as opposed to going along with the experience of the other, as in the case of a human being engaging with another human being—is ensured by the fact that the animal is not worldless, even though it does not have a world in the same way as a human being (i.e., it is incapable of accessing its meaningfulness as such).

The question of animality is thus raised in the light of man's relationship with the world, which is to say, from a comparative perspective: man, who cannot go along with an animal's accessing of its own world, can instead transpose himself into its environmental circle. Therefore, we can only grasp animality in privative terms with respect to our own mode of accessing the world, which is to say, through a negative interpretation: animality amounts to the lack of humanness (along much the same lines, Aristotle states that plants lack eyes).[7] This accessibility, therefore, at the same time coincides with a form of impenetrability. Clearly, all this cannot mean a difference in value or rank between human and animal life, as some authors have all too rashly argued (Derrida 1987; Derrida and Wills 2002). "On the contrary – Heidegger states – life is a domain which possesses a wealth of openness with

[6] In the Zollikon Seminars, Heidegger observes: "Why doesn't an animal speak? Because it has nothing to say" (p. 87).

[7] Dastur (1995) has clearly highlighted how Aristotelian *steresis* constitutes the foundation of the privative interpretation.

which the human world may have nothing to compare" (GA 29-30 1983, p. 371–372).

In other words, no primacy or privilege can be assigned to man by comparison to other modes of being, since the difference is due to man's relationship with the world. The relationship with the world is what ontologically determines man and, comparatively, the stone ("worldless") and the animal ("poor in world"). This distinctive (pre-logical) openness to the world of man—whereby, in relating to it (as a whole), he also relates to his own being—provides the foundation for man's ipseity and hence freedom.

In the closing pages of *The Fundamental Concepts of Metaphysics*, Heidegger writes: "this 'as a whole', the world, admits precisely the manifestness of manifold beings in the various contexts of their being – other human beings, animals, plants, material things, artworks, i.e. everything we are capable of identifying as beings" (2001, p. 353). It is precisely here that we meet that "something unthought of his own" which shaped Eugen Fink's path, down to his development of a phenomenological cosmology. Already during Heidegger's lecture course—as the letter to Patočka reveals—Fink had grasped that for man openness to oneself is a consequence of the fact of being subject to the world and of keeping within it. This is neither a mythical world nor an objective nature, but rather a universe understood as a connection which governs all things, as a manifestation of the entity in its wholeness.

This perspective is the source of the question of human freedom as the "freeing of the gaze" (Tardivel 2011), which leads Fink to adopt a non-metaphysical reading of the Allegory of the Cave and sets Patočka on the path to reclaim Plato—the path of negative Platonism (Patočka 1953).[8] Patočka, who shared the same research horizon as Fink from his early days, sets the movement of life within the context of this cosmology intertwined with Heidegger's ontology. He describes this intertwining in a letter addressed to Fink's widow on 23 September 1976: "Heidegger's late ontology has grown increasingly cosmological, while Fink's ontology has significantly ontologized itself" (quoted in Dastur 2013, p. 14). It is according to this perspective that we should also view the last, remarkable seminar on Heraclitus (GA 15 1986) which Heidegger held with Fink in Freiburg and which sealed his university career.

Another letter, addressed to Campbell on 28 October 1961, provides some clues for better grasping both Fink's position with regard to early philosophers and the point where Patočka's thought meets Fink's cosmology: "Fink has drawn attention to the fact that for early thinkers down to Aristotle, movement primarily coincides with the *cosmogonic* movement, i.e. the movement through which the world constitutes itself starting from chaos (to resolve itself at a new level)" (quoted in Jacquet 2011, p. 397). This conceptualization of the world as movement allows Patočka to frame existence and life more generally—embodied by a body which takes shape

[8] Marion Bernard (2011) observes: "It is not just man who has the power to bring things to manifestation: he himself is subject to the general movement of individuation which discloses things and determines them in relation to one another independently of all human bringing to manifestation" (p. 367).

again and again by becoming anchored to what manifests itself—within a world which is itself grasped as it takes shape.[9] The living being in all its forms, then, understood as the movement of life, coincides with the unique occurrence of that cosmological movement which defines the world as such as movement: "Movement is that which allows things to be, that which allows at the same time that their being is alive, that it is a something in the order of life, a unity, a comprehensible sense, a path 'from... toward...'" (1988, p. 129).

10.4 The Uniqueness of Life

The single occurrence of the animal as a unique movement that cannot be reduced to any mechanism—an occurrence that takes shape through an ever-actual relationship with the movement of the world which constitutes it—finds expression in one of the many examples of niche construction provided by evolutionary and developmental biology: that of the common earthworm, which is tied to the soil and transforms it in order to live within it (Turner 2002; Odling-Smee et al. 2003).[10]

When one sets out to determine whether the earthworm belongs to the class of land creatures, freshwater creatures, or sea creatures, the most perplexing thing is the fact that, although this annelid lives in the earth, it is structurally very poorly adapted to coping with life on land. The earthworm has a kidney-like filtering apparatus—to maintain the proper hydroelectrolytic balance within its body—that is suited for life in a freshwater environment. One crucial question thus emerges in relation to this strange creature. According to standard evolutionary theory, when the aquatic ancestors of terrestrial earthworms evolved toward their new environment, they ought to have developed kidney-like organs suited to terrestrial rather than freshwater life—organs capable of favoring water conservation. Selective pressure, in other words, ought to have selected an adaptive match between the organism and its environment.

The evolutionary path taken by the earthworm instead followed a different trajectory: the maintenance of a hydroelectrolytic balance was achieved by behavioral means, adapting the soil to the requirements of the worm's freshwater physiology. Oriented by this need, earthworms operate according to three basic patterns (Turner 2002). First of all, they dig tunnels in the soil and consolidate their walls by making them more compact and leaving a trail of mucus behind to stabilize them. The worms swallow the soil they dig, digesting organic matter and bacteria in their

[9] Patočka thus makes it possible to conceive animality starting from the universe, according to "a sensory sympathy with the world": a sympathetic mode of living whereby the animal is "outside the boundaries of the organism, its private being, precisely because the cosmos itself is already more than a *res extensa*" (Patočka 1995, p. 61).

[10] It is perhaps by virtue of the fact that they are rooted to the soil in a different way from plants that worms—be they earthworms or sea worms—have held such great appeal for biology. We have already mentioned the importance of the *Sipunculus nudus* for Uexküll and Magnus. Before his death, Darwin himself published the studies on the earthworm he had been conducting for over 40 years.

intestinal tract. The excrements they produce are fecal pellets that, once dry, form large and durable casts capable of absorbing water. Finally, the worms return to the soil surface and collect leaves or other rotten organic material, dragging it into their tunnels in order to either consume it immediately or store it and eat it later on. The general outcome of this constant movement is the transformation of the soil, which becomes much more porous, thereby increasing its infiltration rates and hydraulic capacities, making them several hundred percent higher compared to soil with no earthworms. The final stage of this transformation of the soil is the creation of a layer—a soil horizon—which is capable of holding water and has a high degree of humidity, sufficient air space to make oxygen abundantly available, and warmer surface temperatures. In such a way, earthworms construct for themselves a semi-aquatic niche in the soil in which they live and which functions as an accessory organ of water balance.

Instead of restructuring their internal physiology, earthworms have changed the physiology of the soil they inhabit, significantly contributing to the transmission of modified natural selection pressures to descendant generations. On the other hand, this sort of "external physiology" (Turner 2002)—which results from the adaptive modification of the environment and with respect to which the conventionally defined "internal physiology" is in each case enacted—is governed by structures and devices inside the integumentary boundary of the organism, which corresponds to a change of the structure and chemistry of the soil within the limits that a given kind of earth is capable of ensuring.

The intertwining of ontology and cosmology mentioned by Patočka now becomes even more evident. On the one hand, our recognition of life is based on our mode of access to it, and this is the foundation of privative zoology and its method; on the other hand, the life to which we have access, our existence, is only one mode of manifestation of the motility of life—of life more generally conceived as motility. It is starting from this dimension that the unity of the life of living beings, of man and animals, becomes accessible, along with the difference between them. The two modes of being never enter into conflict with one another: to defend the humanness of man (difference) is not to deny that it is only one of many forms of manifestation of the essence of life (unity).

In this sense, the fluctuation between an ontology which bases the study of animality on man and his possibilities to access it, and a cosmology which binds life to Earth and reveals our existence to be one of the many modes in which life itself unfolds (the movement of the world), enables us to grasp the living being beyond the constructivism and reductionism that shape biology. If we concretely access our experience through the actualization of a relationship which formal indication articulates as sense of content, sense of relation, and sense of enactment, then it is once again this formal schematism which enables us to grasp the reality of each animal: for, in all likelihood, this can only be accessed as an enactment. And it is precisely because life and life alone—in all of its manifestations—amounts to enactment that man in his humanness can be described as a living being.

In the enactment of motility, each life form defines its own relationship with the world, its individuality, thereby constantly making up for that essential incompleteness which distinguishes each living being, while at the same time determining its possibilities of individuation. Within the framework of the motility of the world—understood as a cosmic ground, as "the pre-existing, all-encompassing and undefined setting of all individuation" (Patočka 1988, p. 132)—the living being, be it a man or an animal, distinguishes itself by enacting a specific relationship with this whole and with the individual entities inhabiting it. The process of individuation takes shape as a mode of relating to the world as a whole and, at the same time, in virtue of the relationship itself, as a distinguishing of oneself from the whole itself. Thus the movement of life coincides with the orienting of oneself in relation to those things of the world which address the living being and which each living being meets according to its possibilities. Man and animal are therefore different modes of enacting the motility of life: different ways in which life becomes a phenomenon starting from its embodiment and hence from the belonging of the body to the world.

Patočka thus grasps the phenomenizing living of animals in the light of what he describes as the process of cosmic dis-connection of man, which is to say, of the distance which man creates with respect to the world and which is only bridged by death.[11] We here find an echo of Fink's view of Plato and the *Chorismos* mystery (Plato's well-known distinction between the sensible world and that of the ideas; see Dastur 2013). What we have here is Patočka's negative Platonism: horizons of freedom are disclosed starting from our distance with respect to the way in which we are absorbed in things, from our disconnection from the universe; and the individuation of the animal is grasped according to the same register (i.e., that of cosmic disconnection).[12] What we have here is the relationship between the logos and the pre-predicative manifestness of the being in its wholeness, understood as the occurrence of a free "being held toward," which lets itself be bound in one way or another, in accordance to this or that entity. Heidegger describes this fundamental relationship as being free in an originary sense (GA 29-30 1983, p. 497). The organism—each living organism—finds its individuality by enacting it as a relationship with the world and hence by differentiating itself with respect to it.[13]

[11] As in Demeter's cycle, the earth gives birth to all things and all things return to it.

[12] This perspective constitutes the background to the musings on the logos and freedom we find in the 1929 seminar and which have regrettably been overshadowed by the wretched choice that Heidegger made in 1933. (For a serious analysis of this decision, which was quite foreign to the kind of media attention-grabbing and coarse showiness that marked the publication of the *Black Notebooks*, see Patočka (1990).)

[13] Jonas (1967, 1997) also grasped life as an active separation from the exteriority of the world: a continuous act that each individual constantly renews (through a metabolism always giving itself some new matter). However, Jonas failed to realize that the different degrees of distance exhibited by men, animals, and plants are nothing but different ways of becoming rooted in the "exterior world," different ways of making up for the indigence at the heart of life. Jonas does not realize that even photosynthesis coincides with the motility of life, like the earthworm's digging of tunnels and man's growing old; rather, he understands motility as the movement of an animal correlative to the disconnection of animal life from the world. Hence, he essentially conceives of animals by reference to plants, as Barbaras (2008) notes: "L'animal est. un vegetal qui a perdu le contact et qui doit donc trouver le moyen de s'approprier ce qui n'est. plus en continuité avec lui" (p. 219).

10.5 Organisms and Evolution

By proceeding along this path, we come across the life sciences and, in particular, one of the key points in extended evolutionary synthesis (EES): the extended organism (Turner 2000a; Laland et al. 2014b).[14] This new perspective assigns each organism, in its uniqueness, an effective role in relation to its own development and evolution. The crucial aspect which establishes this conceptualization on a different level, compared to standard evolutionary theory (SET), is precisely the fact that it grasps the organism and its activities as co-determining factors for evolutionary changes. In other words, the taking shape of the organism over the course of its development goes hand in hand with a transformation of the structures of its ecosystem: through its actions the organism modifies the spectrum of natural selection pressures in its own selective environment (Odling-Smee 2010). The organism is no longer regarded as a transient carrier of selfish genes, less still as a system enclosed within an envelope—be it skin or any other integumentary apparatus—that delimits it against the environment, thereby enabling its distinction in the following Boolean terms: 0–1, non-organism-organism (Turner 2002). The envelope which, as a tangible barrier between inside and outside, ensures the distinctness of the organism with respect to the environment also defines its endurance over time as an autonomous unit.

The new vision shatters this separation insofar as it considers the environment within which the organism lives and reproduces itself as an integral part of the organism itself, of its biology and physiology. It is an integral part of the biology of the organism because the active transformation of the milieu counts as a consequence of the expression of genes and hence as a phenotypic trait; it is an integral part of the physiology of the organism because the creation of environmental niches can be instrumental in managing flows of matter and energy in ecosystems and hence through organisms (Odling-Smee and Turner 2011; Turner 2002, 2004a).

So in altering the environment in which the organism lives, its activity partially modifies the selective pressure. Oding-Smee and Turner write: “Adaptation, by which we mean the complementarity of organism and environment, now becomes a two-way street – not only do organisms adapt to their environments, they also ‘adapt’ their environments to themselves” (p. 281, 2011).[15] This modification of the selective pressure can be passed down to descendent generations through the transmission of organism-built structures (ecological inheritance), which can give rise to a stable process that codirects genetic evolution. One telling example is the selection of alleles for adult lactose absorption in human adults (Odling-Smee et al. 2003; Scott-Phillips et al. 2014; Laland et al. 2010).

---

[14] Turner’s conception of the extended organism implies that structures used and built by animals, and other environmental modifications made by them, are actually extensions of their internal physiology, which harness and control the flow of energy, thereby imparting an “extended physiology” to external environments.

[15] However, niche-constructing effects are not always adaptations (Scott-Phillips et al. 2013).

In most of the world's population, the ability to digest lactose disappears in childhood. In some populations, however, lactase activity, known as lactose tolerance, persists into adulthood. High lactose digestion capacity in adults across cultures correlates strongly with whether a certain culture has a history of milk drinking and dairy farming.

The historical hypothesis put forward by Holden and Mace (1997) to account for this difference among populations was that dairying created the selection pressures that drove alleles for lactose tolerance to high frequency. Their analysis also indicated that the dairying came first, approximately 8000–6000 years ago, and that the strong selection of the lactose tolerance was favored subsequently in the setting of dairy farming. The absence of the allele in DNA extracted from early Neolithic Europeans, which suggests that 7000–8000 years ago the allele was absent or at a low frequency (Bersaglieri et al. 2004), supports the hypothesis that the increased frequency of the lactose-tolerance allele was selected through cultural practices. It is possible that early Neolithic humans exposed themselves to a strong selection pressure for lactose tolerance by drinking fresh milk, a practice that was transmitted by milk drinkers to their children, engendering within a few hundred generations a significant fitness advantage for lactose-tolerant individuals (Laland et al. 2010). Although the cultural transmission of an environment-modifying human activity provides a second inheritance system which may persist for enough generations to codirect human genetic evolution, as in the case of lactose tolerance, in humans cultural changes can of course occur independently of genetic change.

A certain emotional phenotype, for example, such as the depressive personality disorder, may be reproduced by the following generation if a certain kind of habitat favoring the re-enactment of this configuration is transmitted and reconstructed (Arciero Bondolfi 2009; Bondolfi et al. 2015). This view would appear to be confirmed by the sociological research on depression conducted within the framework of the research program of the Tavistock Clinic. In a first seminal study conducted in 1978 in Camberwell, London, Brown and Harris analyzed the role of psychosocial factors contributing to depression in a group of depressed female patients.

This study identified triggering elements—significant recent losses—and factors of vulnerability (the failure of a trusted relationship with a partner, the loss of a mother before the age of 11, having three or more children under the age of 15, and joblessness); their mutual interaction was found to be statistically significant for the onset of depression. In a subsequent study, which provided a more detailed examination of the data emerging from the first, Harris et al. (1986) suggested that the lack of adequate parental care (negligence or indifference) that precedes or follows loss represents an important determinant of morbidity, to the point of making the individual twice as likely to develop depression in the course of his or her adult life.

In a replication of the study conducted by Harris et al. (1986), by examining the history of the female patients under observation, Bifulco et al. (1987) noted that, among the significant characteristics related to lack of care and subsequent depression, an unwanted pregnancy was listed, which was associated with emotional relationships devoid of intimacy and with unreliable partners. In the stories analyzed, lack of adequate parental care was seen to increase the risk of young women embarking on premature marriages devoid of any mutual support. One of the most common factors leading to marriage was unwanted pregnancy, a frequent cause of problems in the relationships examined: extramarital affairs, financial hardship, and the structuring of a context characterized by lack of caring. It is evident that similar conditions of parental care, strongly marked by the absence of any form of support, fostered a tendency in the offspring to repeat the cycle. The individual disposition toward depression was thus passed on from one generation to the next through the inheritance (i.e., reproduction) of specific conditions necessary for the development of the tendency. Therefore, the mechanism at the basis of the disorder is engendered and maintained through the generational link: by selecting a series of environmental factors, this predisposes the existing conditions to the selective pressure which will channel the possibility of replication of the sequence in the offspring.

By centering the research on the organism, and its actual development within the environment that it inhabits, this new evolutionary perspective offers an original view of the living being: it frees the living being from that ontology of death which, through the notion of organization, shaped early biology and—especially from Bichat onwards (Schiller 1978)—the conception of biological individuality by confining the organism within a body. Hence, this new way of conceiving the organism no longer implies an integral immanence—organization—supported by the closure of the living being's operations upon itself: an absolute, timeless domain, which remains the same at different (later) points of space time, down to death.

Biological individuality is instead marked by a fundamental incompleteness, whereby it is always already open to the world and subject to it; the motility of life enacts again and again this relationship, which for each living being takes shape as an act of individuation through the choice of life possibilities. This process, therefore, coincides, not with the preservation of a mode of organization which is already given compared to the environmental perturbations threatening its integrity, but rather with the constant search for a unity that is always taking shape, of an identity whose attainment is always postponed, or—as evolutionary biology suggests—of an organism that, through its actual enactment, modifies the environment in which it lives, partially orienting the process of natural selection. Clearly, as Heidegger had already grasped, the difference between man and other organisms has to do with the former's openness to (i.e., a priori capacity to access) the totality of the world as such, which gives itself as a horizon within whose framework all meaningfulness as such takes shape, and with the consequent need to symbolically express what manifests itself in prereflective experience.

10.6 Naturalizing Phenomenology?

The natural scientist investigating the access to experience and the phenomena of life offered by phenomenology is faced once again—and from a different perspective—with the question which Varela sought to answer in relation to Husserl's approach: can the science of experience be naturalized? The problem which emerges, albeit from a different angle, is whether it is possible to conceive of a phenomenally oriented epistemology starting from the conception of an extended organism (something bodily not limited to the body), which in the case of human beings coincides with an openness to the world as a whole, to the world as such. In being enacted over and over again, this openness modifies not only the individual's relationship with the world but also the way in which the individual's future is imagined and his story is re-enacted. Is it possible, then, to practice a science of experience, to study experience through the investigative tools of natural science, by starting not from an a priori model or from a model practically created within a lab—as in Varela's case—but rather from the actual irreducibility of the experience of being alive?

If the bodying forth of each person—and hence the process of becoming an organism which natural science seeks to retrace—is determined by the way in which each person relates to what he or she encounters, then clearly the signatures that the human being engenders and preserves cannot be interpreted as though they were something absolute. These signatures point to a "whither," namely, the meaningfulness that the person encounters (*Gahaltssinn*) and which determines the constitution of the body as a mode of enactment, as well as a "whence," namely, those tendencies and motivations, those existential possibilities (*Bezugssinn*), that are enshrined within the flesh and are fulfilled through the encounter (*Vollzugssinn*), renewing life's access to itself in the situation at hand and thereby updating the overall cohesion of the sense of self (*Zeitigungsinn)*.

The signature of experience upon an organism, including at a neural level, can only be grasped within the framework of the unfolding of individual life in a world that becomes the world which reveals itself to me, which meets me, and which I choose to encounter, in the light of the immanent historicity of life. Hence, the traces of oneself cannot be grasped as the mere collection of data pertaining to a garden-variety population according to an a priori model. To understand them it is not enough to cross a given statistical threshold—the correct P! The latter can only reveal to what extent and in what way the criterion of the certainty of the datum as a marker of truth is uncertain and uninformative.[16] Even the unfading traces left in the flesh by death, the last traces of a life which the clinical pathologist puts together, adumbrate the last breath exhaled: the signature of the end which the body preserves constitutes the indication of a final encounter with the world, of the way in which the individual experienced the world just before death and, more generally, of the lived life which the deceased body points to.

As we have seen, the solution proposed by Varela's neurophenomenology is to modelize experience through experience itself. The research subjects' experience is disciplined in a laboratory according to the rules governing a given reflective practice, i.e., meditation, so as to grasp the cerebral responses this practice elicits. First-person experience is therefore molded according to the 'impersonality' of the

[16] "A P value of 0.05 does not mean that there is a 95% chance that a given hypothesis is correct. Instead, it signifies that if the null hypothesis is true, and all other assumptions made are valid, there is a 5% chance of obtaining a result at least as extreme as the one observed" (Baker 2016, Nuzzo 2014).

shared procedures orienting those engaging in the practice, enabling the subject to distinguish the various experiences engendered by the very exercising of the practice. When the great monk Ajahn Chā (2011) describes the various stages of meditation, which indicate the emergence of different levels of experience after years of shared meditation practice, he stresses the fact that the drawing of these distinctions is the result of a disciplined methodology ensured by the exercising of the practice.

It is worth pausing briefly to consider only the first stage of the state of concentrated calm resulting from meditation practice (*samādhi),* to grasp the manifestation of the five different phases it comprises. When practicing *samādhi* the attention will be fixed on the in-breath and out-breath at the nose tip or the upper lip. The mental factor by which one's attention is applied to the chosen meditation object is called *vitakka*, or "lifting up." When the contemplation of the breath at the nose tip is fixed, this sustained thought is called *vicāra.* "We use *vicāra* to contemplate the various sensations which arise. When *vicāra* becomes gradually more scattered we once again 'lift' our attention with *vittaka"* (p. 89 Vol. 2). Therefore, *vitakka* and *vicāra* work together until *vicāra* becomes gradually more subtle. When the contemplation is right, a feeling of well-being arises as an emergent state. This condition, called *pīti*, is accompanied by a feeling of pleasure, *sukha*, and by a state of meditative absorption, *ekaggatā.* Ajahn Chah notes that this practice, known as "mindfulness of breathing," must be pursued as continuously as possible in order for it to bear fruit. He continues: "Don't meditate for a short time one day and then in one or two weeks, or even a month, meditate again. This will not bring results" (p. 129, Vol 2).[17] First-person and third-person accounts of this distinctive experience that is practiced and recorded in the form of dynamical neural signatures in a laboratory are provided by the practitioner and researcher respectively, thereby establishing a generative circle between the way in which the disciplined experience is personally lived and the bodying forth of the body—the transformation of the experience grasped by the researcher into a trace. The practice generates experience and experience generates the practice: this reciprocal determination is highlighted through the search for generative passages that bear witness to mutually effective bonds (Depraz 2006).

However, if we move beyond the study of reflective practices, which is inevitably confined to a relationship with the world which the practice disciplines in a self-referential way, and enter into the actual lives of human beings, how can the emergence of the neurological signature, the bodying forth of the body, be grasped in a laboratory? If the body becomes such according to its various modes of relating to—and finding itself and being emotionally disposed with respect to—the meaningfulness it encounters, how can we access individual experience?

[17] This rigor is certainly a far cry from the slapdash approach preached and practiced by many mindfulness therapists. While the further investigation of this practice is of great interest for psychology as a science, the attempt to reduce psychology and psychotherapy to it reflects a serious misunderstanding of the very foundations of psychology.

10.7 The Co-productivity of Psychology and the Natural Sciences

From the perspective of the natural scientist, it is necessary to distinguish phenomena through a method capable of disciplining the study of personal experience, without reducing it either to a priori rules for observation or to the rules of a mode of production governed by a given practice. Rather, each phenomenon is to be grasped in view of what the person regards as meaningful given his own developmental trajectory or life story. This is the domain which phenomenology discloses as an avenue for the articulation, comparative examination, and joint application of those perspectives which psychology and the natural sciences have to offer. In other words, a co-productivity between the kind of analyses of personal experience that psychology ensures through the method of formal indication, and the investigations of organismic responses to these experiences that the positive sciences conduct according to their own particular methodology, can only be founded on the tension between the existential determination of the body and the embodiment of existence which phenomenology makes manifest.

On the one hand, then, the precondition for selecting the individuals to be studied is to re-enact the experience of the other faithfully to the situations, in such a way as to re-enact their prereflective meanings and thereby obtain from the very life of the other the categories through which he understands himself (this is what psychology, as a science, ought to contribute through its relationship with the natural sciences). On the other hand, the natural sciences ought to formulate hypotheses and offer some evidence concerning the way in which the body becomes such, starting from its particular mode of being, from those categories through which individual life gains access to itself and which have already been clarified by psychological research.

A kind of psychology focusing on the person in his or her uniqueness must therefore grasp meaning starting from life itself—not destroy the vitality of life by tracing it back to the "simplest elements," identified according to the a priori determined by the theoretical approach. Psychology, which may be regarded as the science of personal experience, thus takes shape as a research conducted on individual modes of enacting one's sense of reference with respect to the manifestation of the world—understood as an investigation of the primary modes of access (interpretative categories) which the body engenders and preserves as living traces. The demarcation of the research domain of psychology, carried out by psychology itself, at the same time coincides with the demarcation of a field of inquiry—the coming into being of personal experience—which cannot be understood by the neurosciences. Clearly, this does not mean that the investigations conducted by neuroscientists in the field of psychology are useless or unwarranted as a means to establish a dialogue between different scientific perspectives. However, within this context it is important to stress once more what the neurosciences are incapable of understanding: the mode of being of a person in its uniqueness. Hence, asserting the autonomy of psychology means not only clarifying what lies outside the field of investigation of the neurosciences but also delineating the specific task and horizon of psychology as a science and, with it, the space of collaboration, mutual engagement, and research—i.e., of mutual enlightenment—which it is capable of establishing with the neurosciences and with the life sciences more generally.

In this sense, the first-person methodology which enables the natural sciences to interpret dynamical organismic signatures, starting from the personal account which the researcher co-develops with an individual, or which the clinician co-develops with a patient, emerges as the method of psychology. On the other hand, the categories thus distinguished—the different ways in which individual life accesses itself—must be understood as modes of bodying forth of the body starting from those possibilities which corporeality actualizes in its relationship with the entity as a whole and which it preserves as a living trace. So if what is corporeal is always defined by the way in which each person situates himself with respect to this or that thing, those modes of being through which life understands itself (i.e., categories) determine each organ and hence the physiology and pathology of each organ. The description/explanation of the bodying forth of the body which the natural sciences offer in mechanistic terms—e.g., in terms of a conventionally defined "internal physiology" governed by structures and devices inside the integumentary boundary of the organism—amounts to nothing more than a mechanistic reduction of life unless it is combined with a unique generation of meaning, the positum of psychology, which has to do with the way in which the individual lives in the world: in other words, unless it is associated with that "external physiology," as J. Scott Turner describes it, according to the evo-developmental view illustrated in Extended Phenotype (2000b, 2002, 2004a, 2004b), which results from the organism-adaptive modification of the environment. Even a corpse on a dissecting table, then, like a mummy in an Egyptian tomb, bears traces linking it to its lived experience. And it is precisely on this co-productivity between psychology and the natural sciences that phenomenology sheds light, which represents the privileged avenue for the clinician and the researcher.

10.8 Co-productivity and Light on the Illness of an Individual: Cardiology

The phenomenological approach reveals a co-productivity between the determination of the corporeal in the relationship with the meaningful contexts according to which it is enacted (the field of study of psychology) and the bodying forth of the body, which, starting from the different ways in which an individual relates to the situation at hand, always reconfigures itself with respect to the latter, according to a mode of finding oneself and of becoming disposed that the flesh preserves and renews over the course of the individual's existence (the object of the natural sciences).

We can appreciate the significance of this perspective by considering a particularly dramatic phenomenon that cardiologists are all too familiar with: sudden unexpected death (SUD). Sudden cardiac death, which is now recognized in all its various forms as a medical problem of major epidemiological importance, accounts for 50% of all cardiovascular death or 25% of all fatalities (Oppenheimer et al. 1990). Sudden death is an "electrical accident" caused by fatal cardiac arrhythmias (Cheung and Hachinski 2000) that may be triggered by intense emotional states caused by disruptive life events and may complicate various neurological conditions (stroke, epilepsy, subarachnoid hemorrhage, etc.).

There is clear evidence to suggest that a series of sudden deaths—being frightened to death ("voodoo death"), sudden death during a natural catastrophe, sudden death during asthma attacks, sudden death during grief after a major loss, sudden death during panic attacks, sudden death from mental stress, sudden death related to sporting events, sudden death during war, sudden death from personal danger or threat of injury, and sudden death following acts of terrorism (Engel 1971; Samuels 2007; Leor et al. 1996; Meisel et al. 1991; Wilbert-Lampen et al. 2008)—are associated with severe emotional activation. The connection between the nervous system and the cardiopulmonary system is provided by the autonomic nervous system, which plays a key role in modulating the neuro-cardiac axis and determines how a person responds to certain triggers (Vaseghi and Shivkum 2008). Notably, stimulation of the sympathetic branch increases heart rate and myocardial contractility exerting facilitatory effects on function, whereas the stimulation of the parasympathetic branch decreasing heart rate and contractility exerts inhibitory effects.

As revealed by Child and his colleagues (Child et al. 2014), subjects with normal ventricles may experience changes in hemodynamic indices consistent with enhanced sympathetic activity. On the other hand, an exaggerated autonomic stimulation associated with the emotional reaction triggered by life-threatening stressors will have both sympathetic and parasympathetic effects, possibly precipitating myocardial dysfunction. There is solid evidence to suggest that the higher risk of cardiovascular events (e.g., myocardial infarction) is strongly associated with acute experiences of anger, anxiety, sadness, grief, and stress (Edmondson et al. 2012; Mittleman et al. 1995). Indeed, it is well known that negative emotion results in a release of catecholamines, increase in sympathetic input, and decrease in parasympathetic tone (Rozanski et al. 1988; Burg et al. 1993; Verrier et al. 1987).

Thus, alterations in autonomic function may play a pathogenic role in the development of cardiovascular morbidity and mortality and may also be important in the pathogenesis of SUD. Experimental data provide evidence (Barger et al. 1961; Raab et al. 1964) that increasing the sympathetic nervous activity or infusing catecholamines in many different types of laboratory animals can result in the production of neurocardiac lesions, namely, myofibrillar degeneration: a unique form of myocyte injury characterized by hypercontracted sarcomeres, dense eosinophilic transverse bands, and an interstitial mononuclear inflammatory response (Baroldi 1975).These histopathologic changes experimentally reproduce in animals lesions that are morphologically identical to those seen in patients who experienced sudden death.

In one famous study, Cebelin and Hirsch (1980) reported myofibrillar degeneration of the hearts of 15 victims of a homicidal physical assault, who had died a few hours or days after the assault, yet without sustaining internal injuries or trauma to the brain or viscera. The authors interpreted the histopathological data collected over 29 years in a manner consistent with acute "stress cardiomyopathy" and hence as evidence for the mediating but also potentially lethal effect of catecholamines. They found evidence of myofibrillar degeneration in the myocardial tissue and suggested the lethal potential of exaggerated sympathetic activation through its effect on the heart. In addition to introducing the term "stress cardiomyopathy," Cebelin and Hirsch's paper opened up a topic with wide-ranging implications for clinical medicine: the relation between heart function and emotional stress. The puzzle became even more intriguing in the 1990s (Sato et al. 1990a) with the identification of a new syndrome that, following Cebelin and Hirsch's lead, is now referred to as stress cardiomyopathy and which roughly presents itself with chest pain that mimics acute myocardial infarction and a transient left ventricular systolic dysfunction precipitated by acute emotional or physical stress (Kawai et al. 2007; Bybee et al. 2004; Prasad et al. 2008).

Without delving into the diagnostic details, which are still being debated, the most striking aspect of the syndrome is its transient yet at the same time dramatic quality. An important work by Wittstein et al. (2005) helps clarify its features, at the threshold between pathology and physiology. In about half the patients studied by Wittstein, the cardiac dysfunction episode was precipitated by the news of an unexpected death, while the remainder experienced a variety of causes of emotional activation, from surprise to fear. The fact that a central role is played here by exaggerated sympathetic stimulation, associated with a parasympathetic alteration, is also shown by the different plasmatic levels of catecholamines: these were two to four times higher in the 13 patients with stress cardiomyopathy than in the 7 patients with a Killip class III myocardial infarction (i.e., pulmonary edema) and remained markedly elevated even a week after the onset of symptoms.

The most distinguishing aspect of this symptomatological temporariness in terms of the organ-specific pathology, however, would appear to be the transient histology and hence reversibility of the neurocardiac damage. An endocardial biopsy conducted on five patients from Wittstein's study—with pheochromocytomas and subarachnoid hemorrhages—reveals the same histopathological characteristics associated with catecholamine excess that, in a more intense and widespread form, distinguishes neurocardiac lesions in the cases of sudden death syndrome studied by Cebelin and Hirsch. What we find here is precisely a direct myocyte injury caused by catecholamine excess and which histologically manifests itself—as in lab animals—in a form of myocyte damage characterized by hypercontracted sarcomeres, dense eosinophilic transverse bands, and an interstitial mononuclear inflammatory response (Wittstein et al. 2005). It thus seems as though there exist various levels of actualization and manifestation of the dynamics that give the brain-heart axis shape and are mediated by the sympathetic system: from the normal level recorded by Child to the transient toxicity affecting Wittstein's patients to the cardiac death described in Cebelin and Hirsch's autopsies.

This diversity would appear to be correlated with different levels of cardiac responsivity. The evidence in this direction comes from certain studies suggesting either a "differential distribution of catecholamine receptors in the myocardium," such that, for example, the apex has more receptors (Owa et al. 2001), or a "variation in responsiveness of receptors" with apical receptors being more susceptible to the catecholamine surge (Mori et al. 1993). Moreover, an increased sympathetic nerve density was found around blood vessels and areas of myocardial damage of patients with a history of ventricular arrhythmias when compared with patients without such a history, suggesting that sympathetic neural remodeling might facilitate ventricular arrhythmias (Oppenheimer 2006; Cao et al. 2000).

Differences from individual to individual, as opposed to simply organ-specific differences, come prominently into play when we take into account the multicentric study by Eitel et al. (2011), which shows that, across a wide population, only two thirds of patients had a clearly identifiable preceding stressor. Therefore, although the intensity of the trigger may be dependent on clinical risk factors that increase myocardial susceptibility to sympathetic stimulation, some people are more susceptible to the effects of ANS activation, irrespective of whether any structural abnormality is present (Child et al. 2014, Buckley and Shivkumar 2016).

If, as the available data suggest, the modes of becoming emotionally engaged with the things we encounter also correspond to a certain taking shape of bodily dynamics and in particular of the way in which the sympathetic system in each case—and over time—actualizes the brain-heart axis and if, in other words, the coming into being of the body in relation to occurrences is preserved within certain dynamics and constantly inclines us toward things, how are we to understand the gradient between "sympathetic normalcy" and the sympathetic storm through which clinical manifestations of various degrees of gravity take shape? Why does this phenomenon occur and why are some people more susceptible to it, as Buckely asks in his editorial on the subject (Buckley and Shivkumar 2016)?

If the body becomes such in each case starting from the possible modes of relating to things—corporeality—through which it positions *itself* in the world, and if these modes correspond to different ways of finding oneself and becoming emotionally disposed, then emotivity, which can affect cardiac functionality in many ways—to the point of killing an individual, as in the case of sudden cardiac death—can only be understood in the light of the story (and not just the medical history) of the person who experiences such emotivity. Clearly, a form of medicine willing to embrace this perspective should account for how each person's way of living life, as he or she can, leads to the emergence and development of certain disorders.

The relation between cardiovascular events and self-perception becomes particularly evident in those studies highlighting the close link between cardiac events and chronic emotional states such as frequent anger and hostility, chronic anxiety, or deep sadness (Brotman et al. 2007; Barefoot and Schroll 1996; Eaker et al. 2004; Matthews et al. 2004; Kubzansky et al. 2001; Kuper et al. 2002), as well as in those studies focusing on the comorbidity between SMC and mood disorders associated with depression and anxiety (Wittstein 2012; Mudd et al. 2007; Regnante et al. 2009).[18] Collectively, these findings not only support the mechanism of the direct effect of catecholamines on the myocardium but also—and most importantly—suggest that individual myocardial susceptibility varies according to

[18] There is a consistent link between chronic emotional or psychosocial stress and coronary atherosclerosis (and atherosclerotic risk factors) (Brotman et al. 2007).

one's modes of emotional engagement and hence according to how each person lives his or her own life. On what bases do these different dispositions take shape? What is the relation between the way in which the emotional world of the person takes shape and the way in which cardiac functionality is structured? What situations lead to the emergence of those negative emotions that support a certain emotional engagement which is detrimental to the person? In what way, starting from what meaningful horizons and in what periods of the individual's existence, did these situations become structured over time? These questions all concern the individual.

10.9 Co-productivity: Some Experimental Guidelines for the Neurosciences and Psychology

The corporeal, which emerges within the framework of our relationship with the world, modifies and transforms itself according to particular emotional dispositions in relation to the meaningful contexts within which it realizes itself. By eliciting changes in the sympathetic-parasympathetic balance, these conditions can negatively affect the cardiovascular system both acutely, as in the case of the news of an unexpected death, and chronically, as in the case of recurrent negative emotional states. As already noted, emotional states integrate the way in which we perceive ourselves in relation to what we are open to and can therefore be understood as categories that articulate the way in which life understands itself within certain existential contexts or *milieus*. Not only that, but it is starting from these modes of being-in-the-world that the body becomes such.

Certain modes of being-in-the-world, then, lead to the emergence of a cardiac susceptibility that is mediated by the autonomic nervous system and which presents a viscerotropic representation in the insular cortex. In turn, this cortex is closely connected to the limbic system and in particular the amygdala and ACC. The modulation operated by the autonomic nervous system on the neurocardiac axis thus reflects a major influence that the brain, i.e., the insular cortex, has on the cardiac structure and function. Indeed, clinical observations and experimental evidence indicate that the insula, an important hub of central cardiovascular control, plays a pivotal role in producing the cardiovascular responses to emotional stimuli (Oppenheimer and Cechetto 1990; Oppenheimer et al. 1992; Oppenheimer 2006; Oppenheimer and Cechetto 2016).

In other words, the meaningfulness that the individual encounters elicits a different insular involvement and therefore a certain cardiac effect, which is clearly also related to the different forms of emotional engagement and personal stability achieved by the individual up until then. Life changes, existential turning points, and the emergence of new positions, or simply the passing of the years according to certain repeated patterns, and in any case the way in which we become bound to the Earth, are characterized by emotional configurations that engender different modes of openness to the meaningful things we encounter. Living is what determines how the body bodies forth.

As we have already seen, different modes of emotioning corresponding to specific ways in which the corporeal dynamic takes shape in relation to major life changes—e.g., the death of a spouse—and starting from them, or simply according to an emotional style that is "channeled" through specific characteristics, e.g., frequent anger and hostility or chronic anxiety, govern the way in which the individual inhabits the world and grasps its meaningfulness, in some cases negatively affecting the cardiovascular system (Brotman et al. 2007). Cardiac disorders, therefore, point to the concrete and historical life of the individual and to its variability. On the other hand, it is only in the light of the existential horizons whose experience is embodied by the body—the very body of the individual patient—that these disorders become more meaningful. It is difficult to see how medicine could offer an adequate treatment without therapeutically cooperating with a science such as psychology.

If the various forms of openness correspond to different modes of insular activation—e.g., in relation to negative emotions—and if its distinct involvement is associated with the individual differences distinguishing the modulation of the autonomic nervous system on the neurocardiac axis, the insular cortex may be seen to exercise a function in dynamically integrating affective and bodily aspects of the inner and outer environments (Craig 2011). Moreover, recent data of ours (Mazzola et al. 2016) indicate that the insular cortex modulates this relationship by affecting those structures that are broadly involved in social perception and contextual information, such as the bilateral superior temporal gyri (STG) (Allison et al. 2000; Adolphs 2003; Singer et al. 2004; Paulus et al. 2005; Xu et al. 2009). Indeed, the reaction in response to others' anger shifts the bi-directional connection between STG and insula from a bilateral to an ipsilateral one. These findings suggest that facing angry people leads to a greater bodily involvement at the expense of a larger focus on the social contextual information processing (Mazzola et al. 2016). Here—as in all things corporeal—the brain, in its taking shape, is determined by the way in which the individual situates himself in his relating to things, in his being-in-the-world.

The dynamically individualized viscero-cerebral signature which the insula preserves, so to speak, and on the basis of which it disposes the organism, takes shape starting from the meaningfulness which the individual encounters and partly chooses, by in each case choosing his possible future and thereby renewing his personal story. It is only within this emotionally situated organismic historicity (constantly in the process of unfolding starting from this or that thing) that it is possible to investigate those individual differences that might shed some light on myocardial susceptibility to sympathetic stimulation. On the other hand, other regions such as the ventromedial prefrontal cortex, orbitofrontal cortex, amygdala, *n. accumbens*, precuneus, etc. can also be active, depending on different sets of circumstances for different individuals. This opens the study of the emotional domain up to the prospect of an individual variability that cannot be reduced to the primacy of individual neuroanatomical structures.

Naturally, this raises further questions regarding the various ways of becoming emotionally engaged and how different emotional positions over the course of one's existence may correspond to different forms of situatedness: different modes of openness to the world and of self-perception, according to which different neural configurations emerge and become stabilized, and along with them different corporeal dynamics (Arciero Bondolfi 2009). In support of this perspective, in a study of ours (Mazzola et al. 2010), we have shown that the insula displays different degrees of involvement as part of the pain matrix in a situation of vicarious pain across subjects chosen on the basis of the categorization of this situatedness, grasped by the life of the subjects themselves according to the two dimensions of existential relatedness: inward and outward.[19] Other studies have sought to identify the "pain signature" by focusing on the central role played by the brainstem and the descending pain modulatory system in the perception of pain under varying circumstances (Tracey and Mantyh 2007).

If the dynamic cerebral signature takes shape starting from the meaningfulness which the individual encounters and partly chooses, by in each case choosing his possible future and thereby renewing his personal story, the neurosciences must engage in the experimental task of grasping these signatures and analyzing their connections. The co-productivity between psychology and the natural sciences, on which phenomenology sheds light, emerges here according to a different angle. Starting from a preestablished distinction in the study subjects of the categories emerging from their experience, psychology can direct the neurosciences toward those sensible meaningful domains targeted by the experiment.[20] The experiment must therefore be capable of eliciting the enactment of such categories.

If we always find ourselves and become disposed starting from a situation, developing an appropriate task in a lab means manipulating the way in which the subjects encounter the meaningfulness that produces the data, generates the neural signatures and thus ensures a particular mode of access. What is required, in other words, is an active form of experimentation, developed according to the same register adopted for the *ars inveniendi* posited by Helmholtz—albeit from an opposite perspective, so to speak. The experiment must be capable of generating an experience and revealing the neural structures associated with it, which can only be understood in the light of the meaningful essences that bind it, and hence within the framework of the configurations of sense identified beforehand. In turn, the dynamic cerebral signatures thus revealed not only raise questions pertaining to the overall coherence and cerebral modelization but, by becoming manifest in the experiment, open up actual experience to further investigations, engendering the kind of co-productivity which represents the privileged avenue for the clinician and the researcher. Therefore, from the point of view of the natural scientist adopting a phenomenally oriented methodology, there are four crucial moments: (1) the formal indication of the phenomenon to be investigated (focusing on things themselves); then, in the light of this phenomenal disclosure, (2) the choice of cohort subjects on the basis of

[19] During the fMRI session, participants watched the faces of partners and strangers, in both painful and neutral situations. The results show that the highly inward prone participants had a greater activation of the left posterior insula while they processed their partners' painful facial expressions, whereas outward participants did not show an involvement of the insula, but a significant involvement of the left precuneus and mPFC.

[20] Clearly, the same indications can also guide psychometric research.

how the concrete and positive content targeted by the formal indication is articulated at an individual level—an investigation that may unfold according to the categories articulating the way in which each individual life understands itself in given ongoing situations and contexts—(3) the development of a task capable of eliciting the execution of a certain tendency or motive and hence of engendering the phenomenon under investigation; and (4) the actual generation of experience through the regeneration of neural signatures (the bodying forth of the body) and the study of neural connections.

Clearly, the privileged avenue just described can also be explored starting from the natural sciences, as studies on stress-induced cardiomyopathy suggest. Let us go over the main points made (Akashi et al. 2010). Clinical-pathological evidence interpreted in the light of the conditions of death of the subjects (Cebelin and Hirsch 1980) suggested the lethal potential of stress through its effect on the heart. These initial data were confirmed by subsequent studies, which concluded that high catecholamine concentrations can directly damage the myocardium. Researchers then identified a reversible ventricular dysfunction that is often associated with coronary spasm (Sato et al. 1990b) and is also associated with emotional stress and hence with the activation of the ANS as a mediator between the insula and the heart. The acknowledgment of the crucial role of the insula (Cheung and Hachinski 2000), and of its different involvement in various emotional states, points to individual differences. The research cannot be advanced any further without addressing this last point through a methodology capable of grasping such differences—a method capable of grasping those categories that articulate the way in which an individual life understands itself in certain situations, so as to engender a space of co-determination between the natural sciences and psychology.

The need for such co-productivity also clearly emerges in the field of clinical psychology, as illustrated, for instance, by the depression diagnosis (Arciero Bondolfi 2009). Since Kraeplin and Kretschmer, the clarification of the limits between ordinary sadness and clinical depression has been a major concern. Much of the controversy has revolved around the issue of whether, and on what basis, a boundary can be fixed in the continuum from the experience of sadness to a major depressive episode. This state of affairs has further been complicated by the fact that the DSM-5 has introduced bereavement-related depression within the field of depressive disorders. Bereavement possesses certain clinical characteristics (e.g., feelings of intense sadness, rumination about loss, insomnia, poor appetite, and weight loss) which make it qualify for a diagnosis of major depression. Therefore, the chief question is on what basis bereavement-triggered and other loss-triggered forms of uncomplicated intense sadness (e.g., no psychotic ideation, no suicidal ideation, no psychomotor retardation, etc.) should be included or excluded from the diagnosis of major depression.

If, as Maj notes, that decision "inevitably requires the exercise of clinical judgment based on what the clinician knows about the individual in question and the individual's cultural norms for the expression of distress in the context of loss" (Maj 2013, p. 90), then a key question concerns the criteria that should guide the clinician in drawing the line between "normal and abnormal emotions." Indeed, the clinician is left to make the decision as to whether the sadness response of the individual is proportional to the triggering situation or not, a decision based on the knowledge that he has of the individual, of his life story, and of his personal experience. How then can the clinician scientifically decide whether the sadness response of the individual is proportional to the triggering situation or not?

Individual experience acquires meaning by dint of the fact that it is found within a configuring framework, the personal story. The ongoing experience is placed within a referential context by the story that it helps to articulate and develop, both in the case where the event falls within the range of expectations that "naturally" derive from the story and in the case where it "takes by surprise" the predictions made regarding the development of the story, thereby generating new expectations. In this respect, each individual story may lead to a state of depression in conjunction with life circumstances capable of initiating a downward spiral: a vicious circle of cognition and sadness. Therefore, it is evident that operational criteria cannot have priority over clinical judgment; but it is equally clear that to decide that the response to a loss is "normal" on the basis of what the clinician knows of the individual and his or her cultural background requires a method. And this can only be ensured by a kind of psychology capable in each case of grasping the prereflective generation of personal meaning, in such a way as to allow the clinician to investigate the uniqueness of the personal story of his or her patients (Bondolfi et al. 2015). Clinical judgment must therefore be founded on the same mode of investigating other people's experience that lies at the basis of the care of self.

References

Adolphs R (2003) Cognitive neuroscience of human social behaviour. Nat Rev Neurosci 4(3):165–178

Akashi YJ, Nef HM, Möllmann H, Ueyama T (2010) Stress cardiomyopathy. Annu Rev Med 61:271–286

Allison T, Puce A, McCarthy G (2000) Social perception from visual cues: role of the STS region. Trends Cogn Sci 4(7):267–278

Arciero G, Bondolfi G (2009) Selfhood, identity and personality styles. John Wiley & Sons, Chichester

Baker M (2016) Statisticians issue warning over misuse of P values. Nature 531(7593):151–151

Barbaras R (2008) Introduction à une phénoménologie de la vie. Vrin, Paris

Barefoot JC, Schroll M (1996) (1996) Symptoms of depression, acute myocardial infarction, and total mortality in a community sample. Circulation 93:1976–1980

Barger AC, Herd JA, Liebowitz MR (1961) Chronic catheterization of coronary artery induction of ECG pattern of myocardial ischemia by intracoronary epinephrine. Proc Soc Exp Biol Med 107:474–477

Baroldi F (1975) Different morphological types of myocardial cell death in man. In: Fleckstein A, Rona G (eds) Recent advances in studies in cardiac structure and metabolism. pathophysiology and morphology of myocardial cell alteration, vol 6. University Park Press, Baltimore

Bernard M (2011) Le monde comme problème philosophique. Les Études philosophiques (3):351–373

Bersaglieri T, Sabeti PC, Patterson N, Vanderploeg T, Schaffner SF, Drake JA, Rhodes M, Reich DE, Hirschhorn JN (2004) Genetic signatures of strong recent positive selection at the lactase gene. Am J Hum Genet 74:1111–1120

Bifulco AT, Brown GW, Harris TO (1987) Childhood loss of parent, lack of adequate parental care and adult depression: a replication. J Affect Disord 12(2):115–128
Bondolfi G, Mazzola V, Arciero G (2015) In between ordinary sadness and clinical depression. Emot Rev 7(3):216–222
Brotman DJ, Golden SH, Wittstein IS (2007) The cardiovascular toll of stress. Lancet 370:1089–1100
Buckley U, Shivkumar K (2016) Stress-induced cardiac arrhythmias: the heart–brain interaction. Trends Cardiovasc Med 26(1):78
Burg MM, Jain D, Soufer R, Kerns RD, Zaret BL (1993) Role of behavioral and psychological factors in mental stress- induced silent left ventricular dysfunction in coronary artery disease. J Am Coll Cardiol 22:440–448
Bybee KA, Kara T, Prasad A et al (2004) Systematic review: transient left ventricular apical ballooning: a syndrome that mimics ST-segment elevation myocardial infarction. Ann Intern Med 141:858–865
Cao JM, Fishbein MC, Han JB, Lai WW, Lai AC, Wu TJ, Czer L, Wolf PL, Denton TA, Shintaku IP, Chen PS, Chen LS (2000) Relationship between regional cardiac hyperinnervation and ventric- ular arrhythmia. Circulation 101:1960–1969
Cebelin M, Hirsch CS (1980) Human stress cardiomyopathy. Hum Pathol 11:123–132
Chā. (2011). The collected teachings of Ajahn Chah, 3 Vol. Amaravati Buddhist Monastery
Cheung RT, Hachinski V (2000) The insula and cerebrogenic sudden death. Arch Neurol 57(12):1685–1688
Child N, Hanson B, Bishop M, Rinaldi CA, Bostock J, Western D et al (2014) Effect of mental challenge induced by movie clips on action potential duration in normal human subjects independent of heart rate. Circ Arrhythm Electrophysiol 7:518–523
Craig AD (2011) Significance of the insula for the evolution of human awareness of feelings from the body. Ann N Y Acad Sci 1225:72–82
Dastur F (1995) Pour une zoologie privative. In: Alter, Revue de Phénoménologie, n° 3. Editions Alter, Paris, pp 281–318
Dastur F (2013) Fink et Patočka: deux conceptions de la liberté. Philosophie 3:13–20
Depraz N (2006) Mettere al lavoro il metodo fenomenologico nei protocolli sperimentali."Passaggi generativi" tra l'empirico e il trascendentale. In: Cappuccio M (ed) Neurofenomenologia. Le scienze della mente e la sfida dell'esperienza cosciente. Bruno Mondadori, Milano
Derrida J (1987) De l'esprit. Galilée, Paris
Derrida J, Wills D (2002) The animal that therefore I am (more to follow). Crit Inq 28(2):369–418
Eaker ED, Sullivan LM, Kelly-Hayes M, D'Agostino RB Sr, Benjamin EJ (2004). Anger and hostility predict the development of atrial fibrillation in men in the Framingham Offspring Study. Circulation; 109: 1267–1271.
Edmondson D, Newman JD, Whang W, Davidson KW (2012) Emotional triggers in myocardial infarction: do they matter? Eur Heart J 34(4):300–306
Eitel I, Knobelsdorff-Brenkenhoff F, Bernhardt P et al (2011) Clinical characteristics and cardiovascular magnetic resonance findings in stress (takotsubo) cardiomyopathy. JAMA 306:277–286
Engel GL (1971) Sudden and rapid death during psychological stress: folklore or folk wisdom? Ann Intern Med 74(5):771–783
Harris T, Brown GW, Bifulco A (1986) Loss of parent in childhood and adult psychiatric disorder: the role of lack of adequate parental care. Psychol Med 16(03):641–659
Heidegger GA 15 1986
Heidegger GA 29-30 (1983)
Heidegger GA 56/57 1999
Holden C, Mace R (1997) Phylogenetic analysis of the evolution of lactose digestion in adults. Hum Biol 69:605–628
Jablonka E, Lamb MJ (2005) Evolution in four dimensions: genetic, epigenetic, behavioral, and symbolic variation in the history of life. MIT Press, Cambridge

Jablonka E, Lamb MJ, Zeligowski A (2014) Evolution in four dimensions, revised edition: genetic, epigenetic, behavioral, and symbolic variation in the history of life, 2nd edn. MIT Press, Cambridge
Jacquet F (2011) Vie et existence: vers une cosmologie phénoménologique. Les Études philosophiques (3):395–419
Jonas H (1967) The phenomenon of life: toward a philosophical biology. Harper & Row, New York
Jonas H (1997) Tecnica, medicina ed etica. Prassi del principio responsabilità, a cura di P. Becchi, Einaudi, Torino
Kawai S, Kitabatake A, Tomoike H (2007) Guidelines for diagnosis of takotsubo (ampulla) cardiomyopathy. Circ J 71:990–992
Kisiel T (1996) L'indication formelle de la facticité: sa genèse et sa transformation. In: Courtine J-F (ed) Heidegger 1919–1929. De l'herméneutique de la facticité à la métaphysique du Dasein. Vrin, Paris
Kubzansky LD, Sparrow D, Vokonas P, Kawachi I (2001) Is the glass half empty or half full? A prospective study of optimism and coronary heart disease in the normative aging study. Psychosom Med 63:910–916
Kuper H, Marmot M, Hemingway H (2002) Systematic review of prospective cohort studies of psychosocial factors in the etiology and prognosis of coronary heart disease. Semin Vasc Med 2:267–314
Laland KN, Odling-Smee J, Myles S (2010) How culture shaped the human genome: bringing genetics and the human sciences together. Nat Rev Genet 11(2):137–148
Laland K, Odling-Smee J, Turner S (2014a) The role of internal and external constructive processes in evolution. J Physiol 592(11):2413–2422
Laland K, Uller T, Feldman M, Sterelny K, Müller GB, Moczek A et al (2014b) Does evolutionary theory need a rethink? Nature 514(7521):161
Leor J, Poole WK, Kloner RA (1996) Sudden cardiac death triggered by an earthquake. N Engl J Med 334(7):413–419
Maj M (2013) "Clinical judgment" and the DSM-5 diagnosis of major depression. World Psychiatry 12(2):89–91
Matthews KA, Gump BB, Harris KF, Haney TL, Barefoot JC (2004) Hostile behaviors predict cardiovascular mortality among men enrolled in the Multiple Risk Factor Intervention Trial. Circulation 109:66–70
Mazzola V, Latorre V, Petito A, Gentili N, Fazio L, Popolizio T et al (2010) Affective response to a loved one's pain: insula activity as a function of individual differences. PLoS One 5(12):e15268
Mazzola V, Arciero G, Fazio L, Lanciano T, Gelao B, Popolizio T et al (2016) What impact does an angry context have upon us? The effect of anger on functional connectivity of the right insula and superior temporal gyri. Front Behav Neurosci 10:109
Meisel SR, Dayan KI, Pauzner H, Chetboun I, Arbel Y, David D, Kutz I (1991) Effect of Iraqi missile war on incidence of acute myocardial infarction and sudden death in Israeli civilians. Lancet 338(8768):660–661
Mittleman MA, Maclure M, Sherwood JB, Mulry RP, Tofler GH, Jacobs SC et al (1995) Triggering of acute myocardial infarction onset by episodes of anger. Circulation 92(7):1720–1725
Mori H et al (1993) Increased responsiveness of left ventricular apical myocardium to adrenergic stimuli. Cardiovasc Res 27(2):192–198
Mudd JO, Kapur NK, Champion HC, Schulman SP, Wittstein IS (2007) Patients with stress-induced (Takotsubo) cardiomyopathy have an increased prevalence of mood disorders and antidepressant use compared to patients with acute myocardial infarction. J Cardiac Fail 13(Supplement 2):S176
Müller GB (2007) Evo–devo: extending the evolutionary synthesis. Nat Rev Genet 8(12):943–949
Nuzzo R (2014) Statistical errors: P values, the 'gold standard' of statistical validity, are not as reliable as many scientists assume. Nature 506(7487):150–153
Odling-Smee J (2010) Niche inheritance. In: Pigliucci M, Müller GB (eds) Evolution: the extended synthesis. MIT Press, Cambridge

Odling-Smee J, Turner JS (2011) Niche construction theory and human architecture. Biol Theor 6(3):283–289
Odling-Smee FJ, Laland KN, Feldman MW (1996) Niche construction. Am Nat 147(4):641–648
Odling-Smee FJ, Laland KN, Feldman MW (2003) Niche construction: the neglected process in evolution (No. 37). Princeton University Press, Princeton
Oppenheimer S (2006) Cerebrogenic cardiac arrhythmias: cortical lateralization and clinical significance. Clin Auton Res 16:6–11
Oppenheimer SM, Cechetto DF (1990) Cardiac chronotropic organization of the rat insular cortex. Brain Res 533:66–72
Oppenheimer S, Cechetto D (2016) The insular cortex and the regulation of cardiac function. Compr Physiol 6(2):1081–1133
Oppenheimer SM, Cechetto DF, Hachinski VC (1990) Cerebrogenic cardiac arrhythmias: cerebral electrocardiographic influences and their role in sudden death. Arch Neurol 47:513–519
Oppenheimer SM, Gelb AW, Girvin JP, Hachinski VC (1992) Cardiovascular effects of human insular stimulation. Neurology 42:1727–1732
Owa M et al (2001) Emotional stress-induced 'ampulla cardiomyopathy': discrepancy between the metabolic and sympathetic innervation imaging performed during the recovery course. Jpn Circ J 65(4):349–352
Patočka J (1953) Le platonism negative in Jan PATOČKA, Liberté et Sacrifice. Écrits politiques, ed. and transl. E. Abrams (Grenoble: Millon, 1990). English Transation 'Negative Platonism". In: Erazim Kohák (ed) Jan Patočka, Philosophy and selected writings. The University of Chicago Press 1989, Chicago, p 175–206, 175
Patočka J (1988) Le Monde naturel et le mouvement de l'existence humaine, traduit par E. Abrams. Kluwer, Dordrecht
Patočka J (1990) Liberté et sacrifice: écrits politiques. Editions Jérôme Millon
Patočka J (1995) Papiers phénoménologiques. Editions Jérôme Millon, Grenoble
Paulus MP, Feinstein JS, Leland D, Simmons AN (2005) Superior temporal gyrus and insula provide response and outcome-dependent information during assessment and action selection in a decision-making situation. Neuroimage 25(2):607–615
Pfennig DW, Wund MA, Snell-Rood EC, Cruickshank T, Schlichting CD, Moczek AP (2010) Phenotypic plasticity's impacts on diversification and speciation. Trends Ecol Evol 25(8):459–467
Prasad A, Lerman A, Rihal CS (2008) Apical ballooning syndrome (Tako-Tsubo or stress cardiomyopathy): a mimic of acute myocardial infarction. Am Heart J 155:408–417
Raab W, Chaplin JP, Bajusz K (1964) Myocardial necroses produced ill domesticated rats and wild rats by sensory and emotional stresses. Proc Soc Exp Biol Med 116:665
Regnante RA, Zuzek RW, Weinsier SB et al (2009) Clinical characteristics and four-year outcomes of patients in the Rhode island takotsubo cardiomyopathy registry. Am J Cardiol 103:1015–1019
Rizzolatti G, Cattaneo L, Fabbri-Destro M, Rozzi S (2014) Cortical mechanisms underlying the organization of goal-directed actions and mirror neuron-based action understanding. Physiol Rev 94(2):655–706
Rozanski A, Bairey CN, Krantz DS, Friedman J, Resser KJ, Morell M et al (1988) Mental stress and the induction of silent myocardial ischemia in patients with coronary artery dis- ease. N Engl J Med 318:1005–1012
Samuels MA (2007) The brain–heart connection. Circulation 116(1):77–84
Sato H, Tateishi H, Uchida T (1990a) Takotsubo-type cardiomyopathy due to multivessel spasm. In: Kodama K, Haze K, Hon M (eds) Clinical aspect of myocardial injury from ischemia to heart failure. Kagakuhyouronsha, Tokyo, pp 56–64
Sato H, Tateishi H, Uchida T et al (1990b) Tako-tsubo-like left ventricular dysfunction due to multivessel coronary spasm. In: Kodama K, Hori M (eds) Clinical aspect of myocardial injury: from ischemia to heart failure [in Japanese]. Kagakuhyoronsha Publishing Co, Tokyo, pp 56–64
Schiller J (1978) La notion d'organisation dans l'histoire de la biologie. Maloine, Paris

Scott-Phillips TC, Laland KN, Shuker DM, Dickins TE, West SA (2014) The niche construction perspective: a critical appraisal. Evolution 68(5):1231–1243

Singer T, Seymour B, O'doherty J, Kaube H, Dolan RJ, Frith CD (2004) Empathy for pain involves the affective but not sensory components of pain. Science 303(5661):1157–1162

Tardivel É (2011) La liberté au principe: essai sur la philosophie de Patočka. Vrin, Paris

Tracey I, Mantyh PW (2007) The cerebral signature for pain perception and its modulation. Neuron 55(3):377–391

Turner JS (2000a) Architecture and morphogenesis in the mound of Macrotermes michaelseni (Sjöstedt)(Isoptera: Termitidae, Macrotermitinae) in northern Namibia. Cimbebasia 16:143–175

Turner JS (2000b) The extended organism. The physiology of animal-built structures. Harvard University Press, Cambridge

Turner JS (2002) A superorganism's fuzzy boundaries. Nat Hist 111(6):62–69

Turner JS (2004a) Extended phenotypes and extended organisms. Biol Philos 19(3):327–352

Turner JS (2004b) Gaia, extended organisms, and emergent homeostasis. In: Schneider SH, Miller JR, Crist E, Boston PJ (eds) Scientists debate gaia. The next century. MIT Press, Cambridge

Vaseghi M, Shivkum K (2008) The role of the autonomic nervous system in sudden cardiac death. Prog Cardiovasc Dis 50(6):404–419

Verrier RL, Hagestad EL, Lown B (1987) Delayed myocardial ischemia induced by anger. Circulation 75:249–254

Volpi F (1996) La Question du Logos dans l'articulation de la facticité chez le jeune Heidegger, lecteur d'Aristote. In: Courtine J-F (ed) Heidegger 1919-1929. De l'herméneutique de la facticité à la métaphysique du Dasein. Vrin, Paris

West-Eberhard MJ (2003) Developmental plasticity and evolution. Oxford University Press, Oxford

Wilbert-Lampen U, Leistner D, Greven S, Pohl T, Sper S, Völker C et al (2008) Cardiovascular events during World Cup soccer. N Engl J Med 358(5):475–483

Wittstein IS (2012) Stress cardiomyopathy: a syndrome of catecholamine-mediated myocardial stunning? Cell Mol Neurobiol 32(5):847–857

Wittstein IS, Thiemann DR, Lima JA, Baughman KL, Schulman SP, Gerstenblith G et al (2005) Neurohumoral features of myocardial stunning due to sudden emotional stress. N Engl J Med 352(6):539–548

Xu J, Gannon PJ, Emmorey K, Smith JF, Braun AR (2009) Symbolic gestures and spoken language are processed by a common neural system. Proc Natl Acad Sci 106(49):20664–20669

The Care of Self and Psychotherapy 11

11.1 Errancy and Therapeutic Care

One of the key themes we have been exploring in this work is the close connection between therapeutic care and sickness on the one hand and freedom on the other hand. We have conceived the experience of freedom as the open possibility of understanding what addresses us, a condition that belongs to man by virtue of his having come into the world, of having being born. We have seen how this experience is the cornerstone on which the distinction between man and animal rests. Whereas the latter is captivated by the enactment of the relationship with the environment within which it lives, man lets the meaningful things he encounters be; in relating to them, man brings them to manifestation and, at the same time, reaches himself precisely through what he causes to appear.

Unlike in the case of animals, this relationship is not bound by necessity: man is free. This means that in relating to things in view of the world in its wholeness—a world which co-appears as a background beyond all objects, yet always together with them—man discloses the domain of sense and its problematic quality as he realizes his own experience over and over again. The experience of the world is therefore primordial, as Patočka clearly notes: "I grasp the world as a phenomenon, but at the same time I see that, while I can replace and modify any worldly entity, I can never do the same with the world as such, the world as a whole" (Patočka 1995, p. 237). This world in its totality, which includes the sphere of objects as such, is inaccessible to animals, who are bound to things on account of their instinct.

From this perspective, ipseity manifests itself as the very possibility of the experience of freedom: as an ever-unfulfilled condition which enables us to grasp the meaningfulness that we encounter starting from a horizon, the world, which withdraws as we approach it, constantly regenerating our incompleteness, want, and precariousness. We thus once again grasp the motility of life as an opening up to (*Bezugssinn*) what addresses us (*Gehaltssinn*) and actually realizes itself again and again (*Volzugssinn*) as the manifestation of oneself and appearance of the world. Clearly, the subject neither creates the world nor "forms" it; rather, in contributing

G. Arciero et al., *The Foundations of Phenomenological Psychotherapy*,
https://doi.org/10.1007/978-3-319-78087-0_11

to the appearance of the world, ipseity relates to itself and comes to manifestation by virtue of that appearance which it conditions. In this sense, every phenomenon coincides with the appearance of the world and the manifestation of an ipseity (Patočka 1995).

Patočka's contribution here is crucial. He establishes the actualization of life and hence the very experience of freedom on corporeality, understood as motility. Patočka takes one decisive step further compared to Heidegger's ontology by giving the latter a dimension which it has itself been incapable of developing. Patočka, that is, explains motility as a corporeality which, in constantly determining itself with respect to what occurs (existential determination), brings the world to manifestation and reaches itself through that which it causes to appear, thereby establishing its substrate (embodiment of existence). Thus corporeality, and with it existence, can only be understood starting from our belonging to the world, conceived as our relationship with the world, with all the possibilities it offers, as disclosed by the motility of life in its enactment. Patočka writes: "Primordial freedom *(la liberté originaire*) is something which I am in the purest sense of the term, [but] which I never possess" (Patočka 1988, p. 113)

In the light of this way of conceiving the corporeal, we have regarded sickness as the modification of this relationship: as an alteration of specific abilities to enact our being-there with respect to what we encounter and hence as a restriction of our freedom. Sickness is therefore understood as a stifling of the motility of human life in its relationship with the meaningfulness which addresses it: a limitation that brings suffering. From this perspective, there is no difference in principle between physical and psychological sickness.[1] On the other hand, we have seen how therapeutic care coincides with the restoration of those possibilities of openness which enable the motility of life to expose the world to the complexity of sense and ipseity to access the latter in its wholeness. In actualizing the possibilities it chooses, the motility of life thus finds a place in the world, causing it to appear, and at the same time it reveals itself in its finitude. As already stated at the beginning of this work, the difference between men takes shape from the orientation of their way of life.

The questioning of this phenomenon, which coincides with both the disclosure of things of the world and the revelation of oneself, the investigation of the nature of this dazzling and marvelous truth, lies at the origin of that ancient tradition that, according to sources stretching as far back as Pythagoras, constitutes the foundation of philosophy: the care of the soul. Indeed, if the world and things show themselves for what they are to the soul, and through the soul, then in order to understand the soul it is necessary to take care of it. This praxis coincides with that know-how that was originally evoked by the term *sophia* and which Pythagoras had sought to teach the tyrant Leon (see Chap. 1). From the dawn of the ancient Greek world, the care of self coincided with contemplation, *theorein*, a practice which, by transforming one's way of life, seeks to grasp those entities that cannot be other than what they are. Thus the early physiologists turned their restless eyes to the heavens, to

[1] The differences between the two become evident when we analyze the mechanisms at work, which are the focus of specific sciences.

discover the changeless divinity of the stars and cleave to it as though to God's own body.

It was in the light of this practice of contemplation, spread by Pythagoras, that Socrates burst onto the public stage in Athens, to show that he who wishes to take care of the common good must possess an enduring kind of wisdom in order to act according to justice. To correct that *errancy* which shapes the destiny of men, the soul—which loses itself in the situations and affairs of everyday life—must constantly examine itself: hence the care of the soul, as it is expounded in the *First Alcibiades*, which consists in directing to oneself that form of contemplation which the Ancients had directed to the starry heavens.

What we have here is the famous Platonic *epistrophe*, the philosopher's conversion, which Hadot outlines as follows: "The philosopher is himself converted because he knew to turn his gaze away from the shadows of the sensible world and turn it towards the light which emanates from the idea of the Good" (p. 176, 1988).[2] Through a movement that leads it back to itself, and which corresponds to a theoretical life practice, the soul accomplishes a constant transformation of itself, retaining only what resists change. In searching for the enduring truth within changing things, the soul only recognizes what is stable, changeless, and divine in itself, while distancing itself from all those affections, desires, and yearnings that bind it to such things. By turning back upon itself, the soul finds and, at the same time, recognizes what it is (*anamnesis*). This is the key theme of the *Alcibiades*: taking care of one's own soul as though it were an ontologically autonomous reality. Patočka states: "We can only understand the soul and grasp its essence by practicing this care, by devoting ourselves to it" (1998, p. 121).

Tireless analysis, laying siege through dialogue, the care of self as the constant making of a fresh start in speech based on a single certainty, knowing that we do not know: such is Socrates' method, whose concrete application inevitably produces a detachment from things, a freeing up of the gaze and, with it, perpetual dissatisfaction with the sensible realm—a discontent which leads us to call ourselves into question again and again.[3] According to Patočka, this experience of transcendence, devoid of any positive content, coincides precisely with the experience of freedom: "Those who have undergone this experience know that objects are not all there is. The experience of freedom is precisely that of the nothingness of the real world, left to itself" (p. 95, 1990). This view constitutes the cornerstone of a non-metaphysical reading of Plato: Patočka's negative Platonism.

If the soul, then, coincides with its own movement (*to hauto kinoun—Phaedrus* 245c–246a), the care of the soul must coincide with the transforming and

[2] On the different forms taken by *epistrophe* over the course of the development of ancient thought, down to its articulation in a Christian context, see Foucault 2001.

[3] This is the crux of the *Apology*, where Socrates explains to the people of Athens—before whom he defends himself against the charge of corrupting the young—that what the oracle of Delphi meant by presenting him as the wisest of all men was that human wisdom is of little or no account. "Therefore I am still even now going about and searching and investigating at the god's behest anyone, whether citizen or foreigner, who I think is wise; and when he does not seem so to me, I give aid to the god and show that he is not wise" (23 b-c).

reorienting of such movement. As already noted at the beginning of this work, the difference between men takes shape starting from the orientation of the motility of life. Therefore, although in the case of most people the soul is usually engrossed in everyday pursuits and worries, those who love the soul can modify it by taking care of it. This corresponds to the *conversion* of one's orientation upon itself and hence to the reestablishment of a dialogue with oneself. In ancient thought, this reversion to oneself is achieved through the search for and identification of the universal, shared truths to be found in all men. Clearly, this emphasis on interiority has nothing to do with the notion of the centrality of the subject emphasized in modern thought (Hadot 2002, 2014).

The theoretical attitude, therefore, concretely translates into a conduct which the philosopher embodies through his way of life, in accordance with what may be described as his "objective" knowledge. This crucial element is what distinguishes philosophy as care of the soul. In this sense, self-knowledge stems from the constant shaping of oneself: self-knowledge requires the care of self, and the care of self goes hand in hand with the capacity to grasp the truth within oneself (*parrhesia*). This remained the essence of philosophy throughout antiquity (Hadot 2002).

11.2 "Learning How to Live" and "Learning How to Die": The Theory and Technology of the Soul

As we have seen (Chap. 1), it was Aristotle who discovered that the contemplative life is one of the possible ways of articulating the movement we have been describing. While maintaining that the loftiest way of life is the theoretical one, Aristotle extended his investigation to factical life in its motility, understood as "that which happens or may happen to someone" at any moment. But whereas Aristotelian ethics vanished in the cracks of history, only to resurface centuries later, theory—reduced to succinct and easily applicable precepts—continued to guide that constant act of focusing one's attention upon oneself that distinguishes philosophical living in the Hellenistic and Roman periods. Domanski beautifully sums it up as follows: "After Plato, from the metaphilosophical reflection of the Hellenistic philosophers to Plotinus, a *communis philosophorum opinio* has it that in order to be a true philosopher, it is not enough to know how one must lead one's life, but it is crucial to live in full accordance with this knowledge" (1996, p. 11).

According to every ancient school, the philosopher's way of life is marked by a tension toward that kind of stable knowledge, the exercise of which, *theorein*, far from being only a meditative practice, must also coincide with the concrete enactment of a philosophical conduct, through an accordance between the discourse on truth, the *logos*, and one's lifestyle, *bios*. This view is effectively summed up by two formulas expressing the same essential tension: "learning how to live" and "learning how to die" (Hadot 1986).

Clearly, the peak of this tension, which coincides with the ideal behavior of the wise man and which underpins both the above maxims, corresponds to the perfect self-identity which the theoretical attitude enables the philosopher to maintain throughout his life, even unto death, as witnessed by Socrates. Dying, like living, implies the practice of detachment, an exercising of self-mastery designed to free oneself at every moment from the impermanent, from egotism, so as to participate in the divine essence.

While Socrates had argued that "the true philosophers practice dying" (*Phaedo* 67e), many centuries later the last great Stoic, Marcus Aurelius, by combining an awareness of the present moment with that of the possibility of death, was to state: "Your way of life achieves perfection when you spend each day as if it were your last.... Let death be before your eyes every day, and you will never have a base thought or an excessive desire" (quoted in Hadot 2002–1998, p.19).[4] What emerges here is the underlying theme which appears to shape the relationship with oneself in the ancient world, from the earliest physiologists described by Aristotle (see Chap. 1, p. 14) down to the late Roman Empire: freeing the soul of all passions through the constant reorienting of one's perspective, so as to live each moment in the light of the essence of what endures.

All philosophical schools sought to release man from that impermanence of things and the world which, by generating uncontrollable affections, prevents the individual from leading a happy life. The strategy they adopted consisted in a sort of progressive purification, designed to foster a stable relationship with the soul through the detection of permanent, universal, and "objective" elements within it. Through such practices, philosophers sought precisely to defuse what happens or might happen to someone.

The neutralizing of every factical instant in the light of the continuous present, ensured by theoretical practice, both anchors the world to the certainty of what is changeless and provides the basis for man to access his temporal existence starting from the eternally present time. To grasp the truth within oneself, therefore, is to grasp man's being as an essential presence that lies beyond the historical mutability to which existence is de facto exposed. This view is the preestablished understanding of man which has guided the development of Western thought, as an ontological background which has fully survived from antiquity to the contemporary age.

For a number of reasons we will have to gloss over here and which have to do with the transformation of the exercising of the care of self (Foucault 2008, 2016), the emphasis on the present moment became particularly strong in Stoic, Epicurean, and Neoplatonic practices of vigilance, which were invariably marked by the emotional neutralization of events according to a series of rules and principles to be applied to the ongoing situation. In order to govern the effects of a certain negative occurrence, the event is resignified by grasping and assimilating it in the light of the maxims, aphorisms, and precepts—but also theories and concepts—developed by philosophical masters (e.g., Epicureans) and according to which each school has educated its disciples, imparting a particular training to them. Thus in Hellenistic schools, the spiritual master presents the beginner with a résumé in the form of individual maxims and guiding principles to be memorized: "for example, *The*

[4]This acute emphasis on the present moment (*prosoche*), achieved by paying attention to each instant (Hadot 1988), makes this spiritual exercise akin to mindfulness meditation techniques.

Fourfold Medicine (Tetrapharmakos)[5] and the *Principal Propositions (kyriai doxai)* among the Epicureans, and the *Little Handbook* of Epictetus among the Stoics, which was even used, along with the Pythagorean *Golden Poem*, by the Neoplatonists in elementary education" (Hadot 1986, p. 451).

What emerges here, once again, is that underlying tension between a theory and its enactment through the way of life of the philosopher, which—as we have repeatedly noted—shaped ancient philosophy and was transmitted as a life practice within philosophical communities. As we will see, this structure endured as the foundation of the "philosophy" of the Church Fathers, who rejected ancient philosophy—particular as a way of life—while preserving its theoretical component (Bardy 1949; Domański 1996).

The famous spiritual exercises extending from Pythagoras to Ignatius of Loyola (Rabbow 1954; Hadot 1995, 2002, 2014) consist precisely in the development of practices designed to transform one's relationship with oneself—the movement of the soul toward itself—according to the theoretical principles of a given school. The aim which all these schools share is to free the soul from worries and from the contextuality of existence by directing it toward the vision of what endures. If man is sad, this is because his passions bind him to the things of the word; hence, the path to happiness has to do with detachment from such passions and their affections upon us. The exercises in question promote a transformation of such kind, freeing the I from human miseries through various regulated practices.[6]

Hadot has summed up this common process of detachment of the I from all passions with remarkable clarity in relation to Stoicism and Epicureanism: "Much the same thing can be said for Stoicism, with the help of the distinction between what does and does not depend on us; we can reject all that is alien to us, and return to our true selves. In other words, we can achieve moral freedom. Finally, the same also holds true for Epicureanism. By ignoring unnatural and unnecessary desires, we can return to our original nucleus of freedom and independence, which may be defined by the satisfaction of natural and necessary desires" (1995, p. 103).

Access to the knowledge of oneself, and hence of those universal truths to be found within the soul, is governed by techniques that refer to the theory which produced them. An ongoing negative situation can thus be neutralized by shifting one's attention in accordance with given guiding principles. Likewise, by being oriented by the same principles, it is possible to anticipate or recall a given negative situation, in such a way as to make it constantly present, through a practice that both reproduces the event and defuses it, making the individual ready to accept it. With a device of this sort, there is not much difference between anticipating negative

[5] "The gods are not to be feared; Death is not to be dreaded; What is good is easy to acquire; What is bad is easy to bear" (Hadot 2002–1995, p. 123)

[6] These practices may be seen to be founded on the element which distinguishes the double path (see Ch. 1) that is disclosed by the care of the soul—i.e., the path of what is changeless and stable and of what flows—and which according to Patočka constitutes the polarity of its movement (1998).

circumstances in order to learn how to manage them, should they (newly) present themselves, as the Stoics do (*anticipatio malorum*), and training oneself to recall past pleasures in order to focus only on the pleasant aspects of the present, as the Epicureans do. The aim is self-transformation and hence the release of oneself from worldly bonds. The freeing of the I from human miseries, in other words, is made possible by "the permanent transformation of true discourse, firmly fixed in the subject, into principles of morally acceptable behavior" (Foucault 2001–2005, p. 32).

One remarkable document which sheds light on this specific aspect in relation to the Epicurean school is a short treatise by Philodemus, a philosopher who lived at Herculaneum, in Italy, around 70 BC. The device, or rather the knowledge, which brings about the neutralizing of the event is conceived by Philodemus as a *techne*—presumably in accordance with a tradition directly transmitted by Epicurus to Metrodorus and Zeno of Sidon. In his treatise "On True Discourse" (*Peri parrhesia*)—a work discovered in the Villa of the Papyri at Herculaneum, the fragments of which have been "restored to life" through the erudite efforts of M. Gigante (1969)—Philodemus shows how true discourse coincides with a way of life, with a conduct designed to ensure the attainment of wisdom and happiness. This practice adopted by the wise man and philosopher (Fr. 1) enables him to understand and face ongoing situations through an argumentative procedure designed to lead the listener (the disciple) to acquire such practice and attain serenity.

As a method of conjecture, in Philodemus' treatise *parrhesia* is associated with medicine (Frr. 63, 64 and 69). Like a physician, the Epicurean master first identifies his disciple's ill and then administers his medicine: he applies *parrhesia* to free the disciple from his illness and thereby purify his soul. In turn, by resorting to *parrhesia,* the disciple will not hesitate to confess his shortcomings to his master, so as to be cured. *Parrhesia* is learned and perfected by practicing it.

The fact of interiorizing and fixing truths transmitted by others—as in Stoic or Epicurean practice—is certainly something quite different from the movement of identification of, and conversion toward, the proper and immutable realities behind things that was prescribed by Plato and which had first been introduced by the Pythagorean examination of conscience. It also differs from the effort to distinguish between the will of the Lord and the darkness of sin we find in Christian philosophy. Nonetheless, in all these cases, the theoretical attitude is what governs one's relationship with oneself and hence orients the range of practices mediating this relationship, as well as one's relationship with one's guiding master. In this respect, if philosophy coincides with the care of the soul and its passions, the inner transformation it requires must be based on the use of a therapeutic tool. Foucault quite rightly referred to such tools as techniques or technologies of the self that "permit individuals to effect by their own means or with the help of others a certain number of operations on their own bodies and souls, thoughts, conduct, and way of being, so as to transform themselves in order to attain a certain state of happiness, purity, wisdom, perfection or immortality" (Foucault 1988, p. 18). These are a range of exercises designed for the progressive search of the changeless truth that lies within oneself—and not in the starry heavens above.

11.3 The Master, the Disciple, and the *Techne*

The acquisition of practices designed to shape the specific relations individuals have with themselves is clearly linked to two essential aspects: on the one hand, the desire and capacity of the novice to change his own life and, on the other hand—and in close connection to the former aspect—the novice's relationship with his spiritual director as one capable of guiding the disciple in his journey toward truth, just as a physician is capable of guiding his patient on the path to healing.

With regard to the first aspect, in his Seventh Letter Plato writes:

> Ought not the doctor that is giving counsel to a sick man who is indulging in a mode of life that is bad for his health to try first of all to change his life, and only proceed with the rest of his advice if the patient is willing to obey? But should he prove unwilling, then I would esteem him both manly and a true doctor if he withdraws from advising a patient of that description, and contrariwise unmanly and unskilled if he continues to advise. (330c–330d)

Like the physician, the spiritual guide can only treat those who are willing to change their lives—this being the precondition for healing. Plato had learned this the hard way, through his dangerous misadventure with the tyrant of Syracuse.

In the Hellenistic age, the issue of the relationship between master and disciple emerges in Philodemus' text, which discusses the therapeutic action of *parrhesia* and its fallibility. Philodemus refers to those young people who are still in need of a diagnosis or who resist treatment: the *a-therapeutoi*. Unlike in other schools, however, in the Epicurean one, there are no passions, mistakes, or faults that cannot be overcome. The disposition toward mutual goodwill, not only between the wise man and the novice, but among all disciples, is the precondition for the exercising of this therapeutic function (col.V a). Imitating Epicurus, the Epicurean master thus acts like the physician who will treat even patients deemed to be untreatable, based on a range of symptoms.

On the other hand, we have Serenus, who is neither ill nor in good health but turns to Seneca for help. Although Serenus already leads a simple and frugal life, in accordance with Stoic precepts, whenever the splendor of luxury dazzles him, he is seized by doubts and second thoughts. Much the same occurs in Serenus' intellectual life, where he sometimes loses his simplicity and becomes ambitious when searching for words. The instability of Serenus' right will (according to Stoic rules) is what leads him to seek help from Seneca. The latter immediately grasps the ill affecting the young man: his diagnosis is that Serenus is in need of some tranquility of soul. Through examples, metaphors, and the presentation of various different cases, Seneca shows Serenus that tranquility of soul is to be acquired day after day through the careful and constant care of one's wavering soul. This care corresponds to an existential hygiene designed to combine theoretical and practical teachings in such an effective way that nothing will ever undermine the stability, steadfastness, and *tranquillitas* of the soul.

A few centuries later, in the context of Eastern Christian monasticism, the reflections on the willingness and capability to engage in the care of self put forward by Plato, Philodemus, and Seneca find an echo in a letter which Nilus of Sinai (d.

430–450) addressed to a layman by the name of Faustianus. The monk writes: "Your soul is far from God. How can you ask me to send you edifying words, if you mock holy things, as profane men do? What good is it to speak to a dead ear, incapable of hearing?" (quoted in Hausherr 2012, p. 316). Nilus' message is obvious: it is impossible to guide along the path to perfection someone unwilling to walk; it is impossible to treat someone who has no intention of receiving any treatment.

Along much the same lines, in his *First Conference*, Cassian informs us that when he and his friend Germanus asked the abbot Moses to take them under his guidance, the elderly ascetic questioned them and put them to the test. Cassian writes that the abbot, who was renowned for his rigorous austerity of soul, refused to heed "the requests of sluggish and lukewarm postulants" (2000, p. 62). The abbot required a firm and well-articulated conviction on the part of those wishing to take up those practices leading to the purity of soul and contemplation of God under his guidance.[7]

What we find, then, are various different ways of conceiving the possibility of operating upon a disciple's life, which reflect different guiding theories. In any case, the relation between master and disciple always plays a central role. Cassian's *First Conference* also brings this second aspect into relief: the importance of the disciple's relationship with his spiritual guide for his own formation and training.

This theme, which is most certainly of Pythagorean origin, crucially influenced the development of the care of self down to the modern age. Socrates clearly formulates it in the *First Alcibiades*. After providing some examples to illustrate the need to live virtuously, "with your eyes turned on what is divine and bright" (134 d), Socrates states: "before getting virtue, to be governed by a superior is better than to govern, for a man as well as a child" (135b). Again, in a passage of the *Phaedrus*—within the context of that defense of memory against the written word which we previously referred to (274b–275c)—Plato puts the following words into Socrates' mouth: "in my opinion, serious discourse about them is far nobler, when one employs the dialectic method (*dialektike techne*) and through knowledge (*episteme*) plants and sows in a fitting soul words which are able to help themselves and him who planted them, which are not fruitless, but yield seed from which there spring up in other minds other words capable of continuing the process for ever, and which make their possessor happy, to the farthest possible limit of human happiness" (276e–277a).

The hierarchical relation between teacher and student, on the basis of which the care of self takes shape, here emerges as an endless dialogue across successive generations: a living exchange based on an unequal relation which is established between an individual; the teacher, who plants the *logos* into the soul of another individual; and the disciple (Griswold 1996). The care of self, then, presupposes a relationship with another person who is capable of determining its orientation.

[7] From the fifth century onward, the spread of Augustinianism—the view that grace is necessary for salvation and that it is a gift of God—throughout monastic communities marked the gradual disappearance of the voluntaristic perspective that it is possible to ensure one's own salvation through spiritual exercises (Foucault 2001).

The same theme runs through Philodemus' short treatise, which provides detailed indications as to how the method of *parrhesia* should be applied in order for true discourse—and hence knowledge—to be transferred from master to disciple according to an unbroken chain that stretches all the way back to Epicurus. In Fr. 45, Philodemus writes: "we shall admonish others with great confidence, both now and when those {of us} who have become offshoots of our teachers have become eminent. And the encompassing and most important thing is, we shall obey Epicurus, according to whom we have chosen to live."

Philodemus' text is a real treasure trove of information on the Epicurean way of managing the relation between a young person and his spiritual guide. For instance, Fr. 71 gives advice on how to behave if the disciple refuses to speak; Fr. 73, if he disobeys. Fr. 87 provides some suggestions on how to get the better of recalcitrant young men; Fr. 93, on how to encourage weak ones. Fr. 22 explains how to guide young people with a strong personality and stresses the importance of choosing the right moment (*kairos*) to act. Fr. 7 discusses the wise man's use of frankness to this end; Fr. 20 how to treat with moderate words and forgive mistakes. Frr. 13, 32, 61, and 66 deal with the exercising of the art of admonishment, Fr. 26 with the use of irony, and Fr. 25 with how to encourage goodwill among young people through the fostering of true discourse. Fr. 10 discusses the free use of *parrhesia* as means to induce young people to open up; Fr. 67 the need for the wise man to swiftly assist those who have fallen prey to error. Finally, Fr. 43 discusses the use of *sympatheia* as a means to foster the development of an individual's character.

A few centuries later, in the Imperial age—by which time the care of self had become detached from concerns about the public good—Galen (AD 129–201) redefined the relationship with a wise man by justifying the need for it on the basis of an element which distinguishes every human being: self-love. Galen writes: "As Aesop says, we have two sacks suspended from our necks; the one in front is filled with the faults of others; the one behind is filled with our own. This is the reason why we see the faults of others but those which concern ourselves remain invisible to us" (Galen 1963, p. 3). Self-love prevents us from seeing our own ills. What emerges here is the theme of self-deception, which acquired a different significance in Christian thought and was ultimately inherited by modern psychology and psychoanalysis (Goleman 1996; Pahulus 2007).

Given the impossibility of judging our own errors, and hence the need for someone else to grasp them, Galen recommends that we find someone to help us. As the right person to take care of our soul, he recommends someone who shows no love or hate toward us—i.e., a stranger—and who has already given ample proof of impartiality in his life and will therefore be capable of revealing our errors to us without hesitation.

The relationship of friendship or closeness between wise man and disciple that distinguished the schools inspired by the Socratic tradition was replaced by a neutral relationship (Foucault 2016): by the figure of an external supervisor who speaks the truth (*parrhesia*), so as to point out any errors in our actions due to passions. In such a way, it will be possible to perform a diagnosis of the "passions and errors of the soul," to quote the title of Galen's treatise.

For Galen, who displays a Stoic sensibility, since error has to do with the passions of the soul, treatment starts when the wise man, by speaking the truth, attempts to do away with them—at least to some extent. This operation proves difficult at first; progress comes slowly and brings only modest results. Gradually, however, by repeating the same exercises over and over again, the passions lose their strength and are brought under control, as it becomes easier for us to restrain ourselves without the need for a supervisor. This is the path that leads to healing.

It is important to note that, both in the Ancient Greek and in the Hellenistic world, and later in the Roman one, the connection between grasping the truth within oneself and one's relationship with the guide directing this process is governed by a form of knowledge that takes the form of a *techne* (as clearly illustrated by Philodemus' treatise): a form of knowledge—oriented according to the theoretical principles of one's school—that can only be transmitted if the disciple is willing to change the way in which he leads his life, by shaping it after that of the master imparting such knowledge. In the context of Epicureanism, for instance, *parrhesia*, as a *techne*, is regarded as "an ethical practice that leads in different – albeit non-scientific – ways to attain a philosophical education and moral perfection; a technique which, starting from a clear philosophical view of what is useful in life, may be applied to young people by the wise Epicurean teacher" (Gigante p. 63). This regulated knowledge is progressively assimilated by concretely embodying it, until one has become his own master. Hence the emphasis in the ancient world on the need for the philosopher's way of life to be consistent with the principles of his philosophy.

Christianity was to mark a significant turning point within this process of transmission.

11.4 The Church Fathers' *Praxis*

It has often been emphasized that Christian spirituality may be seen as the fulfillment of ancient philosophy, since it inherited so many themes and practices from the latter (see, e.g., Foucault (2001) and Hadot (1988, 1995)). An important article by G. Bardy (1949) provides a succinct overview of the main authors and works through which this passage took shape. While this is hardly the best avenue to explore such a vast topic, it is important to note a key fact that marks this continuity: just as in the ancient world philosophers would validate their principles by their way of life, the assimilation of Christianity to philosophy developed according to this trajectory, as an attempt to change the individual's way of life. This process is described as a *praxis* by the Church Fathers and traditionally applies to the ascetic life—withdrawal from the world, hermitic living, fasting, vigils, toil, and poverty.

"The Greek Fathers – Leclercq writes – refer to the monastic life as 'philosophy according to Christ' and 'the only true philosophy', or more simply as 'philosophy': this term, which indicates the practical discernment of the value of things and the vanity of the world, which must be renounced, applies to the monk who undertakes this renouncement" (p. 221, 1952). What is the relationship between renunciation and philosophy?

We here touch upon a crucial point. The monastic life, "the philosophy of Christ," is achieved through sacrifice: renunciation of the self and of life in this world is a precondition for purity of heart and the contemplation of God (*theoria*). Foucault clearly highlighted this by distinguishing the third–/fourth-century Christian conception of conversion (*metanoia*) from the *epistrophe* of the Platonists (the redirecting of one's attention from the world of shadows to that of eternal truths) and of the Hellenistic and Roman ages (the reorienting of one's attention from what does not depend on us to what does), stressing that the former conception marked a break, a discontinuity, a dramatic turning point (Foucault 2001, 2012). Self-renunciation, the pivotal experience of the cross, lies at the very heart of the process of the Christian transmission of philosophy. Just like the *techne* guiding ancient Greek, Hellenistic, and Roman philosophy, this experience requires a master and is framed within the context of the disciple's relationship with him.

Starting from self-sacrifice, from daily martyrdom, the individual seeks the truth about himself. Self-renunciation is also the distinguishing feature of the relationship between the monk and his spiritual father. In other words, while in the case of one's relationship with oneself renunciation implies a constant effort to detach oneself from the world through the mortification of the flesh and vigilance over the thoughts it elicits, in the case of the relationship between the novice and his guide, the former utterly forgoes his will, putting himself completely in the hands of his spiritual director. In this respect, obedience and renunciation of the world go hand in hand (Hausherr 2012).

In severing his ties with the world as far as possible, and withdrawing into a "boundless solitude" (Cassian, coll. I), into the "harsh sands" (coll. XXIV), the monk entrusts himself to another person, his spiritual father, by whom he allows himself to be governed. For the abbot, the work on his relationship with the novice even comes before—and is more essential than—the work on the content of the latter's thoughts. This relationship forces the monk to face the primordial conflict between pride and humility that cost man his fall from Eden.

This explains the relevance of that famous apothegm—taken up with a different nuance by Cassian among other examples of obedience (1989, Book IV, 23)—in which Father John, who had withdrawn into the Scetis desert, spent 3 years watering a dry stick planted in the sand with his bucket, as his spiritual father had asked him to do until the branch would bear fruit. The water was so far away that John would leave in the evening and only return in the morning (p. 231).

Along much the same lines, in one of his letters Barsanuphius writes about obedience: "He who wants to be a monk must in no way have any will of his own. Christ our Lord taught us this when He said: 'I came down from heaven, not to do mine own will' (John 6: 38). (...) So you must obey in everything, even if it should seem to you that what is ordained is not without sin. The Abba who ordains you to do it will bear your sin and will have to answer for you" (quoted in Hausherr 2012, p. 199).

So while the novice owes unconditional obedience to his abbot, the latter in turn is at his disciple's service, since he takes full responsibility for the latter's eternal salvation. Quoting Gregory the Theologian, Christodulus states: "The rule of all spiritual direction is to always disregard one's own interest for the sake of others" (quoted in Hausherr 2012, p. 140). The spiritual father renounces his own self in order to bear his disciple's burden.

This device centered on sacrifice, which defines the mutual relation between the novice and his abbot according to a given way of life and which is applied in compliance with the teachings of a master, Christ, who has renounced his own self in order to save mankind as a whole, once again falls within the Greek tradition of *techne*. The renunciation which the monk practices and learns day after day engenders an experience of the self and hence a possibility to access (*theoria*) a life that is not of this world. This experience is associated with a particular mode of living (based on solitude, fasting, chastity, vigils, toil, etc.) which is shared with one's spiritual guide and takes shape within one's relationship with the latter.[8]

On the other hand, the truly innovative element, which fourth- and fifth-century monasticism introduced and which, as noted by Foucault, was destined to have a crucial influence on the history of Western subjectivity, is the fact that the monk is expected to establish the connection between self-renunciation and the truth about himself which this produces through an ongoing confession: frank discourse about oneself takes shape within the framework of the aforementioned relationship of submission and obedience to one's spiritual father (Foucault et al. 1981; Foucault 2001, 2005, 2009, 2012, 2013, 2016). The fact of renouncing life in this world and forsaking oneself, the desires of the flesh and one's own will, by humbly and patiently accepting to submit to the will of another person, thus goes hand in hand with the verbalization of one's own existence to one's abbot; and this amounts precisely to the experience of self-sacrifice. In his last apothegm (38), Anthony the Great clearly outlines the form and content of this relationship: "If he is able to, a monk ought to tell his elders confidently how many steps he takes and how many drops of water he drinks in his cell, in case he is in error about it" (The sayings of the desert Fathers, 1975).

11.5 Confession and the Motility of the Soul

What is the purpose of disclosing one's own experience, the truth about oneself, to another person, when one has renounced being oneself? What aspect of this self-annulment is confessed? What truth value can be found in the cogitations that this suppression of the self elicits, and whose manifestation constitutes a key theme in the practice of the philosophy of Christ?

[8] In the Eastern tradition, just as in the Classical age, the contemplative life (*theoria*) is conceived as a more lofty mode of living marked by the inceptive union of man with God upon this earth; after death, it will constitute the very essence of beatitude.

The full weight of the theme of renunciation only comes to be felt in the light of the Judeo-Christian account of creation. Before his expulsion from Eden, man lived in the contemplation of God (*theoria*), since he had been endowed with all virtues by Him: "God created mankind in His own image" (Gen. 1: 27). But when man tasted the fruit of the tree, transgressing God's command, and was driven out of Paradise, he fell from his natural and original condition and found himself in a state contrary to nature: under the violent sway of all the passions that surfaced within his soul when he turned away from virtue out of love for worldly things (and not according to the pleasure principle!). Man became a slave to passions, and only a few individuals enlightened by natural law—the great patriarchs—knew God; the multitudes lived in sin. In order to correct this ill and help His creature to revert to its original condition, God gave Moses the Tables of the Law. "But the ill was not corrected" (Dorotheus of Gaza 1993, I/3). Hence, God sent his Son who, having "been made one of us, like us in all things except sin," released mankind from its bondage to sin and passion, making it free once more. To purify the heart from evil dispositions, the Son of God gave man the following commandments. "You have heard that it was said, '*You shall not commit adultery*.' But I tell you that *anyone who looks at a woman lustfully has already committed adultery with her in his heart*" (Mt. 5: 27–28). "You have heard that it was said to the people long ago, '*You shall not murder*, and anyone who murders will be subject to judgment.' But I tell you that anyone who is angry with a brother or sister will be subject to judgment" (Mt. 5: 21–22) (I/6). Christ gave man the possibility of distinguishing between good and evil, by showing him how he falls into sin. At the same time, He gave man the most suitable strategy to curb his bad inclinations and purify himself from passion: renouncing his own will and accepting to humbly obey His commandments. This quest for the fullness of virtue, oriented by Christ's teachings about the love of God and boundless charity toward others, coincides with apatheia, the lack of pathos. This is an impassioned impassivity—not a form of spiritual dullness—that is constantly exposed to the risk of falling (into sin) and must be sought in the completeness of one's relationship with God, which is always in the making. It coincides with the constant enactment of a relationship of love that brings together God and one's neighbor through the same love.

Dorotheus (1993) grasps this relationship in the following terms: "Imagine that the world is a circle, that God is the center, and that the radii are the different ways human beings live. When those who wish to come closer to God walk towards the center of the circle, they come closer to one another at the same time as to God. The closer they come to God, the closer they come to one another. And the closer they come to one another, the closer they come to God" (VI/78). Love lies at the root of the great theme of humbleness which runs through the apothegms of the Desert Fathers and which is encapsulated in Evagrius Ponticus' saying "Just as pride is the root of all evil, humility is the cause of all good" (quoted in I/7). Humbleness is the source of apatheia. The humbleness of the monk takes shape through his direct relationship with God, since it is precisely within the framework of this relationship that he constantly experiences both the limiting of his own I and the action of divine grace. Therefore, apatheia—understood as the fullness of virtue—is connected to humbleness and hence to love. John Climacus (2007) lays particular emphasis on this link between impassivity (*apatheia*), which occupies step 29 of his *Ladder of Divine Ascent*, and love (*agape*), which is located on the following and last step. "Love, by reason of its nature, is a resemblance to God, as far as that is possible for mortals; in its activity it is inebriation of the soul; and by its distinctive property it is a fountain of faith, an abyss of patience, a sea of humility" (XXX, 3.1156B).

What must be treated, then, is the postlapsarian deviation of the natural yearning for God toward worldly things, by restoring man's orientation in accordance with his nature. In the Fathers' view, the personal movement that guides the soul toward external things, toward sensible realities, is a state against nature which we must free ourselves from or an illness to be treated. This is the underlying theme which shapes the way of life of the Eastern Fathers.

Dorotheus of Gaza makes the above point quite clear when commenting upon Paul's statement "the world has been crucified to me, and I to the world" (Gal. 6: 14). To restore the original human condition, it is not enough for the monk "to flee the world and inhabit the wilderness, to fast, sleep on the naked earth, hold vigils and engage in other mortifications, completely forsaking his homeland, wealth and possessions" (Dorotheus, I/13). It is not enough to crucify the world: the monk must struggle against his own will, desires, pleasures, and passions; he must renounce the very inclination toward material things by crucifying himself for the sake of the world. Along much the same lines, in explaining what makes a monk, John Climacus writes: "Being a monk is synonymous with constant violence to one's nature and incessant guarding of one's senses. Being a monk is synonymous with a purified body, a pure mouth and an enlightened mind" (I – 633C, 10).

For the Fathers, therefore, man's original perfection can be restored by freeing the soul from its errors and passions. The latter are merely the product of the disobedience of the first man; as such, they are foreign to the nature of the soul. What lies at the root of the passions or rather vices—that Evagrius fixed in a list of eight (1998 cc 6–14) which become conventional in Orthodox ascesis (Larchet 1997)—is *philautia*, egoistic self-love.[9] Therefore, the monk's aim is purity of heart, as abbot Moses tells Cassian and Germanus (Coll.I). Solitude, fasting, the keeping of vigils, the reading of Scripture, and poverty (renunciation) are all practices directed toward this goal: "the purity and tranquillity of our mind," to ensure that "it may always be fixed on divine realities and God" (Coll. I, 7).

The soul, fettered by its indulgence in the passions, which divert its motility, moves away from its natural movement toward God and no longer distinguishes between good and evil: it can no longer tell the difference between the yearning for God and the yearning for the apparent goods that please the senses (passions). Maximus the Confessor writes: "Evil is nothing but the unreasonable movement of the natural faculties toward something else than their end, according to an erroneous judgment. By end I mean the Cause of beings, toward which all things are naturally inclined" (quoted in Larchet 1996, p. 189).

[9] The eight passions are gluttony, lust, avarice, anger, sadness, acedia, vainglory, and pride.

The possibility of deception thus surfaces at the heart of the motility of the soul—which is always extremely mobile (*aeikinetos kai polykinetos*, Coll. VII, 4)—precisely because the soul, which cannot help yearning for something, mistakes carnal enjoyment for spiritual enjoyment. Here we encounter a key point in the history of Western thought. In the motility of the soul, in the *kinesis* which Aristotle articulates according to its various modes of actualization, there lies the possibility for illusion which a 1000 years later was to feed the evil genius of Descartes' Meditations.

Athanasius of Alexandria writes:

> The soul is by nature mobile, so even though it may have turned away from what is good, it does not lose its mobility. It moves, then, no longer according to virtue or so as to see God, but imagining false things; it makes a novel use of its power, abusing it as a means to the pleasures it has devised, since it is after all made with power over itself. For it is able, as on the one hand to incline to what is good, so on the other to reject it; but in rejecting the good it of course entertains the thought of what is opposed to it, for it cannot at all cease from movement, being, as I said before, mobile by nature. (quoted in Larchet 1997, p. 72)

In other words, if in the movement proper to it the soul deviates from contemplation, it inevitably falls into worldly things and hence toward worldly passions. Thus one's thoughts will tend to be dominated by the flesh; and since it is impossible to stop the motility of the soul, and since the latter can conceal evil thoughts, in addition to saintly and spiritual ones, it is up to man to distinguish between the two.

In the conferences we find two key passages in which the motility of the soul is thematized. The first is in the conversation which Cassian and Germanus have with abbot Moses (Coll. I, 15–17); here the theme is used to introduce the problem of the discernment (*discriminatio*—Coll. II) between thoughts coming from God and ones inspired by the devil. It is in this context that abbot Moses famously resorts to two metaphors in order to illustrate the nature of thoughts: that of the miller separating the wheat from the chaff and that of the money changer telling genuine coins from counterfeit ones.[10]

The motility of the soul is then thematized a second time in the conversation with abbot Serenus (Coll. VII), where it is seen to lie at the heart of the possibility of directing one's thoughts to God or "to the earth and flesh" (Coll. VII, 4). In this dialogue a new metaphor is used to shed light on the tension between the disorders and vices of our soul and our good thoughts: that of the spiritual centurion (Mt. 8, 9), who governs evil suggestions by struggling against the flesh and extinguishing the passions, while welcoming good suggestions that "take captive every thought to make it obedient to Christ" (2 Cor. 10:5). Like the centurion of the Gospel, the monk will only feed his inclination toward the good, doing away with all worldly desires, guided by his observance of Christ's teachings.

[10]These two metaphors may be traced back to Epictetus (**Discourses** I, 20, 7–9), with the crucial difference that, for the latter, the purpose of guarding one's representations is to determine whether they stem from one's own will or not.

11.6 The Melodic Unit of Eastern Monasticism

Abbot Moses and abbot Serenus seem to suggest the same strategy for leading the motility of the soul—which is normally slave to countless desires—back to the stability of contemplation: fasting, vigils, contempt for all worldly things, and anchoritic living are not enough, since "the mother, guardian, and mediator of all virtues" is the capacity to discern. This—Cassian informs us (Coll. II, 2)—was the conclusion that Anthony the Great reached after an all-night discussion with many wise elderly monks. These monks had gathered from all corners of the Thebaid to discuss with Anthony what virtues and observances could safeguard the monk against the deception of the Evil One.

Addressing Cassian and Germanus' queries, abbot Moses goes on to explain that:

> True discretion is only secured by true humility. And of this humility the first proof is given by reserving everything (not only what you do but also what you think), for the scrutiny of the elders, so as not to trust at all in your own judgment but to acquiesce in their decisions in all points. (Coll. II, 10)

This view is echoed by abbot Serenus: "we shall be taught by the authority of those who in converse with the Lord say in all confidence: 'My soul hath stuck close to Thee'" (Coll. VII, 6).

The monk, therefore, learns to guard and discern precisely the motility of his soul in its various manifestations, the *cogitationes (logismoi)* of the Latin Father, in order to lead it toward the stability of divine contemplation.[11] Maximus the Confessor describes this rest of motility as "rest in perpetual movement and movement in rest" (quoted in Larchet 1997, p. 667; see too Maximus the Confessor 2014, *Ambigua* 7 and 15). This inner activity (*theoria*) is the good which the monk's way of life has in view (Coll. I, 7–8; Coll. XIV).

What emerges here is the need to speak the truth about oneself, along with the examination of conscience this entails—a subject to which Foucault devoted many memorable pages. The examination of conscience[12] therefore focuses on those representations, images, desires, suggestions, and impulses of one's interiority *(logismoi)* that at any moment might deviate the motility of the soul by "persevering in guarding our thoughts" (Coll. VII, 6). What is required is a careful, ongoing, and permanent relationship with one's own *cogitationes* that, like the money changer evoked by abbot Moses, must ensure the veracity of the thought that is being guarded.

But could this vigilance, this first-hand assessment of authenticity, not prove fallacious, deceptive, and illusory itself? One example would be the case of abbot John of Lycus, who despite his exhaustion decided to extend his fasting for 2 more days. When he was finally about to break his fast, the devil appeared before him in the

[11] "Perfect impassivity—Evagrius states—emerges in the soul after its victory over all the demons that oppose the practical life" (1998, cc.58).

[12] For a detailed history of the examination of conscience, see Hasso Jaeger (1959).

guise of an Ethiopian and told him that it was he who had appointed that labor for him. The old man, despite his perfection in the virtue of discretion, regretted the fact that under the pretense of abstinence the devil had deceived and misled him (Coll. I, 21). Hence the need to humbly submit one's *cogitationes* to the control of elders, fellow brethren or one's *abba*, for confession is a requirement even for a monk of consummated virtue. Whereas in Greek philosophy obedience to one's master ends when the disciple becomes a master himself, as Foucault notes "in Christian obedience, there is no end, for … one obeys in order to be obedient, in order to arrive at a state of obedience" (2004).

This urgency to keep watch over the motility of the soul by verbalizing its content—so as to grasp (*discretio*) whether a cogitation comes from those virtues which by nature reside in the soul of man or whether it originates from the passions which capture that motility—is supported by a light that provides some guidelines for discernment: the testimonies of Scripture, to which the soul of a genuine master is bound. In other words, God's help is crucial for orienting one's discretion. Cassian writes: "the Author of our salvation teaches us what we ought not merely to think, but also to acknowledge in everything that we do" (I.C. XII, 17).

The quest for the truth about oneself, aimed at rediscovering those (hidden) natural virtues that resided in man before the Fall, here meets the testimony of Scripture, which inspires their discernment. Confession to one's *abba*, or brother, verbalization to another person who is capable of acting as a guide—not through vain dreams or disputations, but by showing one the path for attaining purity of heart (I.C. XII,15)—links the certainty regarding the truth of one's own *cogitationes* to the truth of Scripture. In order to make sure that one's own *cogitatio* is not illusory, it is necessary to submit it to one's abbot for interpretation: is it a cogitation stemming from the virtues naturally residing in the soul, or does it spring from the passions that capture the motility of the soul? This is the suspicion that lurks behind all thought that lies concealed within the motility of the soul.

Besides, is it not precisely at the heart of this motility that temptation takes shape? Temptation: that constant disquietude which fills a Christian's soul, that restless worry (*Bekümmertsein)* which leads not to contemplation, but to the constant questioning of oneself. "For the Christian life there is no certainty; constant uncertainty is the distinguishing feature of things that are of a fundamental significance in actual life". The uncertain is not accidental, but rather necessary. This necessity is no logic or natural requirement. In order to see things clearly here, one must retrieve one's own life and its enactment (GA 60 1995, p. 105).[13] Larivée and Leduc (2001/2) quite rightly note: "It seems evident that many aspects of the notion of care which Heidegger identifies in Augustine can easily be traced back to the ancient philosophical tradition of self-care (*epimeleia heautou*): the call to vigilance, the exhortation to self-knowledge and the denouncing of all forms of self-forgetfulness and self-flight which scatter our unity 'across so many beings and so many things'" (p. 40).

[13] This distinctive interpretation of the motility of Christian life represents the core of Heidegger's commentary on Augustine's *Confessions* (GA 60 1995).

Confession, therefore, concerns the capacity to verbally express the experience of oneself without concealing anything—an experience which emerges starting from the renunciation of one's worldly self. At the same time, it also has to do with the capacity of "genuine masters" to discern the truth by separating the wheat from the chaff in the light of Scripture. As Foucault states, confession lies at the threshold between the hermeneutics of self and textual interpretation: "One thing is important: the fact that a bond, a fundamental connection, exists between the duty to believe and the duty to discover oneself. Christianity has established a profound link between the text and the self" (Foucault et al. 1981, p. 159).

There is a close connection between the constant interpretative analysis of oneself in view of the restoration of that primordial perfection by now concealed by the desire and passion for worldly things, foreign to the soul—a stain, an impurity, and a filth to be removed (Rm 13:13; II Cor 12:21; Eph 4:19)—and the textual hermeneutics that guides the process of discernment, the interpreter's *discretio*. The close link established by Christianity between the text and the self is ensured by an interpreter, who listens to the spoken word in order to discern its good or bad origin in the speaker, in the light of Scripture, and thereby lead the novice's self-examination toward a resonance between the natural perfection of the soul and the content of the text: toward a sort of reflection of the truth of the text within oneself, and of oneself within the truth of the text, mediated by another person, the abba, who is capable of attuning the testimony of Scripture to the original revelation of the novice's conscience. This is the "melodic unit" that sprung from Eastern monasticism and which Cassian brought to the West.

This practice would appear to have two implications. The first, immediately evident implication, concerns the tradition of the care of self, as we have already mentioned: in order to gain access to the truth, the subject must transform himself by struggling against the passions in accordance with the teachings of the text. The relationship with oneself becomes an ongoing hermeneutic exercise designed to disclose and modify the self, the ultimate aim being the attainment of a state of freedom from the passions that may ensure the contemplation of God. What is at play here is the pursuit of the kind of impassivity famously described by Evagrius—and reminiscent of Stoic impassivity (apatheia)—which constitutes the first step in the spiritual life, referred to as *praxis* or *praktike* (Origen (1957)[14]:

> One is strolling and sees something; his thought tells him: 'Look there,' but he answers his thought: 'no, I will not look,' and thus he breaks his own will and does not look. Then he meets others who are talking and his thought tells him: 'You, too, speak this word,' and he breaks his will by not speaking it. (...) Thus in breaking his will again and again, he acquires the habit of breaking: after small things, he starts to effortlessly break even big ones, so as to ultimately relinquish his will. (Dorotheus I, 21)

[14]The impassivity of which Evagrius speaks (1970) and which is described even more clearly by John Climacus—who devotes step 29 of his *Ladder of Divine Ascent* to it—certainly cannot be conflated with Stoic apatheia. Whereas the latter is the result of an individual effort, Christian impassivity—as already noted—is the fruit of man's zeal and God's grace. See also Malingrey (1961).

This discrimination of thoughts is accomplished according to the textual exegesis performed by the spiritual father, who guides the novice in his ongoing soul-searching. As already noted, confession actualizes the connection between the truth of one's original conscience, concealed by the passions, and the truth enshrined in the Book through an interpretation designed to engender a mutual resonance between oneself and Scripture. This is the second implication. What we have here is another crucial point for Western thought: through the mutual reflection between the truth of individual conscience and the truth of the text, confession certifies, and gives certainty to, the former in the light of the latter. As Basil (1976) states, it is necessary to accept God's word and conform one's heart, speech, and actions to it. From this perspective, conscience coincides with the natural law that men have "blocked and infringed upon with the rise of sin" and which can be reawakened by heeding the message of the text. Our true conscience is the good conscience which by nature "enlightens the mind and shows the good as distinct from evil," always reminding us of our duty. This is the basis of the experience of shame, which according to the Fathers reveals one's difficulty or reluctance to disclose one's thoughts. But "just as no one can see his own face in muddy water, we reach a stage where we no longer perceive what our conscience tells us, to the point that we think that we no longer have one" (Dorotheus III, 40). Access to the truth about oneself, then, the very act of acquiring knowledge, coincides with one's attunement to this "faculty as warm and bright as a spark," validated by the message of Scripture, as mediated by another person. This melodic unit established by fourth- and fifth-century Eastern monasticism continued to be developed down to Descartes, through various changes made to the practice of confession (see Bossy (1975) and Tentler (1977)).

11.7 Confession and Theoretical Certainty

Foucault mentions Descartes in relation to the history of confession a number of times and from different perspectives. He is followed in this by Hadot, who emphasizes the continuity between Descartes and ancient philosophy (p. 252, 1998). We will focus on this aspect—which Foucault leaves in the background—by referring to a short yet impressive work by Juliusz Domański (1996), which brings together the various meaningful trajectories suggested by Bardy in his seminal 1949 paper, extending them from the apologist Fathers to the fifteenth–sixteenth centuries.

Domanski's work outlines the underlying themes that reveal the theoretical continuity between ancient and Hellenistic philosophy, on the one hand, and that of the Church Fathers on the other hand. In doing so, he notes that the latter radically called into question the practical component of philosophy represented by the art of living, to the point of denying its legitimacy. Things could hardly have gone any differently, given that in the Fathers' eyes the ethics actually applied by the pagan philosophers, the philosophical life, did not have Christ and the Christian saints as its model. Domanski sums up this common opinion of the Fathers as follows:

> Although the philosophers' efforts proved considerably successful in the domain of the knowledge of essential truths and that of the theory of ethics, they failed to personally adopt a behavior that conformed to their theories, a behavior that would have earned the full approval of Christians. (p. 28)

The key point made by Domanski is that in the process of transmission and assimilation of ancient knowledge, the Fathers' judgment on pagan philosophy shattered the compactness of the notion of philosophy, progressively detaching the theoretical component from the practical one. Hence, it is precisely by setting out from this new formula, which runs through Scholasticism down to Erasmus of Rotterdam, that it is possible to understand the cultural context of Suarez's work and hence the dawn of the modern age. Given the complexity of the topic, we will not be discussing this crucial turning point at the heart of late Scholasticism, which led Suarez not only to redefine metaphysics, de facto paving the way for the independent foundation of knowledge, but also—and more interestingly from our perspective—to timidly attempt to call the institution of confession into question.

Foucault was therefore quite right to emphasize Descartes' place in the history of confession, insofar as the fundamental proposition of the Cartesian method, the famous *cogito ergo sum* formula from which all other axioms derive and which is self-evident, at the same time corresponds to the resolution of "the most ancient tradition of Christian spirituality" (Foucault 2013, p. 120): suspicion toward the tempting devil who lurks behind the motility of the soul and whom the monks had faced through their capacity for discernment, their *discriminatio*. Descartes' axiom provides a certainty that not even the most cunning illusion of the Evil One can undermine. In the following passage from his *Second Meditation*, further developing the topic of the exercising of doubt discussed in the *First Meditation*, Descartes writes:

> But there is a deceiver of supreme power and cunning who is deliberately and constantly deceiving me. In that case I too undoubtedly exist, if he is deceiving me; and let him deceive me as much as he can, he will never bring it about that I am nothing so long as I think that I am something. So after considering everything very thoroughly, I must finally conclude that this proposition, *I am, I exist*, is necessarily true whenever it is put forward by me or conceived in my mind. (Descartes and Cottingham 1996, p. 17)

Before the prospect of an evil genius, the certainty of Scripture is replaced by that of self-foundation. What men now have in common is the structure of their minds, their faculty of reasoning (Arendt 1958).

The undermining of the authority of the text with respect to the truth about oneself coincides with the establishment of a new relation between the truth of reason and the truth about oneself. Foucault repeatedly stresses this point.

According to one perspective, that of confession, the certainty of not being deceived rests on the capacity for discrimination of one's *abba* who, guided by

Scripture, is like the money changer who tells genuine coins from counterfeit ones (*discretio*). According to the other perspective, that of a new philosophical praxis, certainty starts from the independent foundation of knowledge: the thinking I at the basis of all thought. In other words, it is possible to govern the motility of the soul without submitting to any power other than that of one's own reason and of its laws. Based on this unquestionable and absolutely certain principle, which severs the link between the text and the truth about oneself, while establishing reason as the foundation of truth, Descartes integrates Galileo's mathematical project on nature: the mathesis. Hence, in his famous letter to Mersenne of 10 January 1634, Descartes et al. (1991) comments on the sentencing of Galileo (22 June 1633), stating that he is ready to set fire to all his papers or at any rate not to show them to anyone. "I must confess – Descartes writes to his friend – that if that claim (i.e. that the Earth moves) is false, then so are all the foundations of my philosophy. ... And that claim is so closely bound to all the parts of my system that it could not be removed without making the rest highly deficient."

The mathematical schematization of nature developed by Galileo outlined a uniform domain within which all things (being mobile) were determined according to the spatiotemporal coordinates of motion. "*Mathesis* – Heidegger writes – is based on such a claim, i.e., the application of a determination of the thing which is not experientially derived from the thing and yet lies at the base of every determination of the things, making them possible and making room for them" (GA 41 1962, p. 82). This domain requires a method which makes it possible both to investigate how, given certain coordinates, each thing shows itself and to come up with a unitary measurement to determine reality.

Galileo was forced to resign from his chair at Pisa because, in defiance of the visual experience of the fact that it does not take the same time for two falling objects of different weight to reach the ground, he continued to uphold the principle that all bodies fall at the same speed and that differences in their fall are due to the resistances they encounter. What guided his observations was the principle that, in the absence of any attrition, bodies fall through equal distances in equal times. "Both Galileo and his opponents – Heidegger states – saw the same 'fact'. But they interpreted the same fact differently and made the same happening visible to themselves in different ways" (GA 41 1962, p. 82). This point is taken up again by Feyerabend (1975) in his key text *Against Method*, which presents the famous episode of the proof of the usefulness of the telescope as a means to observe heavenly bodies provided by Galileo before an audience of 24 renowned professors. While these scholars acknowledged the effectiveness of the instrument in relation to terrestrial objects, none of them was able to appreciate its effectiveness as a means to observe heavenly bodies. Indeed, the fundamental scheme of nature, the mathesis, through which Galileo perceived the movement of things, was not shared by these renowned professors, who inevitably saw the same things in a different light. As anyone who has ever used a microscope knows, any change in one's way of interpreting appearances will alter one's actual perceptions and sensations, engendering wholly new experiences (Arciero 2006).

What does mathesis—which at every moment provides the certainty of the being of a given mobile body (according to a method applicable to the determination of all things)—have to do with the Cartesian certainty of the self-foundation of the thinking I? Heidegger masterfully illustrates this transition by showing that, once the independent foundation of knowledge has been established—the "I am" ensuring the certainty of this foundation, through which all things can be determined—Descartes can base Galileo's project on this first certainty of the I and, at the same time, newly conceptualize the problem of the being of each entity: the central problem in metaphysics.[15] Thus the subject who thinks—the ever-present mode of the "I am"—and who thinks according to the principles of reason, uses categories to determine all the other things (objects, angels, and swans) he encounters. It is on these fundamental axioms, the principles of the I, that Descartes establishes his whole metaphysics (which is therefore entirely shaped by the essential problem of the subject-object relation). Through mathesis, it is modeled after the principles of reason.

By assigning a central role to rationality, metaphysics came to inquire about the Creator (*theologia rationalis*) and the *ens creatum*: the world (*cosmologia rationalis*) and man (*psychologia rationalis*). Within this perspective, which presents the same ontological categories as ancient philosophy—but not the same practical aspect—Descartes conceptualized the "I am" as a res cogitans: as a self-subsistent, ever-present entity that constitutes the foundation of all *cogitationes*. Between it and reality, the res extensa, there lies a chasm, while the relation between the two—the famous issue of the subject-object relation—became the center focus of modern philosophy and science.

This ever-presence of Descartes' "I am" underlies what we have described as the unthought debt of science toward Kant. Starting from this perspective, Kant was to introduce within the natural sciences a conception of the organism tied to the notion of permanence (Huneman 2007): the permanent presence of a substrate, whereby what changes can only be explained in the light of what endures unchanged.

This new position of the subject, who discovers the certainty of truth in self-awareness, establishes within the rational conscience the preconditions for accessing both the truth about nature (as in the case of the natural sciences) and the truth about oneself. Here the relation between the truth of Scripture and the truth about oneself is replaced by the relation between the truth of reason and the truth about oneself—a relation mediated by the act of knowledge-acquisition and the theory which guides it. As we have seen, this seed took root in theoretically oriented forms of psychology and psychotherapy, including psychoanalysis.

However, this "Cartesian moment" also marks a dramatic transition—the endpoint of an ancient trajectory which the Church Fathers had newly articulated—from a mode of accessing the truth that requires the subject to implement some changes in himself, to put himself to the test in order to ensure the possibility of grasping the truth, to a mode of access whereby the access to the truth about oneself

[15] We can read Marion's *Descartes* (1999) and Courtine's *Suarez* (1990) from the new perspective opened up by this transition.

actualized in the act of knowledge-acquisition is fixed by the very principle guiding such act. The former mode requires the person wishing to access himself to accomplish some transformations and to change his position, so as to engender a new disclosure of the truth about himself through this work on himself. As a consequence of the person's change of position with respect to himself, this new access in turn produces some rebound effects that modify the present being of the subject, creating a sort of virtuous circle.

The other mode of grasping the truth about oneself simply requires the subject to reflect upon himself, guided by rational principles that establish the very preconditions for the acquisition of knowledge, without any need for the subject to call himself into question. Foucault writes: "It is there, if you like, that the notion of knowledge of an object comes to substitute itself for the notion of access to truth" (p. 169, 2001). Self-knowledge thus replaces the care of self, ushering in the modern age[16].

It is within the framework of this contrast that we have viewed the difference between phenomenological psychotherapy and all forms of theoretically oriented psychotherapy. We have described its course of treatment as a journey of self-transformation: a transformation which each person must take charge of through the sustained effort of working on himself with the support of a therapist, "a well-meaning friend with a passion for healing." In this sense, the care of self requires the word of the other: the speech of another person which, in order to prove therapeutic, must be capable of revealing the truth about oneself.

We here meet Foucault's analysis (Scarpat 2001; Foucault 2008, 2009, 2013, 2016) of *parrhesia*—understood as "the discourse of truth which the subject is likely and able to speak about himself, which may be, for example, avowal, confession, or examination of conscience" (p. 5, 2009)—and of the relation between the use of frank discourse (*parrhesia*) and the care of self. With a highly significant interpretative gesture, in his last lecture course at the Collège de France, Foucault linked the two dialogues of Plato's that most emphasize the relation between care of self and the practice of *parrhesia*: the *Alcibiades* and the *Laches*. While in the *Alcibiades* Socrates' *parrhesia* leads to the discovery of the soul as the ontological foundation of all rules of conduct, and hence of ethics and politics, through an ongoing and constantly renewed search for the truth, in the *Laches* the care of self is seen to coincide with the concrete realization of a way of life, with an effective mode of living. "So in one case – Foucault states – we have a mode of giving an account of oneself which leads to the *psyche*, and which, in doing this, marks out the site of a possible metaphysical discourse. In the other case, we have a giving an account of oneself, an 'accounting for oneself,' which is directed towards *bios* as

[16] "One of the most persistent trends in modern philosophy since Descartes and perhaps its most original contribution to philosophy - Hannah Arendt writes - has been an exclusive concern with the self, as distinguished from the soul or person or man in general, an attempt to reduce all experiences, with the world as well as with other human beings, to experiences between man and himself ...World alienation, and not self-alienation as Marx thought, has been the hallmark of the modern age" (1958, pag. 254).

existence, towards [a] mode of existence which is to be examined and tested throughout its life" (2009, p. 148). That of Foucault is a bold interpretative gesture, which identifies two unambiguously Pythagorean themes in Plato, reflecting a more fundamental, more archaic divergence—which had already surfaced within the Pythagorean school, probably starting from a common source—between the mathematical way of gaining an understanding of the essence of things (*mathematikoi*) and a rather schematic and rudimentary conception that relies on rules to be followed and a mode of applying them, so as to enact a certain style of living (*akousmatikoi*) (Kahn 2001). It is these two main strands of the care of self that the *Alcibiades* and the *Laches* seem to distinguish and which, according to Foucault, are brought together in the *Seventh Letter*. Discussing some key passages of the letter (341b–342a), Foucault writes: "But someone who has to subject himself to the test of philosophy must 'live with' or, let's use the word, 'cohabit' with it" (2009, p. 238–239). This "living with," this union, this being present together (*synousia*) of philosophy and the philosopher, is sustained by an exercise, by a life practice: the work, toil, duty, and effort of relating to oneself by following the philosophical path—"And a lengthy and arduous labor must be maintained along this way" (2009, p. 233)

Within this context, Foucault (1994) proposes an aesthetics of existence, understood as "the formation and development of a practice of self whose objective was to constitute oneself as the worker of the beauty of one's own life" (p. 1490). In such a way, with the idea of the development of one's own life as a personal work of art through an ongoing and constantly renewed care of self, Foucault bridges the fundamental divergence in Western philosophy between the "being of the soul" and the "style of living." Through this effort to reclaim the philosophical tradition, Foucault makes another striking gesture that enables him to link aesthetics and ethics: he intertwines the idea of turning one's life into a beautiful work (aesthetics) with the practices of self, with the notion of lifestyle, with the way in which each person de facto leads his life (ethics). This leap of Foucault's thought inevitably brings to mind the essential dynamic between the portrait of oneself and one's actual way of acting that governs the life of Dorian Gray.

But if man takes care of himself through the way in which he relates to the world, the way in which he concerns himself with and worries about the things he encounters, thereby engendering the truth about himself, then the aesthetics of existence is only one of the many possible ways in which human life can realize itself, only one of the possible testimonies to it. In this sense, the care of self is designed to ensure a possession of oneself that must be enacted over and over again by heeding that call—that summoning—stemming from the mere fact of existing in the world with other people. It is an effort to inquire, to call things into question, to retrieve control, and to appropriate oneself; it is an effort to *interpret* one's own life which takes the form of a constant attempt to change oneself "bare-handed," so to speak—with no safety net—in view of one's own choices, while remaining faithful to one's own freedom and taking responsibility for it.

Establishing phenomenological psychotherapy on a new foundation, we have placed it within the great tradition of the cure of the soul, which since Pythagoras has always concerned the transformation of one's way of life, as the precondition for opening up to the truth about oneself.[17]

11.8 Back to the Beginning

The starting point of the present work was Cicero's account of Pythagoras' voyage to Phlius and meeting with the tyrant Leon, to whom the wise man explained that philosophers—unlike ordinary men—practice contemplation: *theorein*. Based on this essential frequency, we have explored the various harmonics that compose the care of self, according to a trajectory extending from Classical antiquity to the modern age. Having reached the end of our journey, we now return to Pythagoras, to reclaim the care of self from a different ontological position and place psychotherapy within this tradition. We will therefore provide a new access, which is nothing but a regrasping inclusion of previous interpretations (GA 61 1985, p. 163)—the retrieval of the unsaid—and which points back to the beginning of the text, by once again touching upon Pythagoras.

Pythagoras was the first to realize that assimilating oneself to the divine, and hence following God as a source of harmony, necessarily means following a life practice, where a central role is played not just by underlying precepts but also by silence: "for – to quote Iamblichus (XVII, 72) – the subjugation of the tongue is the most difficult of all victories".[18] Indeed, the dialogue of the soul with itself to which the Eleatic Stranger gives the name of thought (*dianoia*—*Soph*. 263 and 3–5) is conducted without any words (*aneu phone*).[19] This unspoken dialogue (*dialogos aneu logou*), this dialogue that lies beyond logos and is not subject to it, constitutes the avenue for philosophy and its very core. It is in silence that human being-there is summoned to engage in a mute dialogue, so as to fulfill its duty and responsibility to bear witness to its own existence day after day.

> ξυνὸν δέ μοί ἐστιν,
> ὁππόθεν ἄρξωμαι· τόθι γὰρ πάλιν ἵξομαι αὖθις. Parmenides fr. 5 (DK)

> ...It is the same to me
> From what place I should begin, for to that place I shall come back again.

---

[17] It is in the light of this perspective that we read the first note to § 42 of *Sein und Zeit*: "The point of view adopted in the present existential analytic of being-there in relation to care came to the author's attention during his attempts to develop an interpretation of Augustinian (Graeco-Christian) anthropology, with reference to the basic principles of Aristotelian ontology" (§ 42, p. 199 n. 1)

[18] Candidates were required to exercise silence for 5 years before they could be taught any doctrines.

[19] In a note to this passage in his brilliant Italian translation of the *Sophist,* Arangio-Ruiz points out that the term *dianoia*—describing an unspoken dialogue (*dialogos aneu logou*)—was chosen precisely in order to highlight its nature of a silent debate.

References

Arciero G (2006) Sulle tracce di sé. Bollati Boringhieri, Torino

Arendt H (1958) The human condition. University of Chicago Press, Chicago

Bardy G (1949) "Philosophie" et "philosophe" dans le vocabulaire chrétien des premiers siècles. Revue d'ascétique, Toulouse

Bossy J (1975) The social history of confession in the age of the Reformation. Trans Roy Hist Soc (Fifth Series) 25:21–38

Cassiano G (1989) Le istituzioni cenobitiche. Bresseo di Teofilo, Abbazia di Praglia

Cassiano G (2000) Conferenze ai monaci. Città Nuova, Roma

Courtine JF (1990) Suarez et le système de la métaphysique. Presses universitaires de France, Paris

Descartes R (1998) Reguale ad directionem ingenii. Rules for the direction of the natural intelligence. A bilingual edition of the Cartesian treatise on method. Editions Rodopi, Amsterdam

Descartes R, Cottingham J (1996) René Descartes: meditations on first philosophy: with selections from the objections and replies. Cambridge University Press, Cambridge

Descartes R, Cottingham J, Stoothoff R, Murdoch D (1991) The philosophical writings of Descartes: Volume 3, The correspondence, vol 3. Cambridge University Press, Cambridge

di Cesarea B (1976) Il battesimo. In: Paideia, Brescia

di Gaza D (1993) Insegnamenti spirituali. Città nuova, Rome

Domański J (1996) La philosophie, théorie ou manière de vivre?: les controverses de l'Antiquité à la Renaissance, vol 18. Saint-Paul

Evagrius P (1970) The Praktikos, chapters on prayer, trans. John Eudes Bamberger. Cistercian Publications, Spencer, MA

Feyerabend P (1975) Against method: outline ofan anarchistic theory of knowledge. Humanities Press, Atlantic Highlands

Foucault M (1994) Dits et écrits II, 1976-1988. Gallimard, Paris

Foucault M (2001) L'herméneutique du sujet. Cours au Collège de France 1981-1982. Gallimard-Seuil, Paris

Foucault M (2004) Sécurité, territoire, population: cours au Collège de France, 1977-1978. Gallimard, Paris

Foucault M (2005) Discorso e verità nella Grecia antica. Donzelli editore

Foucault M (2008) Le gouvernement de soi et des autres: Cours au Collège de France (1982-1983), vol 1. Seuil, Paris

Foucault M (2009) Le Gouvernement de soi et des autres II: Le Courage de la vérité. Gallimard, Paris

Foucault M (2012) Du gouvernement des vivants: cours au Collège de France: 1979-1980. F. Ewald, A. Fontana, & M. Senellart (Eds.). Gallimard, Paris

Foucault M (2013) L'origine de l'herméneutique de soi. Conférences prononcées à Darmouth College 1980, Vrin

Foucault M (2016) Discours et vérité. Vrin, Paris

Foucault M, Brion F, Harcourt BE (1981) Mal faire, dire vrai: Fonction de l'aveu en justice-cours de Louvain, 1981. Presses universitaires de Louvain, Louvain

Galen (1963) On the passions and errors of the Soul, translated by Paul W. Harkins, with an Introduction and Interpretation by Walter Riese. Ohio State University Press, Athens

Gigante M (1969) Ricerche filodemee. Macchiaroli Editore, Napoli

Giovanni C (2007) La scala del paradiso. Paoline, Milano

Goleman D (1996) Vital lies, simple truths: the psychology of self deception. Simon and Schuster, New York

Griswold CL Jr (1996) Self-Knowledge in Plato's Phaedrus. Penn State University Press, Pennsylvania

Hadot I (1986) The spiritual guide. In: Armstrong AH, Armstrong AA (eds) Classical Mediterranean spirituality: Egyptian, Greek, Roman, vol 15. Routledge, London

Hadot P (1988) Exercices spirituels et philosophie antique. Deuxième édition revue et augmentée. — Paris: Etudes Augustiniennes - Philosophy as a Way of Life. Spiritual Exercises from Socrates to Foucault, edited with an Introduction by Arnold I. Davidson, translated by Michael Chase. Basil Blackwell, Oxford/Cambridge, 1995
Hadot P (1995). Qu'est-ce que la philosophie antique? (eng. Trans.) P. Hadot, What is ancient philosophy?, transl. by Michael Chase, Belknap Press, Cambridge, 2002
Hadot P (2002) Exercices spirituels et philosophie antique. Albin Michel
Hadot I (2014) Sénèque: direction spirituelle et pratique de la philosophie. Vrin, Paris
Hart JG (1995) Husserl and Fichte: with special regard to Husserl's lectures on "Fichte's ideal of humanity". Husserl Stud 12(2):135–163
Hausherr I (2012) Padre, dimmi una parola. La direzione spirituale nell'antico Oriente, trad. di Romano Baldelli, Edizioni Scritti Monastici-Abbazia di Praglia, Bresseo di Teolo (PD) 2012
Heidegger GA 41 (1962)
Heidegger GA 60 1995
Heidegger GA 61 1985
Huneman P (ed) (2007) Understanding purpose: Kant and the philosophy of biology, vol 8. University Rochester Press, Rochester, NY
Jaeger H (1959) L'examen de conscience dans les religions non-chrétiennes et avant le christianisme. Numen 6(Fasc. 3):175–233
Kahn CH (2001) Pythagoras and the Pythagoreans: a brief history. Hackett Publishing, Indianapolis
Larchet JC (1996) La divinisation de l'homme selon saint Maxime le Confesseur. Cerf, Paris
Larchet JC (1997) Thérapeutique des maladies spirituelles. Cerf, Paris
Larivée A, Leduc A (2001/2) Saint Paul, Augustin et Aristote comme sources gréco-chrétiennes du souci chez Heidegger. Philosophie (2):30–50
Leclercq J (1952) Pour l'histoire de l'expression 'philosophie chrétienne'. Mélanges de Science Religieuse 9:221–226
Malingrey AM (1961) "Philosophia" Étude d'Un Groupe de Mots Dans la Littérture Grecque, des Présocratiques au Ive Siècle Après J.-C. Klincksieck, Paris
Marion JL (1999) On Descartes' metaphysical prism: the constitution and the limits of onto-theology in Cartesian thought. University of Chicago Press, Chicago
Maximus the Confessor (2014) On difficulties in the Church Fathers, The Ambigua, Nicholas Constas (ed. and trans.). Harvard University Press, Cambridge
Origen (1957) The song of songs: commentary and homilies. Translated and annotated by RP Lawson. (Ancient Christian Writers, 26). Pp. vi+ 385. Longmans, Green, London
Patočka J (1988) L'homme et le monde. In: Qu'est-ce que la phénoménologie?, a cura di E. Abrams. Millon, Grenoble
Patočka J (1990) Liberté et sacrifice: écrits politiques. Editions Jérôme Millon
Patočka J (1995) Papiers phénoménologiques. Editions Jérôme Millon, Grenoble
Patočka J (1998) Platone e l'Europa. Vita e Pensiero, Milan
Paulhus DL (2007) Self-deception. In: Baumeister RF, Vohs KD (eds) Encyclopedia of social psychology. Sage, Thousand Oaks, pp 189–190
Rabbow P (1954) Seelenfuhrung: Methodik der Exerzitien in der Antike. Kôsel-Verlag, München
Scarpat G (2001) Parrhesia greca, parrhesia cristiana, vol 130. Paideia, Brescia
Tentler TN (1977) Sin and confession on the eve of the reformation. Princeton University Press, Princeton

Bibliography

Abraham TH (2003) Integrating mind and brain: Warren S. McCulloch, cerebral localization, and experimental epistemology. Endeavour 27(1):32–36

Adolphs R (2003) Cognitive neuroscience of human social behaviour. Nat Rev Neurosci 4(3):165–178

Agamben G (2002) The open. Animal and man. Stanford University Press. (Italian Edition) L'aperto: L'uomo e l'animale, vol 118. Bollati Boringhieri, Torino

Akashi YJ, Nef HM, Möllmann H, Ueyama T (2010) Stress cardiomyopathy. Annu Rev Med 61:271–286

Alessandrelli M (2015) L'ordito Della Vita: Esperienza Del Divino Nella Poetica Di Aristotele. Lexicon Philosophicum: Int J Hist Texts Ideas 3

Allison T, Puce A, McCarthy G (2000) Social perception from visual cues: role of the STS region. Trends Cogn Sci 4(7):267–278

Anderson H (1997) Conversation, language and possibilities: a postmodern approach to practice. Basic Books, New York

Anderson H (2003) Postmodern social construction therapies. In: Sexton T, Weeks G, Robbins M (eds) Handbook of family therapy. Brunner-Routledge, New York, pp 125–146

Anderson T, Ogles BM, Patterson CL, Lambert MJ, Vermeersch DA (2009) Therapist effects: facilitative interpersonal skills as a predictor of therapist success. J Clin Psychol 65:755–768

Arbib MA (2000) Warren McCulloch's search for the logic of the nervous system. Perspect Biol Med 43(2):193–216

Arciero G (1989) From epistemology to ontology: a new age of cognition. Paper presented at the AAAS. S.Francisco

Arciero G (1994) I and Thou: notes on the phenomenology of reciprocity. Paper presented at the meeting "Ethology and Knowledge" -Nov. 1994 University of Siena-Published in "Conoscenza e Etologia" (1996) Melusina ed

Arciero G (2002) Studi e dialoghi sull'identità personale: riflessioni sull'esperienza umana. Bollati Boringhieri, Torino

Arciero G (2006) Sulle tracce di sé. Bollati Boringhieri, Torino

Arciero G, Bondolfi G (2009) Selfhood, identity and personality styles. John Wiley & Sons, Chichester

Arciero G, Guidano VF (2000) Experience, explanation, and the quest for coherence. In: Neimeyer RA, Raskin JD (eds) Constructions of disorder: meaning-making frameworks for psychotherapy. American Psychological Association, Washington

Arciero G et al (2004) Identity, personality and emotional regulation. In: Freeman A, Mahoney MJ, Devito P (eds) Cognition and psychotherapy: second edition. Springer Publishing Company, Berlin

Ardizzi M, Martini F, Umiltà MA, Sestito M, Ravera R, Gallese V (2013) When early experiences build a wall to others' emotions: an electrophysiological and autonomic study. PLoS One 8(4):e61004

G. Arciero et al., *The Foundations of Phenomenological Psychotherapy*,
https://doi.org/10.1007/978-3-319-78087-0

Ardizzi M, Martini F, Umiltà MA, Evangelista V, Ravera R, Gallese V (2015) Impact of childhood maltreatment on the recognition of facial expressions of emotions. PLoS One 10(10):e0141732

Arendt H (1958) The human condition. University of Chicago Press, Chicago

Arendt H (1970) On violence. Houghton Mifflin Harcourt, New York

Arrien, Sophie-Jan. 2009. Natorp et Heidegger: Une science originaire est-elle possible? In Jollivet, S., & Romano, C. Heidegger en dialogue (1912-1930). Rencontres, affinités, confrontations, Paris, Vrin.

Asmundson GJ, Taylor S, AJ Smits J (2014) Panic disorder and agoraphobia: an overview and commentary on DSM-5 changes. Depress Anxiety 31(6):480–486

Atlan H (1974) On a formal definition of organization. J Theor Biol 45(2):295–304

Atlan H (1979) Entre le cristal et la fumée: essai sur l'organisation du vivant. Éditions du Seuil, Paris

Augustine S (1991) Confessions. Translated by Henry Chadwick. Oxford University Press, Oxford

Bachelor A (2013) Clients' and therapists' views of the therapeutic alliance: similarities, differences and relationship to therapy outcome. Clin Psychol Psychother 20(2):118–135

Baker M (2016) Statisticians issue warning over misuse of P values. Nature 531(7593):151–151

Baldwin SA, Wampold BE, Imel ZE (2007) Untangling the alliance-outcome correlation: exploring the relative importance of therapist and patient variability in the alliance. J Consult Clin Psychol 75:842–852

Baltes PB, Lindenberger U, Staudinger UM (2006) Life span theory in developmental psychology. Wiley, New York

Bambach CR (1993) Phenomenological research as destruktion the early Heidegger's reading of Dilthey. Philos Today 37(2):115–132

Bandura A (1997) Self-efficacy: the exercise of self-control. Macmillan, New York

Barbaras R (2007) Le mouvement de l'existence: études sur la phénoménologie de Jan Patočka. Éditions de la Transparence, Chatou

Barbaras R (2008) Introduction à une phénoménologie de la vie. Vrin, Paris

Bardy G (1949) "Philosophie" et "philosophe" dans le vocabulaire chrétien des premiers siècles. Revue d'ascétique, Toulouse

Barefoot JC, Schroll M (1996) (1996) Symptoms of depression, acute myocardial infarction, and total mortality in a community sample. Circulation 93:1976–1980

Barger AC, Herd JA, Liebowitz MR (1961) Chronic catheterization of coronary artery induction of ECG pattern of myocardial ischemia by intracoronary epinephrine. Proc Soc Exp Biol Med 107:474–477

Baroldi F (1975) Different morphological types of myocardial cell death in man. In: Fleckstein A, Rona G (eds) Recent advances in studies in cardiac structure and metabolism. pathophysiology and morphology of myocardial cell alteration, vol 6. University Park Press, Baltimore

Baur M (2003) Kant, Lonergan, and Fichte on the critique of immediacy and the epistemology of constraint in human knowing. Int Philos Q 43(1):91–112

Benoist J (1992) Chair et corps dans les séminaires de Zollikon: la différence et le reste. In: Courtine J-F (ed) Figures de la subjectivité. Éditions du CNRS, Paris

Benveniste E (1966) Problèmes de linguistique générale, 1. Gallimard, Paris

Benveniste E (1974) Problèmes de linguistique générale, 2. Gallimard, Paris

Berg IK, de Shazer S (1993) Making numbers talk: language in therapy. In: Friedman S (ed) The new language of change: constructive collaboration in psychotherapy. Guilford Press, New York, pp 5–24

Bernard M (2011) Le monde comme problème philosophique. Les Études philosophiques (3):351–373

Bersaglieri T, Sabeti PC, Patterson N, Vanderploeg T, Schaffner SF, Drake JA, Rhodes M, Reich DE, Hirschhorn JN (2004) Genetic signatures of strong recent positive selection at the lactase gene. Am J Hum Genet 74:1111–1120

Berti E (2009) Nuovi studi aristotelici, vol 4. Morcelliana, Brescia, p 1

Bertolino A, Arciero G, Rubino V, Latorre V, De Candia M, Mazzola V et al (2005) Variation of human amygdala response during threatening stimuli as a function of 5′ HTTLPR genotype and personality style. Biol Psychiatry 57(12):1517–1525

Beutler LE, Malik M, Alimohamed S, Harwood TM, Talebi H, Noble S et al (2004) Therapist variables. In: Lambert MJ (ed) Handbook of psychotherapy and behavior change, 5th edn. Wiley, New York

Bichat, Marie - François - Xavier (1799) Traité des membranes en général et de diverses membranes en particulier. Richard, Caille et Ravier, Paris

Bifulco AT, Brown GW, Harris TO (1987) Childhood loss of parent, lack of adequate parental care and adult depression: a replication. J Affect Disord 12(2):115–128

Bohart AC, Greaves AL (2013) The client in psychoterapy. In: Lambert M (ed) Bergin and Garfield's Handbook of psychotherapy and behavior change, 6th edn. Wiley, New York, pp 219–257

Bohart, A. C., & Tallman, K. (2010). Clients: the neglected common factor in psychotherapy. In B. Duncan, S. Miller, B. Wampold y M. Hubble (Eds), The heart and soul of change: delivering what works in therapy (2nd ed., pp. 83-111). Washington: American Psychological Association.

Boje DM (1999–2005). Narrative therapy. http://business.nmsu.edu/~dboje/narrativetherapy.html

Bondolfi G, Mazzola V, Arciero G (2015) In between ordinary sadness and clinical depression. Emot Rev 7(3):216–222

Bordin ES (1979) The generalizability of the psychoanalytic concept of the working alliance. Psychotherapy 16(3):252

Boss M (1977) Existential foundations of medicine & psychology. Jason Aronson, Northvale

Bossy J (1975) The social history of confession in the age of the Reformation. Trans Roy Hist Soc (Fifth Series) 25:21–38

Brague R (2009) Aristote et la question du monde: essai sur le contexte cosmologique et anthropologique de l'ontologie. Cerf, Paris

Brain RM, Wise MN, Biagioli M (1999) The science studies reader. Psychology Press, Hove

Brentari C (2015) Jakob Von Uexküll: the discovery of the umwelt between biosemiotics and theoretical biology, vol 9. Springer, Berlin

Brock C (2011) Toys are us: models and metaphors in brain research. In: Choudhury S, Slaby J (eds) Critical neuroscience: a handbook of the social and cultural contexts of neuroscience. Wiley-Blackwell, Oxford, pp 113–134

Bromberg PM (1998) Standing in the space. Analytic Press, Hillsdale

Brotman DJ, Golden SH, Wittstein IS (2007) The cardiovascular toll of stress. Lancet 370:1089–1100

Brown GW, Harris T (1978) Social origins of depression: a study of psychiatric disorder in women. Free Press, New York

Buckley U, Shivkumar K (2016) Stress-induced cardiac arrhythmias: the heart–brain interaction. Trends Cardiovasc Med 26(1):78

Burg MM, Jain D, Soufer R, Kerns RD, Zaret BL (1993) Role of behavioral and psychological factors in mental stress- induced silent left ventricular dysfunction in coronary artery disease. J Am Coll Cardiol 22:440–448

Burke K (1969) A rhetoric of motives, vol 111. University of California Press, Berkeley

Bybee KA, Kara T, Prasad A et al (2004) Systematic review: transient left ventricular apical ballooning: a syndrome that mimics ST-segment elevation myocardial infarction. Ann Intern Med 141:858–865

Cahan D (ed) (1993) Hermann von Helmholtz and the foundations of nineteenth-century science. In: University of California Press, Berkeley, CA

Cahan D (1995) Science and culture: popular and philosophical essays. University of Chicago Press, Chicago

Calvino I (1988) Six memos for the next millennium. Harvard University Press, Cambridge

Canguilhem G (1952) Le vivant et son milieu. In idem. La connaissance de la vie. Vrin, Paris

G. Canguilhem (1966) On the normal and the pathological. D. Reidel Publishing Company, Dordrecht, Boston and London. Il normale ed il patologico (1994) trad. it Einaudi

Cao JM, Fishbein MC, Han JB, Lai WW, Lai AC, Wu TJ, Czer L, Wolf PL, Denton TA, Shintaku IP, Chen PS, Chen LS (2000) Relationship between regional cardiac hyperinnervation and ventric- ular arrhythmia. Circulation 101:1960–1969

Cappuccio M (2006) Neurofenomenologia. Le scienze della mente e la sfida dell'esperienza cosciente. Bruno Mondadori, Milano
Caputo JD (1984) Husserl, Heidegger and the question of a "hermeneutic" phenomenology. Husserl Stud 1(1):157–178
Caputo JD (1988) Radical hermeneutics: repetition, deconstruction, and the hermeneutic project. Indiana University Press, Bloomington
Caputo A (2015). Emozioni come infezioni? La critica heideggeriana al contagio emotivo di Scheler. In Dibattista (Eds.) Emozioni infettive. Mimesis, pp 127–155
Carroll KM, Nich C, Rounsaville BJ (1997) Contribution of the therapeutic alliance to outcome in active versus control psychotherapies. J Consult Clin Psychol 65(3):510
Cassiano G (1989) Le istituzioni cenobitiche. Bresseo di Teofilo, Abbazia di Praglia
Cassiano G (2000) Conferenze ai monaci. Città Nuova, Roma
Cassirer E (1996) The metaphysics of symbolic forms. (Italian Edition) Metafisica delle forme simboliche (2003). Sansoni
Cebelin M, Hirsch CS (1980) Human stress cardiomyopathy. Hum Pathol 11:123–132
Chā. (2011). The collected teachings of Ajahn Chah, 3 Vol. Amaravati Buddhist Monastery
Chalmers DJ (1995) Facing up to the problem of consciousness. J Consc Stud 2(3):200–219
Chantraine P (1977) Dictionnaire etymologique grec. Klincksieck, Paris
Cheung RT, Hachinski V (2000) The insula and cerebrogenic sudden death. Arch Neurol 57(12):1685–1688
Child N, Hanson B, Bishop M, Rinaldi CA, Bostock J, Western D et al (2014) Effect of mental challenge induced by movie clips on action potential duration in normal human subjects independent of heart rate. Circ Arrhythm Electrophysiol 7:518–523
Conti V (2018) Il chi della storia. Ph.D. Dissertation, Pontifical University Gregoriana
Courtine JF (1990) Suarez et le système de la métaphysique. Presses universitaires de France, Paris
Craig AD (2011) Significance of the insula for the evolution of human awareness of feelings from the body. Ann N Y Acad Sci 1225:72–82
Crowell SG (2001) Husserl, Heidegger, and the space of meaning: paths toward trancendental phenomenology. Northwestern University Press, Evanston
Dahan A, Pestre D (2004) Transferring Formal and mathematical Tools from war management to Political, Technological, and social Intervention (1940-1960). In: Lucertini M, Gasca AM, Nicolo F (eds) Technological concepts and mathematical models in the evolution of modern engineering systems. Birkhäuser, Basel, pp 79–100
Dahl G (2010) The two time vectors of Nachträglichkeit in the development of ego organization: significance of the concept for the symbolization of nameless traumas and anxieties 1. Int J Psychoanal 91(4):727–744
Dahlstrom DO (1994) Heidegger's method: philosophical concepts as formal indications. Rev Metaphys 47:775–795
Dahlstrom DO (2001) Heidegger's concept of truth. Cambridge University Press, Cambridge
Damasio AR (1994) Descartes' error: emotion, rationality and the human brain. Putnam, New York
Damasio AR (1999) The feeling of what happens: body and emotion in the making of consciousness. Mifflin Harcourt, Houghton
Damasio AR (2003) Looking for Spinoza: joy, sorrow, and the feeling brain. Mifflin Harcourt, Houghton
Dastur F (1995) Pour une zoologie privative. In: Alter, Revue de Phénoménologie, n° 3. Editions Alter, Paris, pp 281–318
Dastur F (2013) Fink et Patočka: deux conceptions de la liberté. Philosophie 3:13–20
Davies B, Harré R (1990) Positioning: the discursive production of selves. J Theory Soc Behav 20(1):44–63
de Barenne JD (1934) The disturbances after laminar thermo-coagulation of the motor cerebral cortex. Brain 57(4):517–526
de Barenne JD, McCulloch WS (1938) The direct functional interrelation of sensory cortex and optic thalamus. J Neurophysiol 1:176–186
De Barenne JD, McCulloch WS (1939) Physiological delimitation of neurones in the central nervous system. Am J Physiol 127(4):620–628

De Kock L (2011) Some preliminary considerations on Helmholtz's Fichte: towards a naturalized epistemology of constraint? Revista de Estud(i)os sobre Fichte, 2 | 2011
Dennett DC (2004) Freedom evolves. Penguin, London
Depraz N (2006) Mettere al lavoro il metodo fenomenologico nei protocolli sperimentali."Passaggi generativi" tra l'empirico e il trascendentale. In: Cappuccio M (ed) Neurofenomenologia. Le scienze della mente e la sfida dell'esperienza cosciente. Bruno Mondadori, Milano
Derrida J (1987) De l'esprit. Galilée, Paris
Derrida J, Wills D (2002) The animal that therefore I am (more to follow). Crit Inq 28(2):369–418
Descartes R (1998) Reguale ad directionem ingenii. Rules for the direction of the natural intelligence. A bilingual edition of the Cartesian treatise on method. Editions Rodopi, Amsterdam
Descartes R, Cottingham J (1996) René Descartes: meditations on first philosophy: with selections from the objections and replies. Cambridge University Press, Cambridge
Descartes R, Cottingham J, Stoothoff R, Murdoch D (1991) The philosophical writings of Descartes: Volume 3, The correspondence, vol 3. Cambridge University Press, Cambridge
Di Cesare G, Di Dio C, Rochat MJ, Sinigaglia C, Bruschweiler-Stern N, Stern DN, Rizzolatti G (2013) The neural correlates of "vitality form" recognition: an fMRI study. Soc Cogn Affect Neurosci 9(7):951–960
di Cesarea B (1976) Il battesimo. In: Paideia, Brescia
di Gaza D (1993) Insegnamenti spirituali. Città nuova, Rome
Domański J (1996) La philosophie, théorie ou manière de vivre?: les controverses de l'Antiquité à la Renaissance, vol 18. Saint-Paul
Dumouchel P, Dupuy JP (1983) L'auto-organisation: de la physique au politique. Éditions du Seuil, Paris
Dupuy JP (1985) L'essor de la première cybernétique (1943-1953). Cahiers du CREA 7:7–140
Dupuy JP (2009) On the origins of cognitive science: the mechanization of the mind. MIT Press, Cambridge, MA
Düring I (1968) Aristotele. Mursia, Milano
Durst M (2004) Birth and Natality in Hannah Arendt. In: Husserliana A (ed) Does the world exist? vol 79. Springer, Dordrecht, pp 777–797
Dush DM, Hirt ML, Schroeder HE (1983) Self-statement modification with adults: a meta-analysis. Psychol Bull 94(3):408
Eaker ED, Sullivan LM, Kelly-Hayes M, D'Agostino RB Sr, Benjamin EJ (2004). Anger and hostility predict the development of atrial fibrillation in men in the Framingham Offspring Study. Circulation; 109: 1267–1271.
Edmondson D, Newman JD, Whang W, Davidson KW (2012) Emotional triggers in myocardial infarction: do they matter? Eur Heart J 34(4):300–306
Eitel I, Knobelsdorff-Brenkenhoff F, Bernhardt P et al (2011) Clinical characteristics and cardiovascular magnetic resonance findings in stress (takotsubo) cardiomyopathy. JAMA 306:277–286
Elliott R, James E (1989) Varieties of client experience in psychotherapy: an analysis of the literature. Clin Psychol Rev 9(4):443–467
Elliott R, Bohart AC, Watson JC, Greenberg LS (2011) Empathy. In: Norcross JC (ed) Psychotherapy relationships that work: evidence-based responsiveness. Oxford University Press, Oxford
Engel GL (1971) Sudden and rapid death during psychological stress: folklore or folk wisdom? Ann Intern Med 74(5):771–783
Eraclito (1978) Frammenti, a cura di M. Marcovich, La Nuova Italia, Firenze
Erickson EH (1982) The life cycle completed: a review. W. W Norton, New York
Evagrius P (1970) The Praktikos, chapters on prayer, trans. John Eudes Bamberger. Cistercian Publications, Spencer, MA
Feist R, Sweet W (2003) Husserl and Stein. Council for Research in Values and Philosophy, Washington
Festugière AJ (1950) Contemplation Et Vie Contemplative Selon Platon. J. Vrin, Paris
Feyerabend P (1975) Against method: outline ofan anarchistic theory of knowledge. Humanities Press, Atlantic Highlands
Fichte JG, Breazeale D (1998) Foundations of transcendental philosophy (Wissenschaftslehre) nova methodo (1796/99). Cornell University Press, Ithaca

Fichte JG, Neuhouser F, Baur M (2000) Foundations of natural right. Cambridge University Press, Cambridge
Fichte JG, Breazeale D, Zoller G (2005) The system of ethics (p. v399) (tr Daniel Breazeale and Günter Zöller). Cambridge University Press, Cambridge
Fonagy P, Gergely G, Jurist E, Target M (2002) Affect regulation, mentalization, and the development of the self. Other Press, New York
Foucault M (1963) Naissance de la clinique. Une archeologie du regard medical. Quadrige, PUF
Foucault M (1994) Dits et écrits II, 1976-1988. Gallimard, Paris
Foucault M (2001) L'herméneutique du sujet. Cours au Collège de France 1981-1982. Gallimard-Seuil, Paris
Foucault M (2004) Sécurité, territoire, population: cours au Collège de France, 1977-1978. Gallimard, Paris
Foucault M (2005) Discorso e verità nella Grecia antica. Donzelli editore
Foucault M (2008) Le gouvernement de soi et des autres: Cours au Collège de France (1982-1983), vol 1. Seuil, Paris
Foucault M (2009) Le Gouvernement de soi et des autres II: Le Courage de la vérité. Gallimard, Paris
Foucault M (2012) Du gouvernement des vivants: cours au Collège de France: 1979-1980. F. Ewald, A. Fontana, & M. Senellart (Eds.). Gallimard, Paris
Foucault M (2013) L'origine de l'herméneutique de soi. Conférences prononcées à Darmouth College 1980, Vrin
Foucault M (2016) Discours et vérité. Vrin, Paris
Foucault M, Burchell G (2011) The government of self and others: Lectures at the College de France, 1982-1983, vol 7. Macmillan, Basingstoke
Foucault M, Brion F, Harcourt BE (1981) Mal faire, dire vrai: Fonction de l'aveu en justice-cours de Louvain, 1981. Presses universitaires de Louvain, Louvain
Foucault M, Ewald F, Gros F, Fontana A (2009) Le courage de la vérité: le gouvernement de soi et des autres II: cours au Collège de France, 1983-1984. Seuil, Paris
Frank JD (1961) Persuasion and healing. Johns Hopkins University Press. Garner, Baltimore
Frank JD (1987) Psychotherapy, rhetoric, and hermeneutics: implications for practice and research. Psychotherapy 24(3):293
Frank JD (1995) Psychotherapy as rhetoric: some implications. Clin Psychol Sci Pract 2(1):90–93
Frank JD, Frank JB (1991) Persuasion and healing: a comparison study of psychotherapy. JHU Press, Baltimore
Freud S (1905) Jokes and their relation to the unconscious. F. Deuticke, Leipzig
Freud S (1920) Beyond the pleasure principle. Hogarth, London
Freud S (1939) Moses and monotheism. Hogarth, London, pp 7–140
Freud S (1955) The Moses of Michelangelo. In: The standard edition of the complete psychological works of Sigmund Freud, Vol XIII
Freud S, Breuer J (1895) Studies on hysteria. SE, 2. Hogarth, London
Friedländer P. (1928–2004) Platon, 3 vols. Berlin: De Gruyter (Trad. It. Platone Bompiani Ed.)
Friedländer P, Reale G, Le Moli A (2004) Platone. Bompiani, Milan
GA 62 (2005) PIA Phänomenologische Interpretationen zu Aristoteles (Anzeige der hermeneutischen Situation) (Natorp-Bericht)
Gadamer HG (1975) Truth and method (1960). Crossroad, New York
Galen (1963) On the passions and errors of the Soul, translated by Paul W. Harkins, with an Introduction and Interpretation by Walter Riese. Ohio State University Press, Athens
Gallagher S (2000) Philosophical conceptions of the self: implications for cognitive science. Trends Cogn Sci 4:14–21
Gallagher S (2012) In defense of phenomenological approaches to social cognition: interacting with the critics. Rev Philos Psychol 3(2):187–212
Gallese V (2005) Embodied simulation: from neurons to phenomenal experience. Phenomenol Cogn Sci 4(1):23–48

Gallese V (2007a) Before and below 'theory of mind': embodied simulation and the neural correlates of social cognition. Phil Trans R Soc Lond B Biol Sci 362(1480):659–669
Gallese V (2007b) Embodied simulation: from mirror neuron systems to interpersonal relations. Novartis Found Symp 278:3–12
Garfield SL (1995) Psychotherapy: an eclectic-integrative approach. John Wiley & Sons, New York
Garland C (1998) Thinking about trauma. In: Garland C (ed) Understanding trauma: a psychoanalytical approach. Duckworth, London, pp 9–31
Gazzaniga MS (1998) The mind's past. University of California Press, Berkeley
Gethmann CF (1993) Dasein: Erkennen und Handeln: Heidegger im phänomenologischen Kontext, vol 3. Walter de Gruyter, Berlin
Gigante M (1969) Ricerche filodemee. Macchiaroli Editore, Napoli
Ginzburg C (1986) Miti, emblemi, spie: morfologia e storia. G. Einaudi, Torino
Giovanni C (2007) La scala del paradiso. Paoline, Milano
Goddard JC (2003) Autonomie, réduction et réflexivité: la philosophie naturelle de Francisco. J Varela et le projet transcendental Intellectica 36(37):205–225
Goldstein K (1934) Der Aufbau des Organismus. Einführung in die Biologie unter besonderer Berücksichtigung der Erfahrungen am kranken Menschen. Den Haag, Nijhoff. English Translation. The organism: a holistic approach to biology derived from pathological data in man. (1939). American Book Company, New York
Goldstein K (1939) The organism: a holistic approach to biology derived from pathological data in man. American Book Company, New York
Goleman D (1996) Vital lies, simple truths: the psychology of self deception. Simon and Schuster, New York
Greenberg LS, Johnson SM (1986) Affect in marital therapy. J Marital Fam Ther 12(1):1–10
Greenberg LS, Safran JD (1989) Emotion in psychotherapy. Am Psychol 44(1):19
Greenberg RP, Constantino MJ, Bruce N (2006) Are patient expectations still relevant for psychotherapy process and outcome? Clin Psychol Rev 26(6):657–678
Greisch J (1994a) Ontologie et temporalité: esquisse d'une interprétation intégrale de Sein und Zeit. Presses universitaires de France, Paris
Greisch J (1994b) Le phénomène de la chair: un «ratage» de Sein und Zeit. In: Floviral G (ed) Dimensions de l'exister. Etudes de anthropologie philosophique, Bibliothèque Philosophique de Louvain. Peeters, Leuven
Greisch J (1999) Les limites de la chair. In: Olivetti MM (ed) Incarnation. CEDAM, Padua
Greisch J (2000) L'Arbre de vie et l'Arbre du savoir.: Le chemin phénoménologique de l'herméneutique heideggérienne (1919-1923). Cerf, Paris
Grencavage LM, Norcross JC (1990) Where are the commonalities among the therapeutic common factors? Prof Psychol Res Pract 21(5):372
Griswold CL Jr (1996) Self-Knowledge in Plato's Phaedrus. Penn State University Press, Pennsylvania
Gros F, Foucault M (2001) L'Herméneutique du sujet: Cours au Collège de France. (eng. Trans.) 1981–82. 'The Hermenutics of the Subject: Lectures at the College de France. Palgrave Macmillan, New York
Gros-Azorin C (1998) Le phénomène du corps (leib): Une entente participative. Les Études Philosophiques, pp 465–477
Guidano VF (1987) Complexity of the self: a developmental approach to psychopathology and therapy. Guilford Press, New York
Guidano VF (1991) The self in process: toward a post-rationalist cognitive therapy. Guilford Press, New York
Gurevič AJ (1996) La nascita dell'individuo nell'Europa medievale. Laterza, Roma-Bari
Hadot I (1986) The spiritual guide. In: Armstrong AH, Armstrong AA (eds) Classical Mediterranean spirituality: Egyptian, Greek, Roman, vol 15. Routledge, London
Hadot P (1988) Exercices spirituels et philosophie antique. Deuxième édition revue et augmentée. — Paris: Etudes Augustiniennes - Philosophy as a Way of Life. Spiritual Exercises from

Socrates to Foucault, edited with an Introduction by Arnold I. Davidson, translated by Michael Chase. Basil Blackwell, Oxford/Cambridge, 1995
Hadot P (1995). Qu'est-ce que la philosophie antique? (eng. Trans.) P. Hadot, What is ancient philosophy?, transl. by Michael Chase, Belknap Press, Cambridge, 2002
Hadot P (2002) Exercices spirituels et philosophie antique. Albin Michel
Hadot P (2008) La filosofia come modo di vivere. Conversazioni con Jeannie Carlier e Arnold I. Davidson, 3-71. Einaudi
Hadot I (2014) Sénèque: direction spirituelle et pratique de la philosophie. Vrin, Paris
Harris T, Brown GW, Bifulco A (1986) Loss of parent in childhood and adult psychiatric disorder: the role of lack of adequate parental care. Psychol Med 16(03):641–659
Hart JG (1995) Husserl and Fichte: with special regard to Husserl's lectures on "Fichte's ideal of humanity". Husserl Stud 12(2):135–163
Hausherr I (2012) Padre, dimmi una parola. La direzione spirituale nell'antico Oriente, trad. di Romano Baldelli, Edizioni Scritti Monastici-Abbazia di Praglia, Bresseo di Teolo (PD) 2012
Hayek FA (1952) The sensory order: an inquiry into the foundations of theoretical psychology. University of Chicago Press, Chicago
Hayek FA (1967) Rules, perception and intelligibility. In: Hayek FA (ed) Studies in philosophy, politics and economics. Routledge & Kegan Paul, London
Hayek FA (1969) The primacy of the abstract. In: Koestler A, Smythies JR (eds) Beyond reductionism—the Alpbach symposium. Hutchinson, London
Hayek, F. A. (1978). New studies. In Philosophy, politics, economics and the history of ideas. Chicago University of Chicago Press.
Heidegger, M. (1926). On the essence of truth (Pentecost Monday) (TJ Kisiel, Trans.). Becoming Heidegger: on the trail of his early occasional writings, 1927, pp 275–288.
Heidegger GA 15 1986
Heidegger GA 17 1994
Heidegger GA 18 2002
Heidegger GA 2 1977
Heidegger GA 20 1988
Heidegger GA 24 1975
Heidegger, GA 27 (1996)
Heidegger GA 28 1997
Heidegger GA 29-30 (1983)
Heidegger GA 33 1981
Heidegger GA 41 (1962)
Heidegger GA 49 1991
Heidegger GA 56/57 1999
Heidegger GA 58 1992
Heidegger GA 59 1993
Heidegger GA 6.1 1996
Heidegger GA 60 1995
Heidegger GA 61 1985
Heidegger GA 62 2005
Heidegger GA 64 2004
Heidegger GA 89 (n.d.)
Heidegger GA 9 (1976)
Heidelberger M (1993) Force, law, and experiment: the evolution of Helmholtz's philosophy of science. In: Cahan D (ed) Hermann von Helmholtz and the foundations of nineteenth-century science. University of California Press, Berkeley
Hempel CG (1942) The function of general laws in history. J Philos 39(2):35–48
Hermans HJM, Gieser T (eds) (2012) Handbook of dialogical self theory. Cambridge University Press, Cambridge
Hermans HJ, Hermans-Jansen E (2004) The dialogical construction of coalitions in a personal position repertoire. In: Hermans HJ, Dimaggio G (eds) The dialogical self in psychotherapy: an introduction. Routledge, New York

Hoag MJ, Burlingame GM (1997) Evaluating the effectiveness of child and adolescent group treatment: a meta-analytic review. J Clin Child Psychol 26(3):234–246
Holden C, Mace R (1997) Phylogenetic analysis of the evolution of lactose digestion in adults. Hum Biol 69:605–628
Horvath AO, Bedi RP (2002) The alliance. In: Norcross JC (ed) Psychotherapy relationships that work: therapist contributions and responsiveness to patients. Oxford University Press, Oxford
Horvath AO, Greenberg LS (1994) The working alliance: theory, research, and practice, vol 173. John Wiley & Sons, New York
Horvath AO, Luborsky L (1993) The role of the therapeutic alliance in psychotherapy. J Consult Clin Psychol 61(4):561
Horvath AO, Symonds BD (1991) Relation between working alliance and outcome in psychotherapy: a meta-analysis. J Couns Psychol 38(2):139
Horvath AO, Del Re AC, Flückiger C, Symonds D (2011) Alliance in individual psychotherapy. Psychotherapy 48(1):9–16
Hubble MA, Duncan BL, Miller SD (1999) The heart and soul of change: what works in therapy. American Psychological Association, Washington
Hubble MA et al (2010) Introduction. In: Duncan BL, Miller SD, Wampold BE, Hubble MA (eds) The heart and soul of change: delivering what works in therapy. American Psychological Association, Washington
Huneman P (2005) Espèce et adaptation chez Kant et Buffon. In: Theis R, Ferrari J, Ruffin M (eds) Kant et la France-Kant und Frankreich. Olms, Hildesheim
Huneman P (ed) (2007) Understanding purpose: Kant and the philosophy of biology, vol 8. University Rochester Press, Rochester, NY
Huneman P (2008) Métaphysique et biologie: Kant et la constitution du concept d'organisme. Kimé, Paris
Husserl E (2006) Logische Untersuchungen. Ergänzungsband. Zweiter Teil.: Texte für die Neufassung der VI. Untersuchung. Zur Phänomenologie des Ausdrucks und der Erkenntnis (1893/94-1921), vol 20. Springer, Berlin
Husserl E, Hahnengress KH, Rapic S (1991) Ding und Raum, vol 437. Meiner Verlag, Leipzig
Hyppolite J (1959) L'idée fichtéenne de la doctrine de la science et le projet husserlien. Husserl et la Pensée Moderne–Husserl und das Denken der Neuzeit. Nijhoff, The Hague
Imel Z, Wampold B (2008) The importance of treatment and the science of common factors in psychotherapy. In: Brown SD, Lent RW (eds) Handbook of counseling psychology. Wiley, New York, pp 249–266
Jablonka E, Lamb MJ (2005) Evolution in four dimensions: genetic, epigenetic, behavioral, and symbolic variation in the history of life. MIT Press, Cambridge
Jablonka E, Lamb MJ, Zeligowski A (2014) Evolution in four dimensions, revised edition: genetic, epigenetic, behavioral, and symbolic variation in the history of life, 2nd edn. MIT Press, Cambridge
Jacquet F (2011) Vie et existence: vers une cosmologie phénoménologique. Les Études philosophiques (3):395–419
Jaeger G (1944) Paideia. Oxford University Press, New York
Jaeger H (1959) L'examen de conscience dans les religions non-chrétiennes et avant le christianisme. Numen 6(Fasc. 3):175–233
Jaspers K (1919) Psychologie der weltanschauungen. Springer, Berlin
Jaspers K (2013) Psychologie der weltanschauungen. Springer, Berlin
Johansson H, Jansson JÅ (2010) Therapeutic alliance and outcome in routine psychiatric out-patient treatment: patient factors and outcome. Psychol Psychother Theory Res Pract 83(2):193–206
Johnson-Laird PN (2005) Mental models and thought. In: Holyoak KJ, Morrison RG (eds) The Cambridge handbook of thinking and reasoning. Cambridge University Press, Cambridge, pp 185–208
Jonas H (1967) The phenomenon of life: toward a philosophical biology. Harper & Row, New York
Jonas H (1997) Tecnica, medicina ed etica. Prassi del principio responsabilità, a cura di P. Becchi, Einaudi, Torino.
Jones EG, Mendell LM (1999) Assessing the decade of the brain. Science 284(5415):739–739

Kahn CH (1979) The art and thought of Heraclitus. Cambridge University Press, Cambridge
Kahn CH (2001) Pythagoras and the Pythagoreans: a brief history. Hackett Publishing, Indianapolis
Kahneman D (2011) Thinking, fast and slow, New York. Macmillan
Kant I (1786). What does it mean to orient oneself in thinking? In Allen W. Wood (transl and ed) Immanuel Kant, religion and rational theology. Cambridge University Press, Cambridge, 1996
Kant, I. (1793) I progressi della metafisica, tr. it. a cura di P. Manganaro, Bibliopolis, Napoli 1967
Kant I (1889) Immanuel Kants Kritik der reinen Vernunft. Mayer & Müller, Berlin
Kant I (1972) Vorlesungen über Metaphysik und Rationaltheologie. (Metaphysik L1). Gerhard Lehmann
Kant E (1992) On the form and principles of the sensible and the intelligible world [Inaugural Dissertation]. In: Walford D, Meerbote R (eds) Theoretical philosophy, 1755–1770 (The Cambridge Edition of the Works of Immanuel Kant). Cambridge University Press, Cambridge
Kant I (1997) Metaphysik Mrongovius. Lectures on metaphysics. Cambridge University Press, Cambridge
Kant I, Deutsche Akademie der Wissenschaften (RDA). (1972) Gesammelte Schriften: Vorlesungen. Vorlesungen über Metaphysik und Nationaltheologie. W. de Gruyter, Berlin
Kawai S, Kitabatake A, Tomoike H (2007) Guidelines for diagnosis of takotsubo (ampulla) cardiomyopathy. Circ J 71:990–992
Kelly G (2003) The psychology of personal constructs: volume two: clinical diagnosis and psychotherapy. Routledge, London
Kierkegaard S (2013) Kierkegaard's writings, XI: stages on life's way, vol 11. Princeton University Press, Princeton
Kim DM, Wampold BE, Bolt DM (2006) Therapist effects in psychotherapy: a random-effects modeling of the National Institute of Mental Health Treatment of Depression Collaborative Research Program data. Psychother Res 16(02):161–172
Kirk GS, Raven JE, Schofield M (1983) The presocratic philosophers: a critical history with a selcetion of texts. Cambridge University Press, Cambridge
Kirsch I (1985) Response expectancy as a determinant of experience and behavior. Am Psychol 40(11):1189
Kisiel T (1993) The genesis of Heidegger's being and time. University of California Press, Berkeley
Kisiel T (1996) L'indication formelle de la facticité: sa genèse et sa transformation. In: Courtine J-F (ed) Heidegger 1919-1929. De l'herméneutique de la facticité à la métaphysique du Dasein. Vrin, Paris
Kisiel, T. (1999). L'indication formelle de la facticité. Etudes phénoménologiques, 15(29/30), 107-126.
Kisiel T (2000) Heidegger-Lask-Fichte. In: Rockmore T (ed) Heidegger, German idealism, and neo-Kantianism. Humanity, New York
Kisiel T (2002) Heidegger's way of thought: critical and interpretive signposts. A&C Black, London
Kisiel T (2008) On the genesis of Heidegger's formally indicative hermeneutics of facticity. In: Raffoul F, Nelson ES (eds) Rethinking Facticity. State University of New York Press, New York
Koselleck R (2004) Futures past: on the semantics of historical time. Columbia University Press, New York
Krupnick JL, Elkin I, Collins J, Simmens S, Sotsky SM, Pilkonis PA, Watkins JT (1994) Therapeutic alliance and clinical outcome in the NIMH Treatment of Depression Collaborative Research Program: preliminary findings. Psychotherapy 31(1):28
Krupnick JL, Sotsky SM, Simmens S, Moyer J, Elkin I, Watkins J, Pilkonis PA (1996) The role of the therapeutic alliance in psychotherapy and pharmacotherapy outcome: findings in the National Institute of Mental Health Treatment of Depression Collaborative Research Program. J Consult Clin Psychol 64(3):532
Kubzansky LD, Sparrow D, Vokonas P, Kawachi I (2001) Is the glass half empty or half full? A prospective study of optimism and coronary heart disease in the normative aging study. Psychosom Med 63:910–916

Kuper H, Marmot M, Hemingway H (2002) Systematic review of prospective cohort studies of psychosocial factors in the etiology and prognosis of coronary heart disease. Semin Vasc Med 2:267–314

Lachaux JP, Rodriguez E, Martinerie J, Varela FJ (1999) Measuring phase synchrony in brain signals. Hum Brain Mapp 8(4):194–208

Laland KN, Odling-Smee J, Myles S (2010) How culture shaped the human genome: bringing genetics and the human sciences together. Nat Rev Genet 11(2):137–148

Laland K, Odling-Smee J, Turner S (2014a) The role of internal and external constructive processes in evolution. J Physiol 592(11):2413–2422

Laland K, Uller T, Feldman M, Sterelny K, Müller GB, Moczek A et al (2014b) Does evolutionary theory need a rethink? Nature 514(7521):161

Lambert MJ (1986) Some implications of psychotherapy outcome research for eclectic psychotherapy. J Integr Elect Psychother 5(1):16–45

Lambert MJ (1992) Implications of outcome research for psychotherapy integration. In: Norcross JC, Goldstein MR (eds) Handbook of psychotherapy integration. Basic Books, New York, pp 94–129

Lambert M, Bergin AE (1994) The effectiveness of psychotherapy. In: Bergin A, Garfield S (eds) Handbook of psychotherapy and behaviour change, 4th edn. Wiley, New York, pp 143–489

Lambert JJ, Ogles BM (2004) The efficacy and effectiveness of psychotherapy. In: Lambert MJ (ed) Bergin and Garfield's handbook of psychotherapy and behavior change, 5th edn. Wiley, New York, pp 139–193

Larchet JC (1996) La divinisation de l'homme selon saint Maxime le Confesseur. Cerf, Paris

Larchet JC (1997) Thérapeutique des maladies spirituelles. Cerf, Paris

Larivée A, Leduc A (2001/2) Saint Paul, Augustin et Aristote comme sources gréco-chrétiennes du souci chez Heidegger. Philosophie (2):30–50

Lazarus AA (1993) Tailoring the therapeutic relationship, or being an authentic chameleon. Psychotherapy 30(3):404

Leclercq J (1952) Pour l'histoire de l'expression 'philosophie chrétienne'. Mélanges de Science Religieuse 9:221–226

Lenoir T (1988) Social interests and the organic physics of 1847. In: Science in reflection. Springer, Netherlands, pp 169–191

Lenoir T (1994) Helmholtz and the materialities of communication. Osiris 9:184–207

Leor J, Poole WK, Kloner RA (1996) Sudden cardiac death triggered by an earthquake. N Engl J Med 334(7):413–419

Lettvin JY, Maturana HR, McCulloch WS, Pitts WH (1959) What the frog's eye tells the frog's brain. Proc IRE 47(11):1940–1951

Levinas E (1979) Totality and infinity: an essay on exteriority, vol 1. Springer Science & Business Media, Dordrecht

Levinas E (1991) Totality and infinity: an essay on exteriority. Kluwer Academic Publishers, Dordrecht

Livet P (1987) Intersubjectivité, réflexivité et récursivité chez Fichte, Archives de Philosophie, tome 50, cahier 4, octobre-décembre 1987

Lopez-Dominguez V (2010) Political realism in idealism: Fichte versus Hegel and their different versions of the foundation of right. In: Breazeale D, Rockmore T (eds) Fichte, German idealism, and early romanticism, vol 24. Rodopi, Amsterdam

Lorenz K (1966) On aggression, trans Marjorie Latzke. Methuen, London

Luborsky L, McLellan AT, Woody GE, O'Brien CP, Auerbach A (1985) Therapist success and its determinants. Arch Gen Psychiatry 42(6):602–611

Luborsky L, Diguer L, Seligman DA, Rosenthal R, Krause ED, Johnson S et al (1999) The researcher's own therapy allegiances: a "wild card" in comparisons of treatment efficacy. Clin Psychol Sci Pract 6(1):95–106

Luborsky L, Rosenthal R, Diguer L, Andrusyna TP, Berman JS, Levitt JT et al (2002) The dodo bird verdict is alive and well—mostly. Clin Psychol Sci Pract 9(1):2–12

Lutz A, Lachaux JP, Martinerie J, Varela FJ (2002) Guiding the study of brain dynamics by using first-person data: synchrony patterns correlate with ongoing conscious states during a simple visual task. Proc Natl Acad Sci 99(3):1586–1591
Lutz W, Leon SC, Martinovich Z, Lyons JS, Stiles WB (2007) Therapist effects in outpatient psychotherapy: a three-level growth curve approach. J Couns Psychol 54(1):32
MacIntyre AC (1981) After virtue. Notre Dame, Notre Dame Press
MacIntyre A (1984) After virtue: a study in moral theory. University of Notre Dame Press, Notre Dame, IN
MacIntyre A (2013) After virtue. A&C Black, London
Magnus R (1906) Goethe als Naturforscher Vorlesungen. Barth, Leipzig
Magnus R (1925) Croonian lecture: animal posture. Proc R Soc Lond Ser B Biol Sci 98(690):339–353
Magnus R (1926) The physiology of posture: cameron lectures. Lancet 211:531–536
Magnus, R. (1930) The physiological a priori, Lane lectures on experimental pharmacology and medicine, Stanford University Publications, University Series, Medical Sciences 2(3) pp. 97–103, Stanford: Stanford University Press
Magnus O (2002) Rudolf Magnus: physiologist and pharmacologist (1873-1927). Springer Science & Business Media, Dordrecht
Mahoney MJ (1984) Integrating cognition, affect, and action: a comment. Cogn Ther Res 8(6):585–589
Maj M (2013) "Clinical judgment" and the DSM-5 diagnosis of major depression. World Psychiatry 12(2):89–91
Malingrey AM (1961) "Philosophia" Étude d'Un Groupe de Mots Dans la Littérture Grecque, des Présocratiques au Ive Siècle Après J.-C. Klincksieck, Paris
Marcovich M (1978) Eraclito: frammenti. La nuova Italia
Marion JL (1999) On Descartes' metaphysical prism: the constitution and the limits of onto-theology in Cartesian thought. University of Chicago Press, Chicago
Marion JL (2013) Sur la pensée passive de Descartes. Presses universitaires de France
Martin WM (2003) Nothing more or less than logic: general logic, transcendental philosophy, and Kant's repudiation of Fichte's Wissenschaftslehre. Topoi 22(1):29–39
Martin DJ, Garske JP, Davis MK (2000) Relation of the therapeutic alliance with outcome and other variables: a meta-analytic review. J Consult Clin Psychol 68(3):438
Marx W (1961) The meaning of Aristotle's ontology. Martinus Nijhoff, Hague
Marx W (1998) Uber das Märchen vom Ende der Philosophie. Eine Streitschrift für systematische Rationalität. Königshausen und Neumann, Würzburg
Matthews KA, Gump BB, Harris KF, Haney TL, Barefoot JC (2004) Hostile behaviors predict cardiovascular mortality among men enrolled in the Multiple Risk Factor Intervention Trial. Circulation 109:66–70
Maturana HR (1970) Biology of cognition. Biological Computer Laboratory Research Report BCL 9.0. University of Illinois, Urbana.
Maturana H, Varela F (1973) De Máquinas y Seres Vivos: una caracterización de la organización biológica. Editorial Universitaria, Santiago
Maturana HR, Varela FJ (1987) The tree of knowledge: the biological roots of human understanding. New Science Library/Shambhala Publications, Boston
Maximus the Confessor (2014) On difficulties in the Church Fathers, The Ambigua, Nicholas Constas (ed. and trans.). Harvard University Press, Cambridge
Mazzola V, Latorre V, Petito A, Gentili N, Fazio L, Popolizio T et al (2010) Affective response to a loved one's pain: insula activity as a function of individual differences. PLoS One 5(12):e15268
Mazzola V, Vuilleumier P, Latorre V, Petito A, Gallese V, Popolizio T et al (2013) Effects of emotional contexts on cerebello-thalamo-cortical activity during action observation. PLoS One 8(9):e75912
Mazzola V, Marano G, Biganzoli EM, Boracchi P, Lanciano T, Arciero G, Bondolfi G (2014) The In-Out dispositional affective style questionnaire (IN-OUT DASQ): an exploratory factorial analysis. Front Psych 5:1005

Mazzola V, Arciero G, Fazio L, Lanciano T, Gelao B, Popolizio T et al (2016) What impact does an angry context have upon us? The effect of anger on functional connectivity of the right insula and superior temporal gyri. Front Behav Neurosci 10:109
Mazzola V Arciero G, Fazio L, Lanciano L, Gelao B, Popolizio T, Bertolino A, Bondolfi G (2018) An association between emotion-body link disposition and salience—action brain network underlies individual differences in people's anger response
McCulloch WS (1951) Why the mind is in die head. In: Jeffress LA (ed) Cerebral mechanisms in behavior. The Hixon symposium. John Wiley and Sons, New York, pp 42–111
McCulloch WS (1974) Recollections of the many sources of cybernetics. ASC Forum 6(2):5–16
McCulloch WS, Pitts W (1943) A logical calculus of the ideas immanent in nervous activity. Bull Math Biophys 5(4):115–133
Meisel SR, Dayan KI, Pauzner H, Chetboun I, Arbel Y, David D, Kutz I (1991) Effect of Iraqi missile war on incidence of acute myocardial infarction and sudden death in Israeli civilians. Lancet 338(8768):660–661
Merleau-Ponty M (2013) Phénoménologie de la perception. Éditions Gallimard, Paris
Mittleman MA, Maclure M, Sherwood JB, Mulry RP, Tofler GH, Jacobs SC et al (1995) Triggering of acute myocardial infarction onset by episodes of anger. Circulation 92(7): 1720–1725
Mohr DC (1995) Negative outcome in psychotherapy: a critical review. Clin Psychol Sci Pract 2(1):1–27
Mori H et al (1993) Increased responsiveness of left ventricular apical myocardium to adrenergic stimuli. Cardiovasc Res 27(2):192–198
Morris C (1972) The discovery of the individual, 1050-1200, vol 5. University of Toronto Press, Toronto
Mudd JO, Kapur NK, Champion HC, Schulman SP, Wittstein IS (2007) Patients with stress-induced (Takotsubo) cardiomyopathy have an increased prevalence of mood disorders and antidepressant use compared to patients with acute myocardial infarction. J Cardiac Fail 13(Supplement 2):S176
Müller GB (2007) Evo–devo: extending the evolutionary synthesis. Nat Rev Genet 8(12):943–949
Nagel T (1974) What is it like to be a bat? Philos Rev 83(4):435–450
Natorp P (1912) Allgemeine Psychologie nach kritischer Methode, vol 1. JCB Mohr (P. Siebeck), Tubingen
Nelson B, Parnas J, Sass LA (2014) Disturbance of minimal self (ipseity) in schizophrenia: clarification and current status. Schizophr Bull 40(3):479–482
Nightingale AW (2004) Spectacles of truth in classical Greek philosophy: theoria in its cultural context. Cambridge University Press, Cambridge
Norcross JC (2010) The therapeutic relationship. In: Duncan BL, Miller SD, Wampold BE, Hubble MA (eds) The heart and soul of change: delivering what works in therapy. American Psychological Association, Washington
Norcross JC, Lambert MJ (2005a) What should be validated? The therapy relationship. In: Norcross J, Levant R, Beutler L (eds) Evidence-based practices in mental health: debate and dialogue on the fundamental questions. American Psychological Association Press, Washington, pp 209–218
Norcross JC, Lambert MJ (2005b) The therapy relationship. In: Norcross JC, Beutler LE, Levant RF (eds) Evidence-based practices in mental health: debate and dialogue on the fundamental questions. American Psychological Association, Washington
Norcross JC, Wampold BE (2011) Adapting psychotherapy to the individual patient. J Clin Psychol Session 67(2)
Nuzzo A (2006) The role of the human body in Fichte's Grundlage des Naturrechts (1796–97). In: Breazeale D, Rockmore T (eds) Rights, bodies and recognition: new essays on Fichte's Foundations of natural right. Ashgate Publishing, Farnham
Nuzzo A (2010) Phenomenologies of intersubjectivity: Fichte between Hegel and Husserl. In: Waibel VL, Breazeale D, Rockmore T (eds) Fichte and the phenomenological tradition. De Gruyter, New York

Nuzzo R (2014) Statistical errors: P values, the 'gold standard' of statistical validity, are not as reliable as many scientists assume. Nature 506(7487):150–153
Odling-Smee J (2010) Niche inheritance. In: Pigliucci M, Müller GB (eds) Evolution: the extended synthesis. MIT Press, Cambridge
Odling-Smee J, Turner JS (2011) Niche construction theory and human architecture. Biol Theor 6(3):283–289
Odling-Smee FJ, Laland KN, Feldman MW (1996) Niche construction. Am Nat 147(4):641–648
Odling-Smee FJ, Laland KN, Feldman MW (2003) Niche construction: the neglected process in evolution (No. 37). Princeton University Press, Princeton
Okiishi J, Lambert MJ, Nielsen SL, Ogles BM (2003) Waiting for supershrink: an empirical analysis of therapist effects. Clin Psychol Psychother 10(6):361–373
Oppenheimer S (2006) Cerebrogenic cardiac arrhythmias: cortical lateralization and clinical significance. Clin Auton Res 16:6–11
Oppenheimer SM, Cechetto DF (1990) Cardiac chronotropic organization of the rat insular cortex. Brain Res 533:66–72
Oppenheimer S, Cechetto D (2016) The insular cortex and the regulation of cardiac function. Compr Physiol 6(2):1081–1133
Oppenheimer SM, Cechetto DF, Hachinski VC (1990) Cerebrogenic cardiac arrhythmias: cerebral electrocardiographic influences and their role in sudden death. Arch Neurol 47:513–519
Oppenheimer SM, Gelb AW, Girvin JP, Hachinski VC (1992) Cardiovascular effects of human insular stimulation. Neurology 42:1727–1732
Origen (1957) The song of songs: commentary and homilies. Translated and annotated by RP Lawson. (Ancient Christian Writers, 26). Pp. vi+ 385. Longmans, Green, London
Orlinsky D, Grave D, Parks B (1994) Process and outcome in psychotherapy. In: Bergin A, Garfield S (eds) Handbook of psychotherapy and behaviour change, 4th edn. Wiley, New York
Otis L (2007) Mullers Lab. Oxford University Press, Oxford
Owa M et al (2001) Emotional stress-induced 'ampulla cardiomyopathy': discrepancy between the metabolic and sympathetic innervation imaging performed during the recovery course. Jpn Circ J 65(4):349–352
Paley G, Shapiro DA (2002) Lessons from psychotherapy research for psychological interventions for people with schizophrenia. Psychol Psychother Theory Res Pract 75(1):5–17
Patočka J (1953) Le platonism negative in Jan PATOČKA, Liberté et Sacrifice. Écrits politiques, ed. and transl. E. Abrams (Grenoble: Millon, 1990). English Transation 'Negative Platonism". In: Erazim Kohák (ed) Jan Patočka, Philosophy and selected writings. The University of Chicago Press 1989, Chicago, p 175–206, 175
Patočka J (1976) Le monde naturel comme problème philosophique. Martinus Nijhoff, La Haye
Patočka J (1988a) Le Monde naturel et le mouvement de l'existence humaine, traduit par E. Abrams. Kluwer, Dordrecht
Patočka J (1988b) L'homme et le monde. In: Qu'est-ce que la phénoménologie?, a cura di E. Abrams. Millon, Grenoble
Patočka J (1990) Liberté et sacrifice: écrits politiques. Editions Jérôme Millon
Patočka J (1995) Papiers phénoménologiques. Editions Jérôme Millon, Grenoble
Patočka J (1996) Heretical essays in the philosophy of history. Open Court, Chicago
Patočka J (1997) Platone e l'Europa, vol 55, Milan. Vita e Pensiero
Patočka J (1998) Body, community, language, world. Open Court Publishing, Chicago
Patočka J (1998) Platone e l'Europa. Vita e Pensiero, Milan
Patočka J (2003) Il mondo naturale e la fenomenologia, trad. di A. Pantano Mimesis, Milano
Paulhus DL (2007) Self-deception. In: Baumeister RF, Vohs KD (eds) Encyclopedia of social psychology. Sage, Thousand Oaks, pp 189–190
Paulus MP, Feinstein JS, Leland D, Simmons AN (2005) Superior temporal gyrus and insula provide response and outcome-dependent information during assessment and action selection in a decision-making situation. Neuroimage 25(2):607–615
Peirce CS (1955) Abduction and induction. In: Buchler J (ed) Philosophical writings of Peirce. Dover, New York, pp 150–156

Perec G (2009) Life a user's manual. David R Godine, Boston
Petitot J et al (1999) Naturalizing phenomenology: issues in contemporary phenomenology and cognitive science. Stanford University Press, Stanford
Pfennig DW, Wund MA, Snell-Rood EC, Cruickshank T, Schlichting CD, Moczek AP (2010) Phenotypic plasticity's impacts on diversification and speciation. Trends Ecol Evol 25(8):459–467
Pihlström S (2009) Pragmatism and naturalized transcendental subjectivity. Contemporary Pragmatism 6(1):1–13
Pitts W, McCulloch WS (1947) How we know universals the perception of auditory and visual forms. Bull Math Biophys 9(3):127–147
Plutarch (1927) On listening to lectures. In: Babbitt FC (ed) Plutarch's Moralia, vol 1. Harvard University Press, Cambridge
Pöggeler, O. (1989–1995). Heideggers logische Untersuchungen. In Martin Heidegger: Innen- und Aussenansichten: Forum für Philosophie Bad Hamburg (pp 75–100) trad.it. In Poggi S, Tomasello P (eds) Martin Heidegger. Ontologia, fenomenologia, verità. LED Edizioni Universitarie. A cura di. Milano, 1995
Polanyi M (1958) Personal knowledge, towards a post critical epistemology. University of Chicago, Chicago
Pontico E (1998) Trattato pratico sulla vita monastica, Rome. Città nuova
Popper KR (1972) Objective knowledge: an evolutionary approach. Oxford University Press, Oxford
Popper KR, Eccles JC (1986) Self and its brain: an arguments for interactionism. Springer, Berlin
Prasad A, Lerman A, Rihal CS (2008) Apical ballooning syndrome (Tako-Tsubo or stress cardiomyopathy): a mimic of acute myocardial infarction. Am Heart J 155:408–417
Priebe S, Richardson M, Cooney M, Adedeji O, McCabe R (2010) Does the therapeutic relationship predict outcomes of psychiatric treatment in patients with psychosis? A systematic review. Psychother Psychosom 80(2):70–77
Putnam FW (1988) The switch process in multiple personality disorder and other state-change disorders. Dissociation 1(1):24–32
Quesne P (2004) Les recherches philosophiques du jeune Heidegger, vol 171. Springer Science & Business Media, Dordrecht
Raab W, Chaplin JP, Bajusz K (1964) Myocardial necroses produced ill domesticated rats and wild rats by sensory and emotional stresses. Proc Soc Exp Biol Med 116:665
Rabbow P (1954) Seelenfuhrung: Methodik der Exerzitien in der Antike. Kôsel-Verlag, München
Rachman S (1981) The primacy of affect: some theoretical implications. Behav Res Ther 19(4):279–290
Reale, G. (2004) Metafísica di Aritotele. Bompiani.
Regnante RA, Zuzek RW, Weinsier SB et al (2009) Clinical characteristics and four-year outcomes of patients in the Rhode island takotsubo cardiomyopathy registry. Am J Cardiol 103:1015–1019
Richardson WJ (1988) The place of the unconscious in Heidegger. In: Hoeller K (ed) Heidegger & psychology. Review of Existential Psychology & Psychiatry, Seattle
Ricoeur P (1954) Sympathie et respect: phénoménologie et éthique de la seconde personne. Revue de Métaphysique et de Morale 59(4):380–397
Ricoeur P (1983) Narrative time. In: Mitchell WT (ed) On narrative. University of Chicago Press, Chicago
Ricoeur P (1992) Oneself as another. University of Chicago Press, Chicago
Ricœur P (2001) Le juste 2. Ed. Esprit, Paris, p 9
Ricoeur P (2004) In: Blamey K, Pellauer D (eds) Memory, History, Forgetting, trans. University of Chicago Press, Chicago
Ricoeur P (2010a) Time and narrative, vol 3. University of Chicago Press, Chicago
Ricoeur P (2010b) Power and Violence. Theor Cult Soc 27(5):18–36
Riedweg C (2005) Pythagoras. His life, teaching, and influence. Translated from the German by Steven Rendall. Cornell University Press, Ithaca/London

Rizzolatti G, Cattaneo L, Fabbri-Destro M, Rozzi S (2014) Cortical mechanisms underlying the organization of goal-directed actions and mirror neuron-based action understanding. Physiol Rev 94(2):655–706
Robinson LA, Berman JS, Neimeyer RA (1990) Psychotherapy for the treatment of depression: a comprehensive review of controlled outcome research. Psychol Bull 108(1):30
Rogers CR (1957) The necessary and sufficient conditions of therapeutic personality change. J Consult Psychol 21(2):95
Rogers CR (1975) Empathic: an unappreciated way of being. Counsel Psychol 5(2):2–10
Romano, Claude (2009). Le monde animal: Heidegger et von Uexküll. In S. Jollivet et C. Romano (éd.), Heidegger en dialogue 1912-1930. Vrin Paris
Rosenzweig S (1936) Some implicit common factors in diverse methods of psychotherapy. Am J Orthopsychiatry 6(3):412
Rozanski A, Bairey CN, Krantz DS, Friedman J, Resser KJ, Morell M et al (1988) Mental stress and the induction of silent myocardial ischemia in patients with coronary artery dis- ease. N Engl J Med 318:1005–1012
Rubino V, Blasi G, Latorre V, Fazio L, d'Errico I, Mazzola V et al (2007) Activity in medial prefrontal cortex during cognitive evaluation of threatening stimuli as a function of personality style. Brain Res Bull 74(4):250–257
Rudrauf D, Lutz A, Cosmelli D, Lachaux JP, Le Van Quyen M (2003) From autopoiesis to neurophenomenology: Francisco Varela's exploration of the biophysics of being. Biol Res 36(1):27–65
Russiello F, Arciero G, Decaminada F, Corona R, Ferrigno L, Fucci M et al (1995) Stress, attachment and skin diseases: a casecisco Varelaal. J Eur Acad Dermatol Venereol 5(3):234–239
Safran JD, Christopher Muran J (1996) The resolution of ruptures in the therapeutic alliance. J Consult Clin Psychol 64(3):447
Safran JD, Muran JC (2006) Has the concept of the therapeutic alliance outlived its usefulness? Psychotherapy 43(3):286
Safran JD, Muran JC, Eubanks-Carter C (2011) Repairing alliance ruptures. Psychotherapy 48(1):80
Samuels MA (2007) The brain–heart connection. Circulation 116(1):77–84
Sato H, Tateishi H, Uchida T (1990a) Takotsubo-type cardiomyopathy due to multivessel spasm. In: Kodama K, Haze K, Hon M (eds) Clinical aspect of myocardial injury from ischemia to heart failure. Kagakuhyouronsha, Tokyo, pp 56–64
Sato H, Tateishi H, Uchida T et al (1990b) Tako-tsubo-like left ventricular dysfunction due to multivessel coronary spasm. In: Kodama K, Hori M (eds) Clinical aspect of myocardial injury: from ischemia to heart failure [in Japanese]. Kagakuhyoronsha Publishing Co, Tokyo, pp 56–64
Scarpat G (2001) Parrhesia greca, parrhesia cristiana, vol 130. Paideia, Brescia
Scheler MF (1970) The nature of sympathy. Transaction Publishers, London
Schiller J (1978) La notion d'organisation dans l'histoire de la biologie. Maloine, Paris
Schopenhauer A, Payne EFJ (1995) On the basis of morality. Hackett Publishing, Indianapolis
Schore AN (2002) Advances in neuropsychoanalysis, attachment theory, and trauma research: implications for self psychology. Psychoanal Inq 22(3):433–484
Scott-Phillips TC, Laland KN, Shuker DM, Dickins TE, West SA (2014) The niche construction perspective: a critical appraisal. Evolution 68(5):1231–1243
Shapiro DA, Shapiro D (1982) Meta-analysis of comparative therapy outcome studies: a replication and refinement. Psychol Bull 92(3):581
Sheehan T (2014) Making sense of Heidegger: a paradigm shift. Rowman & Littlefield International, London
Sheets-Johnstone M (2011) The primacy of movement, vol 82. John Benjamins Publishing
Siegal ML, Bergman A (2002) Waddington's canalization revisited: developmental stability and evolution. Proc Natl Acad Sci 99(16):10528–10532
Singer T, Seymour B, O'doherty J, Kaube H, Dolan RJ, Frith CD (2004) Empathy for pain involves the affective but not sensory components of pain. Science 303(5661):1157–1162

Sloane RB, Staples FR, Cristol AH, Yorkston NJ, Whipple K (1975) Psychotherapy vs. behavior therapy. Harvard University Press, Cambridge
Smalheiser NR (2000) Walter Pitts. Perspect Biol Med 43(2):217–226
Smith ML, Glass GV, Miller TI (1980) The benefits of psychotherapy. Johns Hopkins University Press, Baltimore
Solomon RC (ed) (2004) Thinking about feeling. Oxford University Press, Oxford, p 76
Solomon M, Siegel DJ (2003) Healing trauma. WW Norton, New York
Stendhal (2002) Romanzi e racconti 2 Vol. I Meridiani. Mondadori
Stolorow RD (2008) The contextuality and existentiality of emotional trauma. Psychoanal Dialogues 18(1):113–123
Stolorow RD (2011) Trauma and human existence: autobiographical, psychoanalytic, and philosophical reflections. Taylor & Francis, New York
Strawson G (1999) The self. In: Gallagher S, Shear J (eds) Models of the Self. Imprint Academic, Charlottesville
Strupp HH, Fox RE, Lessler K (1969) Patients view their psychotherapy. Johns Hopkins Press, Oxford
Tardivel É (2011) La liberté au principe: essai sur la philosophie de Patočka. Vrin, Paris
Tentler TN (1977) Sin and confession on the eve of the reformation. Princeton University Press, Princeton
The sayings of the Desert Fathers: the alphabetical collection (1975) Cistercian Publications, Kalamazoo
Thompson E (2001) Between ourselves: second-person issues in the study of consciousness. Imprint Academic, Charlottesville
Tinbergen N (1968) On war and peace in animals and man. Science. 160(3835):1411–1418
Tracey I, Mantyh PW (2007) The cerebral signature for pain perception and its modulation. Neuron 55(3):377–391
Turner RS (1977) The Ohm-Seebeck dispute, Hermann von Helmholtz, and the origins of physiological acoustics. Br J Hist Sci 10(01):1–24
Turner JS (2000) Architecture and morphogenesis in the mound of Macrotermes michaelseni (Sjöstedt)(Isoptera: Termitidae, Macrotermitinae) in northern Namibia. Cimbebasia 16:143–175
Turner JS (2000b) The extended organism. The physiology of animal-built structures. Harvard University Press, Cambridge
Turner JS (2002) A superorganism's fuzzy boundaries. Nat Hist 111(6):62–69
Turner JS (2004) Extended phenotypes and extended organisms. Biol Philos 19(3):327–352
Turner JS (2004b) Gaia, extended organisms, and emergent homeostasis. In: Schneider SH, Miller JR, Crist E, Boston PJ (eds) Scientists debate gaia. The next century. MIT Press, Cambridge
Uexkull JV (1903) Der biologische Bauplan des Sipunculus. Zeitschr f Biol 44:269–344
Uexküll JV (1909) Umwelt und innenwelt der tiere. Springer, Berlin
Uexküll JV (1926) Theoretical biology, tr. DL Mackinnon. Kegan Paul, Trench, Trubner, London
Van der Kolk BA (1989) The compulsion to repeat the trauma. Psychiatr Clin North Am 12(2):389–411
Varela FJ (1979) Principles of biological autonomy. North-Holland/Elsevier, New York
Varela F (1986) Laying down a path in walking: a biologist's look at a new biology. Cybernetic 2:6–15
Varela F (1987) The science and technology of cognition: emergent directions. In: Roos JL (ed) Economics and artificial intelligence, 1st IFAC international symposium. Pergamon Press, Oxford/New York, pp 1–9
Varela F (1992) Un know-how per l'etica. Laterza, Bari-Roma
Varela F (1996) Neurophenomenology: a methodological remedy to the hard problem. J Consc Stud 3:330–350
Varela FJ (1999) The specious present: a neurophenomenology of time consciousness. In: Petitot J et al (eds) Naturalizing phenomenology: issues in contemporary phenomenology and cognitive science, vol 64. Stanford University Press, Stanford, pp 266–329

Varela F, Shear J (1999a) (eds) The view from within: first-person methodologies in the study of consciousness (Special Issue). J Consc Stud 6 (2–3)
Varela FJ, Shear J (1999b) First-person methodologies: what, why, how. J Consc Stud 6(2-3):1–14
Varela FJ, Thompson E, Rosch E (1991) The embodied mind: cognitive science and human experience. MIT press, Cambridge
Vaseghi M, Shivkum K (2008) The role of the autonomic nervous system in sudden cardiac death. Prog Cardiovasc Dis 50(6):404–419
Vernant JP, Detienne M (1974) Les Ruses de l'intelligence: la mètis des Grecs. Flammarion, Paris
Verrier RL, Hagestad EL, Lown B (1987) Delayed myocardial ischemia induced by anger. Circulation 75:249–254
Volpi F. (1996) Introduzione. In Kant I (1996). Che cosa significa orientarsi nel pensare? Adelphi
Volpi F (1996b) La Question du Logos dans l'articulation de la facticité chez le jeune Heidegger, lecteur d'Aristote. In: Courtine J-F (ed) Heidegger 1919-1929. De l'herméneutique de la facticité à la métaphysique du Dasein. Vrin, Paris
Volpi F (2008) Guida ad Heidegger. Laterza, Roma-Bari
Von Glasersfeld E (1995) Radical constructivism: a way of knowing and learning, Studies in mathematics education series, vol 6. Falmer Press, Taylor & Francis, Bristol
Von Helmholtz H, Cahan D (1995) Science and culture: popular and philosophical essays. University of Chicago Press, Chicago
von Helmholtz H, Southall JPC (2005) Treatise on physiological optics, vol 3. Courier Corporation, North Chelmsford
von Herrmann FW (2013) Hermeneutics and reflection: Heidegger and Husserl on the concept of phenomenology. University of Toronto Press, Toronto
von Uexküll J (1921) Umwelt und Innenwelt der Tiere, 2nd edn. Springer, Berlin
Jakob von Uexküll (1946–2014) Der unsterbliche Geist in der Natur (Italian Edition) L'immortale spirito nella natura (I Timoni)
Von Wright GH (1971) Explanation and understanding. Cornell University Press, Ithaca
Von Wright GH (2004) Explanation and understanding. Cornell University Press, Ithaca
Waddington CH (1957) The strategy of the genes. A discussion of some aspects of theoretical biology. With an appendix by H. Kacser. Allen & Unwin, Sydney
Wampold BE (2001a) Contextualizing psychotherapy as a healing practice: culture, history, and methods. Appl Prevent Psychol 10(2):69–86
Wampold BE (2001b) The great psychotherapy debate: model, methods, and findings. Lawrence Erlbaum Associates, Mahwah
Wampold BE (2005) Establishing specificity in psychotherapy scientifically: design and evidence issues. Clin Psychol Sci Pract 12(2):194–197
Wampold BE (2007) Psychotherapy: the humanistic (and effective) treatment. Am Psychol 62(8):857
Wampold BE (2010a) The basics of psychotherapy: an introduction to theory and practice. American Psychological Association, Washington, DC
Wampold BE (2010b) The research evidence for the common factors models: a historically situated perspective. In Barry L. Duncan, Scott D. Miller, Bruce E. Wampold, Mark A. Hubble (eds) The heart and soul of change: delivering what works in therapy, 2nd Revised edition. American Psychological Association, Washington
Wampold BE (2011) Qualities and actions of effective therapists. American Psychological Association, Washington
Wampold BE, Bhati KS (2004) Attending to the omissions: a historical examination of evidence-based practice movements. Prof Psychol Res Pract 35(6):563
Wampold BE, Brown GSJ (2005) Estimating variability in outcomes attributable to therapists: a naturalistic study of outcomes in managed care. J Consult Clin Psychol 73(5):914
Wampold BE, Ahn H-n, Coleman HLK (2001) Medical model as metaphor: old habits die hard. J Counsel Psychol 48(3):268–273

Wampold BE, Imel ZE, Bhati KS, Johnson-Jennings MD (2006) Insight as a common factor. In: Castonguay LG, Hill CE (eds) Insight in psychotherapy. American Psychological Association, Washington, pp 119–139
Weimer WB (1977) A conceptual framework for cognitive psychology: motor theories of mind. In: Shaw R, Bransford J (eds) Perceiving, acting, and knowing. Erlbaum, Hillsdale
West-Eberhard MJ (2003) Developmental plasticity and evolution. Oxford University Press, Oxford
Westphal CFO (1871) Die Agoraphobie, eine neuropathische Erscheinung. Archiv für Psychiatrie und Nervenkrankheiten 3:138–161, and 219–221
White M, Epston D (1990) Narrative means to therapeutic ends. WW Norton & Company, New York
Whitehead C (2004) Everything I believe might be a delusion. Whoa! Tucson 2004: ten years on, and are we any nearer to a science of consciousness? J Consc Stud 11(12):68–88
Wiesel E (1996) Mémoires 2, Paris. Éditions du Seuil, Paris
Wilbert-Lampen U, Leistner D, Greven S, Pohl T, Sper S, Völker C et al (2008) Cardiovascular events during World Cup soccer. N Engl J Med 358(5):475–483
Wittstein IS (2012) Stress cardiomyopathy: a syndrome of catecholamine-mediated myocardial stunning? Cell Mol Neurobiol 32(5):847–857
Wittstein IS, Thiemann DR, Lima JA, Baughman KL, Schulman SP, Gerstenblith G et al (2005) Neurohumoral features of myocardial stunning due to sudden emotional stress. N Engl J Med 352(6):539–548
Xu J, Gannon PJ, Emmorey K, Smith JF, Braun AR (2009) Symbolic gestures and spoken language are processed by a common neural system. Proc Natl Acad Sci 106(49):20664–20669
Zahavi D (2003) How to investigate subjectivity: Natorp and Heidegger on reflection. Continent Philos Rev 36(2):155–176
Zeleny M (1981) Autopoiesis, a theory of living organizations. North-Holland/Elsevier, New York
Zeller E, Mondolfo R (1961) La filosofia dei greci nel suo sviluppo storico. La nuova Italia, Milano

Index

G. Arciero et al., *The Foundations of Phenomenological Psychotherapy*,
https://doi.org/10.1007/978-3-319-78087-0

MIX
Papier aus verantwortungsvollen Quellen
Paper from responsible sources
FSC® C105338

If you have any concerns about our products,
you can contact us on
ProductSafety@springernature.com

In case Publisher is established outside the EU,
the EU authorized representative is:
Springer Nature Customer Service Center GmbH
Europaplatz 3, 69115 Heidelberg, Germany

Printed by Libri Plureos GmbH
in Hamburg, Germany